T0227734

MONOGRAPHS ON
STATISTICS AND APPLIED PROBABILITY

General Editors

**D.R. Cox, D.V. Hinkley, N. Reid, D.B. Rubin
and B.W. Silverman**

(Full details concerning this series are available from the publisher)

Tensor Methods
in Statistics

PETER McCULLAGH

The University of Chicago
Department of Statistics
Chicago, USA

CRC Press
Taylor & Francis Group
Boca Raton London New York

CRC Press is an imprint of the
Taylor & Francis Group, an **informa** business

First published 1987 by CRC Press
Taylor & Francis Group
6000 Broken Sound Parkway NW, Suite 300
Boca Raton, FL 33487-2742

Reissued 2018 by CRC Press

© 1987 by P. McCullagh
CRC Press is an imprint of Taylor & Francis Group, an Informa business

No claim to original U.S. Government works

This book contains information obtained from authentic and highly regarded sources. Reasonable efforts
have been made to publish reliable data and information, but the author and publisher cannot assume
responsibility for the validity of all materials or the consequences of their use. The authors and publishers
have attempted to trace the copyright holders of all material reproduced in this publication and apologize to
copyright holders if permission to publish in this form has not been obtained. If any copyright material has
not been acknowledged please write and let us know so we may rectify in any future reprint.

Except as permitted under U.S. Copyright Law, no part of this book may be reprinted, reproduced, transmit-
ted, or utilized in any form by any electronic, mechanical, or other means, now known or hereafter invented,
including photocopying, microfilming, and recording, or in any information storage or retrieval system,
without written permission from the publishers.

Trademark Notice: Product or corporate names may be trademarks or registered trademarks, and are used
only for identification and explanation without intent to infringe.

Library of Congress Cataloging-in-Publication Data
McCullagh. P. (Peter), 1952-
　Tensor methods in statistics.
　(Monographs on statistics and applied probability)
　Bibliography : p
　Includes indexes.
　1. Calculus of tensors. 2. Mathematical statistics.
I. Title. II. Series.
QA433.M325 1987　519.5　87-13826
ISBN 0-412-27480-9

A Library of Congress record exists under LC control number: 87013826

Publisher's Note
The publisher has gone to great lengths to ensure the quality of this reprint but points out that some imperfec-
tions in the original copies may be apparent.

Disclaimer
The publisher has made every effort to trace copyright holders and welcomes correspondence from those they
have been unable to contact.

ISBN 13: 978-1-315-89801-8 (hbk)
ISBN 13: 978-1-351-07711-8 (ebk)

Visit the Taylor & Francis Web site at http://www.taylorandfrancis.com and the
CRC Press Web site at http://www.crcpress.com

TO ROSA

Contents

Preface

In matters of aesthetics and mathematical notation, no one loves an index. According to one school of thought, indices are the pawns of an arcane and archaic notation, the front-line troops, the cannon-fodder, first to perish in the confrontation of an inner product. Only their shadows persist. Like the putti of a Renaissance painting or the cherubim of an earlier era or, like mediaeval gargoyles, indices are mere embellishments, unnecessary appendages that adorn an otherwise bare mathematical symbol. Tensor analysis, it is claimed despite all evidence to the contrary, has nothing whatever to do with indices. 'Coordinate-free methods' and 'operator calculus' are but two of the rallying slogans for mathematicians of this persuasion. 'Computation', on the other hand, is a reactionary and subversive word. Stripped of its appendages, freed from its coordinate shackles, a plain unadorned letter is the very model of a modern mathematical operator.

Yet this extreme scorn and derision for indices is not universal. It can be argued, for example, that a 'plain unadorned letter' conceals more than it reveals. A more enlightened opinion, shared by the author, is that, for many purposes, it is the symbol itself, — that 'plain unadorned letter' — not the indices, that is the superfluous appendage. Like the grin on Alice's Cat, the indices can remain long after the symbol has gone. Just as the grin rather than the Cat is the visible display of the Cat's disposition, so too it is the indices, not the symbol, that is the visible display of the nature of the mathematical object. Surprising as it may seem, indices are capable of supporting themselves without the aid of crutches!

In matters of index notation and tensor analysis, there are few neutral observers. Earlier workers preferred index notation partly, perhaps, out of an understandable concern for computation. Many

modern workers prefer coordinate-free notation in order to foster
geometrical insight. The above parodies convey part of the flavour
and passion of the arguments that rage for and against the use of
indices.

Index notation is the favoured mode of exposition used in this
book, although flexibility and tolerance are advocated. To avoid
unfamiliar objects, indices are usually supported by symbols, but
many of the most important formulae such as the fundamental
identity (3.3) could easily be given in terms of indices alone without
the assistance of supporting symbols. The fundamental operation
of summing over connecting partitions has everything to do with
the indices and nothing to do with the supporting symbol. Also,
in Section 4.3.2, where supporting symbols tend to intrude, the
Cheshire notation is used.

Chapter 1 introduces the reader to a number of aspects of
index notation, groups, invariants and tensor calculus. Examples
are drawn from linear algebra, physics and statistics. Chapters
2 and 3, dealing with moments, cumulants and invariants, form
the core of the book and are required reading for all subsequent
chapters.

Chapter 4 covers the topics of sample cumulants, symmetric
functions, polykays, simple random sampling and cumulants of k-
statistics. This material is not used in subsequent chapters. Unless
the reader has a particular interest in these fascinating topics, I
recommend that this chapter be omitted on first reading.

Chapters 5 and 6, dealing with Edgeworth and saddlepoint
approximations, are presented as a complementary pair, though
they can be read independently. Chapter 6 refers backwards to
Chapter 5 at only one point.

The final two chapters are again presented as a pair. Chapter 7, dealing with likelihood functions, log likelihood derivatives,
likelihood ratio statistics and Bartlett factors, makes use only of
material in Chapters 2 and 3. Chapter 8, dealing with ancillary
statistics, makes considerable use of saddlepoint approximations
and Legendre transforms from Chapter 6 and other expansions
given in Chapter 7. The book concludes with a derivation of
Barndorff-Nielsen's formula for the conditional distribution of the
maximum likelihood estimator given a suitable ancillary statistic.

Exercises are an integral part of the book, though results derived as exercises are rarely used in the remainder of the book. One

exception is the Taylor expansion for the log determinant, derived
in Exercise 1.16, subsequently used in Section 8.6. Exercises vary
considerably in length and, I suppose, difficulty. I have decided
against difficulty ratings on the grounds that the composer of the
exercise is least suited to judge difficulty. Results derived in one
exercise are commonly used in the exercises immediately following.

The book is intended mainly for graduate students in statistics and as a reference work for professional statisticians. Readers
should have some familiarity with linear algebra, eigenvalue decompositions, linear models and, for later chapters, with likelihood
functions, likelihood ratio statistics and so on. As soon as students become comfortable using index notation, most of the first
four chapters will be within their grasp. Final year undergradu-
ate mathematics students at U.K. universities can master most of
Chapter 2 and parts of Chapters 3 and 4.

In addition to the topics covered in these eight chapters, the
original plan called for additional chapters on *Riemannian geometry*, *nuisance parameters* and applications to *Bayesian methodol-
ogy*. Because of constraints on length and on time, these topics
have had to be abandoned, at least temporarily.

In the three years that it has taken me to write this book, I
have benefitted greatly from the help and advice of colleagues at
Imperial College, AT&T Bell Labs and University of Chicago. I
am especially indebted to Colin Mallows for numerous discussions
concerning invariants, for debunking hastily conceived conjectures
and for calculations leading to a number of exercises. Allan Wilks
is responsible for relieving me of the arduous task of compiling
by hand the Tables in the Appendix. Others who have read and
commented on parts of the manuscript or who have made useful recommendations short of abandoning the enterprise, include
R. Bahadur, J.M. Chambers, D.R. Cox, M. Handcock, C. Inclan,
V. Johnson, P. Lang, D. Pregibon, N. Reid, L.A. Shepp, I.M. Skovgaard, T.P. Speed, J.M. Steele, S.M. Stigler, J.W. Tukey, P. Vos,
D.L. Wallace, Daming Xu and S.L. Zabell. I am grateful also to
H.E. Daniels and to I.M. Skovgaard for providing copies of unpublished papers on saddlepoint approximation and to O.E. Barndorff-
Nielsen & P. Blaesild for providing unpublished papers on *strings*
and related topics.

In a book of this kind, it is unreasonable to expect all formulae
and all exercises to be free from error however careful the checking.

All formulae have been checked but undoubtedly errors remain. I would be grateful, therefore, if diligent eagle-eyed readers could notify me of any further errors.

The book was typeset partly using the TₑX system, (Knuth, 1986) and partly using the TROFF system. The figures were typeset using the PᵢCTₑX system (Wichura, 1986). I am grateful to M. Wichura for assistance with PᵢCTₑX and to Diana Wilson for typesetting advice.

Finally, it is a great pleasure to thank the typist whose unfailing courtesy, inexhaustible patience and careful attention to detail have made this book possible.

Chicago   P. McCullagh
December, 1986

Financial support for this research was provided in part by NSF Grants No. DMS-8404941 and DMS-8601732.

CHAPTER 1

Index notation

1.1 Introduction

It is a fact not widely acknowledged that, with appropriate choice of notation, many multivariate statistical calculations can be made simpler and more transparent than the corresponding univariate calculations. This simplicity is achieved through the systematic use of index notation and special arrays called tensors. For reasons that are given in the following sections, matrix notation, a reliable workhorse for many second-order calculations, is totally unsuitable for more complicated calculations involving either non-linear functions or higher-order moments. The aim of this book is to explain how index notation simplifies many statistical calculations, particularly those involving moments or cumulants of non-linear functions. Other applications where index notation greatly simplifies matters include k-statistics, Edgeworth and conditional Edgeworth approximations, calculations involving conditional cumulants, moments of maximum likelihood estimators, likelihood ratio statistics and the construction of ancillary statistics. These topics are the subject matter of later chapters.

In some ways, the most obvious and, at least initially, most disconcerting aspect of index notation is that the components of the vector of primary interest, usually a parameter, θ, or a random variable, X, are indexed using superscripts. Thus, θ^2, the second component of the vector θ, is not to be confused with the square of any component. For this reason, powers are best avoided unless the context leaves no room for ambiguity, and the square of θ^2 is written simply as $\theta^2\theta^2$. In view of the considerable advantages achieved, this is a very modest premium to pay.

1.2   The summation convention

Index notation is a convention for the manipulation of multi-dimensional arrays. The elements of these arrays are called *components* if they are functions of selected components of the vector of interest. In the context of parametric inference and in manipulations associated with likelihood functions, it is appropriate to take the unknown parameter as the vector of interest: see the first example in Section 1.4. Here, however, we take as our vector of interest the p-dimensional random variable X with components $X^1, ..., X^p$. In this context, arrays of constants are called *coefficients*. This terminology is merely a matter of convention but it appears to be useful and the notation does emphasize it. Thus, for example, $\kappa^i = E(X^i)$ is a one-dimensional array whose components are the means of the components of X and $\kappa^{ij} = E(X^i X^j)$ is a two-dimensional array whose components are functions of the joint distributions of pairs of variables.

Probably the most convenient aspect of index notation is the implied summation over any index repeated once as a superscript and once as a subscript. The range of summation is not stated explicitly but is implied by the positions of the Thus,

$$a_i X^i \equiv \sum_{i=1}^{p} a_i X^i \tag{1.1}$$

specifies a linear combination of the Xs with coefficients $a_1, ..., a_p$. repeated index and by conventions regarding the range of the index. Quadratic and cubic forms in X with coefficients a_{ij} and a_{ijk} are written

$$a_{ij} X^i X^j \quad \text{and} \quad a_{ijk} X^i X^j X^k \tag{1.2}$$

and the extension to homogeneous polynomials of arbitrary degree is immediate.

For the sake of simplicity, and with no loss of generality, we take all multiply-indexed arrays to be symmetric under index permutation but, of course, subscripts may not be interchanged with superscripts. The value of this convention is clearly apparent when we deal with scalars such as $a_{ij} a_{kl} \omega^{ijkl}$, which, by convention only, is the same as $a_{ik} a_{jl} \omega^{ijkl}$ and $a_{il} a_{jk} \omega^{ijkl}$. For instance, if $p = 2$

and $a_{ij} = \delta_{ij} = 1$ if $i = j$ and 0 otherwise, then, without the convention,

$$a_{ij}a_{kl}\omega^{ijkl} - a_{ik}a_{jl}\omega^{ijkl} = \omega^{1122} + \omega^{2211} - \omega^{1212} - \omega^{2121}$$

and this is not zero unless ω^{ijkl} is symmetric under index permutation.

Expressions (1.1) and (1.2) produce one-dimensional or scalar quantities, in this case scalar random variables. Suppose instead, we wish to construct a vector random variable Y with components $Y^1, ..., Y^q$, each of which is linear in X, we may write

$$Y^r = a_i^r X^i \qquad (1.3)$$

and $r = 1, ..., q$ is known as a *free* index. Similarly, if the components of Y are homogeneous quadratic forms in X, we may write

$$Y^r = a_{ij}^r X^i X^j. \qquad (1.4)$$

Non-homogeneous quadratic polynomials in X may be written

$$Y^r = a^r + a_i^r X^i + a_{ij}^r X^i X^j.$$

Where two sets of indices are required, as in (1.3) and (1.4), one referring to the components of X and the other to the components of Y, we use the sets of indices $i, j, k, ...$ and $r, s, t, ...$. Occasionally it will be necessary to introduce a third set, $\alpha, \beta, \gamma, ...$ but this usage will be kept to a minimum.

All of the above expressions could, with varying degrees of difficulty, be written using matrix notation. For example, (1.1) is typically written as $\mathbf{a}^T \mathbf{X}$ where $\mathbf{a}$ and $\mathbf{X}$ are column vectors; the quadratic expression in (1.2) is written $\mathbf{X}^T \mathbf{A} \mathbf{X}$ where $\mathbf{A}$ is symmetric, and (1.3) becomes $\mathbf{Y} = \mathbf{A}^* \mathbf{X}$ where $\mathbf{A}^*$ is of order $q \times p$. From these examples, it is evident that there is a relationship of sorts between column vectors and the use of superscripts, but the notation $\mathbf{X}^T \mathbf{A} \mathbf{X}$ for $a_{ij} X^i X^j$ violates the relationship. The most useful distinction is not in fact between rows and columns but between coefficients and components and it is for this reason that index notation is preferred here.

1.3   Tensors

The term *tensor* is used in this book in a well-defined sense, similar in spirit to its meaning in differential geometry but with minor differences in detail. It is not used as a synonym for *array*, *index notation* or the *summation convention*. A cumulant tensor, for example, is a symmetric array whose elements are functions of the joint distribution of components of the random variable of interest, X say. The values of these elements in any one coordinate system are real numbers but, when we describe the array as a tensor, we mean that the values in one coordinate system, Y say, can be obtained from those in any other system, X say, by the application of a particular transformation formula. The nature of this transformation is the subject of Sections 2.4 and 3.5, and in fact, we consider not just changes of basis, but also non-invertible transformations.

When we use the adjectives *covariant* and *contravariant* in reference to tensors, we refer to the way in which the arrays transform under a change of variables from the original x to new variables y. In statistical calculations connected with likelihood functions, x and y are typically parameter vectors but in Chapters 2 and 3, x and y refer to random variables. To define the adjectives covariant and contravariant more precisely, we suppose that ω is a d-dimensional array whose elements are functions of the components of x, taken d at a time. We write $\omega = \omega^{i_1 i_2 \ldots i_d}$ where the d components need not be distinct. Consider the transformation $y = g(x)$ from x to new variables $y = y^1, \ldots, y^p$ and let $a_i^r \equiv a_i^r(x) = \partial y^r / \partial x^i$ have full rank for all x. If $\bar{\omega}$, the value of ω for the transformed variables, satisfies

$$\bar{\omega}^{r_1 r_2 \ldots r_d} = a_{i_1}^{r_1} a_{i_2}^{r_2} \ldots a_{i_d}^{r_d} \omega^{i_1 i_2 \ldots i_d} \qquad (1.5)$$

then ω is said to be a contravariant tensor. On the other hand, if ω is a covariant tensor, we write $\omega = \omega_{i_1 i_2 \ldots i_d}$ and the transformation law for covariant tensors is

$$\bar{\omega}_{r_1 r_2 \ldots r_d} = b_{r_1}^{i_1} b_{r_2}^{i_2} \ldots b_{r_d}^{i_d} \omega_{i_1 i_2 \ldots i_d} \qquad (1.6)$$

where $b_r^i = \partial x^i / \partial y^r$, the matrix inverse of a_i^r, satisfies $a_i^r b_r^j = \delta_i^j = a_r^j b_i^r$.

The function $g(.)$ is assumed to be an element of some group, either specified explicitly or, more commonly, to be inferred from the statistical context. For example, when dealing with transformations of random variables or their cumulants, we usually work with the general linear group (1.3) or the general affine group (1.8). Occasionally, we also work with the smaller orthogonal group, but when we do so, the group will be stated explicitly so that the conclusions can be contrasted with those for the general linear or affine groups. On the other hand, when dealing with possible transformations of a vector of parameters, it is natural to consider non-linear but invertible transformations and $g(.)$ is then assumed to be a member of this much larger group. In other words, when we say that an array of functions is a tensor, the statement has a well defined meaning only when the group of transformations is specified or understood.

It is possible to define hybrid tensors having both subscripts and superscripts that transform in the covariant and contravariant manner respectively. For example, if ω^{ij} and ω_{ijk} are both tensors, then the product $\gamma^{ij}_{klm} = \omega^{ij}\omega_{klm}$ is a tensor of covariant order 3 and contravariant order 2. Furthermore, we may sum over pairs of indices, a process known as *contraction*, giving

$$\gamma^i_{kl} = \gamma^{ij}_{klj} = \omega^{ij}\omega_{klj}.$$

A straightforward calculation shows that γ^i_{kl} is a tensor because, under transformation of variables, the transformed value is

$$\bar{\gamma}^{ij}_{klm} = \gamma^{rs}_{tuv}a^i_r a^j_s b^t_k b^u_l b^v_m$$

and hence, summation over $m = j$ gives

$$\bar{\gamma}^i_{kl} = \gamma^r_{tu}a^i_r b^t_k b^u_l.$$

Thus, the tensor transformation property is preserved under multiplication and under contraction. An important consequence of this property is that scalars formed by contraction of tensors must be invariants. In effect, they must satisfy the transformation law of zero order tensors. See Section 1.6.

One of the problems associated with tensor notation is that it is difficult to find a satisfactory notation for tensors of arbitrary order. The usual device is to use subscripted indices as in (1.5) and (1.6), but this notation is aesthetically unpleasant and is not particularly easy to read. For these reasons, subscripted indices will be avoided in the remainder of this book. Usually we give explicit expressions involving up to three or four indices. The reader is then expected to infer the necessary generalization, which is of the type (1.5), (1.6) if we work with tensors but is usually more complicated if we work with arbitrary arrays.

1.4 Examples

In this and in the following chapter, X and Y are random variables but when we work with log likelihood derivatives, it is more appropriate to contemplate transformation of the parameter vector and the terms covariant and contravariant then refer to parameter transformations and not to data transformations. To take a simple example, relevant to statistical theory, let $l(\theta; Z) = \log f_Z(Z; \theta)$ be the log likelihood function for $\theta = \theta^1, ..., \theta^p$ based on observations Z. The partial derivatives of l with respect to the components of θ may be written

$$U_r(\theta) = \partial l(\theta; Z)/\partial \theta^r$$
$$U_{rs}(\theta) = \partial^2 l(\theta; Z)/\partial \theta^r \partial \theta^s$$

and so on. The maximum likelihood estimate of θ satisfies $U_r(\hat{\theta}) = 0$ and the observed information for θ is $I_{rs} = -U_{rs}(\hat{\theta})$, with matrix inverse I^{rs}. Suppose now that we were to re-parameterize in terms of $\phi = \phi^1, ..., \phi^p$. If we denote by an asterisk derivatives with respect to ϕ, we have

$$\begin{aligned} U_r^* &= \theta_r^i U_i, & U_{rs}^* &= \theta_r^i \theta_s^j U_{ij} + \theta_{rs}^i U_i, \\ I_{rs}^* &= \theta_r^i \theta_s^j I_{ij}, & I^{*rs} &= \phi_i^r \phi_j^s I^{ij} \end{aligned} \tag{1.7}$$

where

$$\theta_r^i = \partial \theta^i/\partial \phi^r, \qquad \theta_{rs}^i = \partial^2 \theta^i/\partial \phi^r \partial \phi^s$$

and θ_r^i is assumed to have full rank with matrix inverse $\phi_i^r = \partial \phi^r/\partial \theta^i$. Arrays that transform like U_r, I_{rs} and I^{rs} are tensors, the first two being covariant of orders 1 and 2 respectively and the third being contravariant of order 2. The second derivative, $U_{rs}(\theta)$, is not a tensor on account of the presence of second derivatives with respect to θ in the above transformation law. Note also that the array U^*_{rs} cannot be obtained by transforming the array U_{rs} alone: it is necessary also to know the value of the array U_r. However $E\{U_{rs}(\theta); \theta\}$, the Fisher information for θ, is a tensor because the second term in U^*_{rs} has mean zero at the true θ.

To take a second example, closer in spirit to the material in the following two chapters, let $X = X^1, ..., X^p$ have mean vector $\kappa^i = E(X^i)$ and covariance matrix

$$\kappa^{i,j} = \text{cov}(X^i, X^j) = E(X^i X^j) - E(X^i)E(X^j).$$

Suppose we make an affine transformation to new variables $Y = Y^1, ..., Y^q$, where

$$Y^r = a^r + a^r_i X^i. \tag{1.8}$$

The mean vector and covariance matrix of Y are easily seen to be

$$a^r + a^r_i \kappa^i \quad \text{and} \quad a^r_i a^s_j \kappa^{i,j}$$

where $a^r_i = \partial Y^r / \partial X^i$. Thus, even though the transformation may not be invertible, the covariance array transforms like a contravariant tensor. Arrays that transform in this manner, but only under linear or affine transformation of X, are sometimes called *Cartesian tensors* (Jeffreys, 1952). Such transformations are of special interest because $a^r_i = \partial Y^r / \partial X^i$ does not depend on X. It will be shown that cumulants of order two or more are not tensors in the sense usually understood in differential geometry, but they do behave as tensors under the general affine group (1.8). Under non-linear transformation of X, the cumulants transform in a more complicated way as discussed in Section 3.4.

Tensors whose components are unaffected by coordinate transformation are called *isotropic*. This terminology is most commonly used in Mechanics and in the physics of fluids, where all three coordinate axes are measured in the same units. In these contexts, the two groups of most relevance are the orthogonal group, O, and the orthogonal group with positive determinant, O^+. In either case, δ^i_j, δ_{ij} and δ^{ij} are isotropic tensors. There is exactly one isotropic third-order tensor under O^+ (Exercise 1.22). However, this tensor, called the *alternating tensor*, is anti-symmetrical and does not occur in the remainder of this book. All fourth-order isotropic tensors are functions of the three second-order isotropic tensors (Jeffreys, 1952, Chapter 7). The only symmetrical isotropic fourth-order tensors are

$$\delta_{ij}\delta_{kl} + \delta_{ik}\delta_{jl} + \delta_{il}\delta_{jk}, \quad \delta^{ij}\delta^{kl} + \delta^{ik}\delta^{jl} + \delta^{il}\delta^{jk},$$

$$\delta^{ij}\delta_{kl} \quad \text{and} \quad \delta^i_k\delta^j_l + \delta^i_l\delta^j_k,$$

(Thomas, 1965, Section 7). Isotropic tensors play an important role in physics (see Exercise 1.21) but only a minor role in statistics.

1.5   Elementary matrix theory

For later use, we state here without detailed proof, some elementary matrix-theory results using tensor notation and terminology. Our main interest lies in matrix inverses and spectral decompositions or eigenvalue decompositions of real symmetric matrices. We consider first the tensorial properties of generalized inverse matrices.

1.5.1   *Generalized inverse matrices*

Let ω_{ij} be a symmetric covariant tensor written as Ω using matrix notation. A generalized inverse of Ω is any matrix Ω^- satisfying

$$\Omega\Omega^-\Omega = \Omega, \tag{1.9}$$

implying that Ω^- is the identity on the range of Ω. It follows that $\text{rank}(\Omega^-) \geq \text{rank}(\Omega)$ and also that $\text{rank}(\Omega\Omega^-) \geq \text{rank}(\Omega)$ implying $\text{rank}(\Omega\Omega^-) = \text{rank}(\Omega)$. Post-multiplication of (1.9) by Ω^- gives $(\Omega\Omega^-)(\Omega\Omega^-) = \Omega\Omega^-$, implying that $\Omega\Omega^-$ is idempotent and hence that

$$\text{tr}(\Omega\Omega^-) = \text{rank}(\Omega\Omega^-) = \text{rank}(\Omega). \tag{1.10}$$

This conclusion is independent of the choice of generalized inverse. In addition, if Ω^- is a generalized inverse of Ω and if $\mathbf{A}$ is any full-rank matrix, then $\mathbf{A}\Omega^-\mathbf{A}$ is a generalized inverse of $\mathbf{A}^{-1}\Omega\mathbf{A}^{-1}$. The last result follows directly from the definition (1.9).

Reverting now to tensor notation, we write the generalized matrix inverse of ω_{ij} as ω^{ij}. The inverse can in fact always be chosen to be symmetric, so that this notation does not conflict with our convention regarding symmetry. It follows from (1.10) that, whatever the choice of inverse,

$$\text{rank}(\omega_{ij}) = \omega_{ij}\omega^{ij}.$$

In addition, if ω^{ij} is any generalized inverse of ω_{ij}, then $a_i^r a_j^s \omega^{ij}$ is a generalized inverse of $b_r^i b_s^j \omega_{ij}$, where a_r^i is a full rank matrix with inverse b_i^r. In other words, ω^{ij} is a contravariant tensor.

These arguments are entirely independent of the choice of generalized inverse. Occasionally, however, it is convenient to choose a generalized inverse with the property that

$$\Omega^- \Omega \Omega^- = \Omega^- \qquad (1.11)$$

implying that $\text{rank}(\Omega^-) = \text{rank}(\Omega)$. In other words, Ω is a generalized inverse of Ω^-. The symmetry of the tensor formulae is greatly enhanced if the generalized inverse matrix is chosen to have the property (1.11) for then we need not distinguish between ω^{ij} as a generalized inverse matrix of ω_{ij} and ω_{ij} as a generalized inverse of ω^{ij}.

In fact, it is always possible to choose a generalized inverse with the properties (1.9) and (1.11) and having the additional symmetry property that

$$\Omega^- \Omega = (\Omega^- \Omega)^T, \quad \Omega \Omega^- = (\Omega \Omega^-)^T.$$

Such a generalized inverse is unique and is known as the Moore-Penrose inverse (Rao, 1973, Section 1b). See also Exercise 1.10.

Conditions (1.9), (1.11) are entirely natural whether Ω is the matrix representation of a symmetric covariant tensor, a symmetric contravariant tensor or an asymmetric (1,1) tensor. On the other hand, the symmetry conditions, as stated above, appear to be quite unnatural if Ω is the matrix representation of an asymmetric (1,1) tensor, but otherwise, the conditions seem sensible. The symmetry condition arises naturally if the usual Euclidean inner product with equal weights is used to determine orthogonality: see Kruskal (1975, Section 6). On the other hand, if a weighted inner product is appropriate, as it often is in statistical applications, then the symmetry condition would seem to be inappropriate. See, for example, Exercise (1.11).

1.5.2 Spectral decomposition

Any real symmetric covariant tensor ω_{ij} may be written in the form

$$\omega_{ij} = \sigma_i^r \sigma_j^s \Lambda_{rs} \qquad (1.12)$$

where $\Lambda_{rr} = \lambda_r$, a real number, $\Lambda_{rs} = 0$ for $r \neq s$ and σ_i^r is a real orthogonal matrix satisfying

$$\sigma_i^r \sigma_j^s \delta_{rs} = \delta_{ij}.$$

The values $\lambda_1, ..., \lambda_p$ are known as the eigenvalues of ω_{ij} and (1.12) is known as the eigenvalue decomposition or spectral decomposition of ω. This decomposition implies that the quadratic form $Q = \omega_{ij} x^i x^j$ may be written as $Q = \sum_1^p \lambda_r (y^r)^2$ where $y^r = \sigma_i^r x^i$ is an orthogonal transformation of x. The set of eigenvalues is unique but evidently the representation (1.12) is not unique because we may at least permute the components of y. Further, if some of the eigenvalues are equal, say $\lambda_1 = \lambda_2$, any orthogonal transformation of the components (y^1, y^2) satisfies (1.12).

Under orthogonal transformation of x, ω_{ij} transforms to $\bar{\omega}_{ij} = a_i^k a_j^l \omega_{kl}$ where a_i^k is orthogonal. The spectral decomposition (1.12) then becomes

$$\bar{\omega}_{ij} = (a_i^k \sigma_k^r)(a_j^k \sigma_l^s) \Lambda_{rs} \qquad (1.13)$$

where $a_i^k \sigma_k^r$ is an orthogonal matrix. On comparing (1.12) with (1.13) we see that the set of eigenvalues of ω_{ij} is invariant under orthogonal transformation of coordinates. The eigenvalues are not invariant under arbitrary nonsingular transformation because $a_i^k \sigma_k^r$ is not, in general, orthogonal unless a_i^k is orthogonal.

Consider now the alternative decomposition

$$\omega_{ij} = \tau_i^r \tau_j^s \epsilon_{rs} \qquad (1.14)$$

where $\epsilon_{rr} = \pm 1$ or zero, $\epsilon_{rs} = 0$ for $r \neq s$ and no constraints are imposed on τ_i^r other than that it should be real and have full rank. The existence of such a decomposition follows from (1.12). Again, the representation (1.14) is not unique because, if we write $y^r = \tau_i^r x^i$, then $Q = \omega_{ij} x^i x^j$ becomes

$$Q = \sum{}^+ (y^r)^2 - \sum{}^- (y^r)^2 \qquad (1.15)$$

where the first sum is over those y^r for which $\epsilon_{rr} = +1$ and the second sum is over those components for which $\epsilon_{rr} = -1$. Two orthogonal transformations, one for the components y^r for which $\epsilon_{rr} = +1$ and one for the components for which $\epsilon_{rr} = -1$, leave (1.15) unaffected. Furthermore, the components for which $\epsilon_{rr} = 0$ may be transformed linearly and all components may be permuted without affecting the values of ϵ_{rs} in (1.14). At most, the order of the diagonal elements, ϵ_{rr} can be changed by the transformations listed above.

Under linear transformation of x, (1.14) becomes

$$\bar{\omega}_{ij} = (a_i^k \tau_k^r)(a_j^l \tau_l^s)\epsilon_{rs},$$

so that the matrix rank

$$\omega_{ij}\omega^{ij} = \epsilon_{rs}\epsilon^{rs}$$

and signature,

$$\delta^{rs}\epsilon_{rs} = \sum \epsilon_{rr}$$

are invariant functions of the covariant tensor ω_{ij}. This result is known as Sylvester's law of inertia (Gantmacher, 1960, Chapter X; Cartan, 1981, Section 4).

The geometrical interpretation of Sylvester's law is that the equation

$$x^0 = Q = \omega_{ij}x^i x^j$$

describes a hypersurface of dimension p in R^{p+1} and the qualitative aspects of the shape of this surface that are invariant under linear transformation of $x^1, ..., x^p$ are the numbers of positive and negative principal curvatures. This makes good sense because the effect of such a transformation is to rotate the coordinates and to re-define distance on the surface. The surface, in a sense, remains intact. If we were to change the sign of x^0, the positive and negative curvatures would be reversed.

In the particular case where ω_{ij} is positive definite of full rank, the matrix τ_i^r in (1.14) is known as a matrix square root of ω. For this case, if σ_r^s is an orthogonal matrix, then $\tau_i^r \sigma_r^s$ is also a matrix square root of ω_{ij}. Subject to this choice of orthogonal transformation, the matrix square root is unique.

1.6 Invariants

An invariant is a function whose value is unaffected by transformations within a specified class or group. To take a simple example, let ω_i^r be a mixed tensor whose value under linear transformation becomes

$$\bar\omega_i^r = a_s^r \omega_j^s b_i^j,$$

where $a_s^r b_i^s = \delta_i^r$. In matrix notation, $\bar\Omega = \mathbf{A}\Omega\mathbf{A}^{-1}$ is known as a *similarity* transformation or *unitary* transformation, (Dirac, 1958, Chapter 26). In Cartan's terminology (Cartan, 1981, Section 41), the matrices Ω and $\bar\Omega$ are said to be *equivalent*. It is an elementary exercise to show that $\mathrm{tr}(\bar\Omega) = \mathrm{tr}(\Omega)$, so that the sum of the eigenvalues of a real $(1,1)$ tensor is (a) real and (b) invariant. The same is true of any symmetric polynomial function of the eigenvalues. In particular, the determinant is invariant. For an interpretation, see Exercise 1.13.

The second example is closer in spirit to the material in the following chapters in the sense that it involves random variables in an explicit way. Let $\kappa^{i,j}$ be the covariance matrix of the components of X and consider the effect on $\kappa^{i,j}$ of making an orthogonal transformation of X. By (1.13), the set of eigenvalues is unaffected. Thus the set of eigenvalues is an invariant of a symmetric contravariant tensor, but only within the orthogonal group. Only the rank and signature are invariant under nonsingular linear or affine transformation. Other examples of invariant functions of the cumulants are given in Section 2.8.

These examples pinpoint one serious weakness of matrix notation, namely that no notational distinction is made between a $(1,1)$ tensor whose eigenvalues are invariant under the full linear group, and a $(2,0)$ tensor, invariably symmetric, wwhose eigenvalues are invariant only under the smaller orthogonal group.

The log likelihood function itself is invariant under arbitrary smooth parameter transformation, not necessarily linear. Under nonsingular transformation of the data, the log likelihood is not invariant but transforms to $l(\theta;z)+c(z)$ where $c(z)$ is the log determinant of the Jacobian of the transformation. However $l(\hat\theta;z)-l(\theta;z)$, the maximized log likelihood ratio statistic, is invariant under transformation of the data. Some authors define the log likelihood as the equivalence class of all functions that differ from $l(\theta;z)$ by a function of z and, in this sense, the log likelihood function is an invariant.

To test the simple hypothesis that θ takes on some specified value, say θ_0, it is desirable in principle to use an invariant test statistic because consistency of the observed data with the hypothesized value θ_0 is independent of the coordinate system used to describe the null hypothesis value. Examples of invariant test statistics include the likelihood ratio statistic, $l(\hat{\theta}; z) - l(\theta_0; z)$, and the quadratic score statistic, $U_r U_s i^{rs}$, where i^{rs} is the matrix inverse of $-E\{U_{rs}; \theta\}$.

One of the main reasons for working with tensors, as opposed to arbitrary arrays of functions, is that it is easy to recognize and construct invariants. For example, any scalar derived from tensors by the process of contraction is automatically an invariant. This is a consequence of the tensor transformation properties (1.5) and (1.6). If the arrays are tensors only in some restricted sense, say under linear transformation only, then any derived scalars are invariants in the same restricted sense.

By way of illustration, consider the group of orthogonal transformations and suppose that the array ω^{ij} satisfies the transformation laws of a tensor under orthogonal transformation of X. Since we are dealing with orthogonal transformations only, and not arbitrary linear transformations, it follows that δ^{ij} and δ_{ij} are tensors. This follows from

$$a_i^r a_j^s \delta^{ij} = \delta^{rs}$$

where a_i^r is any orthogonal matrix. Note, however, that δ^{ijk} and δ_{ijk} are not tensors (Exercise 1.19). Hence the scalars

$$\omega^{ij}\delta_{ij}, \quad \omega^{ij}\omega^{kl}\delta_{ik}\delta_{jl}, \dots$$

are invariants. In terms of the eigenvalues of ω^{ij}, these scalars may be written as power sums,

$$\sum_j \lambda_j, \quad \sum_j \lambda_j^2, \dots$$

Another function invariant under nonsingular linear transformation is the matrix rank, which may be written $\omega^{ij}\omega_{ij}$, where ω_{ij} is any generalized inverse of ω^{ij}. However not every invariant function can easily be derived by means of tensor-like manipulations of the type described here. For example, Sylvester's law of inertia states that the numbers of positive and negative eigenvalues of a real symmetric matrix are invariant under nonsingular linear transformation. In other words, if a_i^r is a nonsingular matrix then the sign pattern of the eigenvalues of ω^{ij} is the same as the sign pattern of the eigenvalues of $a_i^r a_j^s \omega^{ij}$. There appears to be no simple way to deduce this result from tensor-like manipulations alone. See, however, Exercise 1.9 where the signature is derived by introducing an auxiliary tensor ω_{ij}^+ and forming the invariant $\omega^{ij} \omega_{ij}^+$.

1.7 Direct product spaces

1.7.1 *Kronecker product*

Suppose that $Y^1, ..., Y^n$ are independent and identically distributed vector-valued random variables, each having p components. Should we so wish, we can regard $\mathbf{Y} = (Y^1, ..., Y^n)$ as a point in R^{np}, but usually it is preferable to consider $\mathbf{Y}$ explicitly as an element of the direct product space $R^n \times R^p$. One advantage of this construction is that we can then require derived statistics to be invariant under one group of transformations acting on R^n and to be tensors under a different group acting on R^p. For example, if $\kappa^{r,s}$ is the $p \times p$ covariance matrix of Y^1, then $\kappa^{r,s}$ is a tensor under the action of the affine group on R^p. Any estimate $k^{r,s}$, say, ought to have the same property. Furthermore, since the joint distribution of $Y^1, ..., Y^n$ is unaffected by permuting the n vectors, $k^{r,s}$ ought to be invariant under the action of the symmetric group (of permutations) on R^n.

Using tensor notation, the covariance matrix of $\mathbf{Y}$ may be written as $\delta^{ij} \kappa^{r,s}$, where indices i and j run from 1 to n while indices r and s run from 1 to p. Both δ^{ij} and $\kappa^{r,s}$ are tensors, but under different groups acting on different spaces. The Kronecker product, $\delta^{ij} \kappa^{r,s}$, is a tensor under the direct product group acting on $R^n \times R^p$.

More generally, if $\omega^{ij} \kappa^{r,s}$ is a tensor under the direct product of two groups acting on R^n and R^p, it may be necessary to compute the matrix inverse or generalized inverse in terms of ω_{ij} and $\kappa_{r,s}$, the generalized inverses on R^n and R^p respectively. It is immediately apparent that $\omega_{ij} \kappa_{r,s}$ is the required generalized inverse and that the inverse is covariant in the sense implied by the positions of the indices.

In matrix notation, the symbol $\otimes$ is usually employed to denote the Kronecker product. However, $\mathbf{A} \otimes \mathbf{B}$ is not the same matrix as $\mathbf{B} \otimes \mathbf{A}$ on account of the different arrangement of terms. No such difficulties arise with index notation because multiplication of real or complex numbers is a commutative operation.

Other kinds of direct products such as the Hadamard product (Rao, 1973, p. 30) do not arise in tensor calculations.

1.7.2 Factorial design and Yates's algorithm

Suppose that one observation is taken at each combination of the levels of factors A, B and C. Denote by Y^{ijk}, the yield or response recorded with A at level i, B at level j and C at level k. We make no claim that Y is, in any useful sense, a tensor. In fact, we occasionally write Y_{ijk} in place of Y^{ijk} where convenient to do so. This is the conventional notation for factorial models, where the indices are ordered according to the factors, and the array is not symmetric under index permutation. Further, unless the factors have equal numbers of levels, the indices have unequal ranges. This application is included here more to illustrate the value of index notation and the summation convention than as an example of a tensor.

Corresponding to each factor, we introduce *contrast matrices*, a_r^i, b_s^j, c_t^k, where the letters i,j,k refer to factor levels and the letters r,s,t refer to factor contrasts. The term 'contrast' is misused here because usually the first column of the contrast matrix is not a contrast at all, but a column of unit values. The remaining columns typically sum to zero and are therefore contrasts in the usual sense. The contrast matrices have full rank and are usually chosen to be orthogonal in the sense that

$$a_r^i a_{r'}^{i'} \delta_{ii'} = a_{rr'} = 0 \quad \text{if } r \neq r'$$
$$b_s^j b_{s'}^{j'} \delta_{jj'} = b_{ss'} = 0 \quad \text{if } s \neq s'$$
$$c_t^k c_{t'}^{k'} \delta_{kk'} = c_{tt'} = 0 \quad \text{if } t \neq t'$$

For example, if A has two levels, B has three ordered levels and C has four ordered levels, it is customary to make use of the orthogonal polynomial contrasts

$$a_r^i = \begin{matrix} 1 & -1 \\ 1 & 1 \end{matrix} \qquad b_s^j = \begin{matrix} 1 & -1 & 1 \\ 1 & 0 & -2 \\ 1 & 1 & 1 \end{matrix} \qquad c_t^k = \begin{matrix} 1 & -3 & 1 & -1 \\ 1 & -1 & -1 & 3 \\ 1 & 1 & -1 & -3 \\ 1 & 3 & 1 & 1 \end{matrix}$$

Hence, the inner products give

$$a_{rr'} = \mathrm{diag}\{2,2\}, \quad b_{ss'} = \mathrm{diag}\{3,2,6\}, \quad c_{tt'} = \mathrm{diag}\{4,20,4,20\}.$$

By convention, contrast matrices are arranged so that rows refer to factor levels and columns to factor contrasts.

Whatever the contrasts chosen, the design matrix $\mathbf{X}$ corresponding to the factorial model $A * B * C$ is just the Kronecker product

$$x_{rst}^{ijk} = a_r^i b_s^j c_t^k.$$

In other symbols, the saturated factorial model is just

$$E(Y^{ijk}) = x_{rst}^{ijk} \beta^{rst} = a_r^i b_s^j c_t^k \beta^{rst},$$

where β^{rst} is the 'interaction' of contrast r of factor A with contrast s of factor B and contrast t of factor C. For instance, with the contrast matrices given above, β^{111} is just the mean and β^{132} is written conventionally as $B_Q C_L$. In other words, β^{132} is a measure of the change in the quadratic effect of B per unit increase in the level of C.

The so-called 'raw' or unstandardized contrasts that are produced by the three steps of Yates's algorithm, are given by

$$b_{rst} = a_r^i b_s^j c_t^k Y_{ijk}. \qquad (1.16)$$

The linear combinations that are implied by the above expression are exactly those that arise when Yates's algorithm is performed in the conventional way.

To derive the least squares estimates of the parameters, we raise the indices of b, using the expression

$$\hat{\beta}^{rst} = a^{rr'} b^{ss'} c^{tt'} b_{r's't'}.$$

If the contrast matrices are each orthogonal, this expression reduces to

$$\hat{\beta}^{rst} = b_{rst}/(a_{rr}b_{ss}c_{tt}),$$

with variance $\sigma^2/(a_{rr}b_{ss}c_{tt})$, no summation intended.

The extension to an arbitrary number of factors, each having an arbitrary number of levels, is immediate. For further discussion, see Good (1958, 1960) or Takemura (1983).

From the numerical analytic point of view, the number of computations involved in (1.16) is considerably fewer than what would be required to solve n linear equations if the factorial structure of $\mathbf{X}$ were not utilized. For example, with k factors each at two levels, giving $n = 2^k$ observations, (1.16) requires nk additions and subtractions as opposed to $O(n^2)$ operations if the factorial structure were ignored. Herein lies the appeal of Yates's algorithm and also the fast Fourier transform, which uses the same device.

1.8 Bibliographic notes

Tensor notation is used widely in applied mathematics, mathematical physics and differential geometry. Definitions and notation vary to some extent with the context. For example, Jeffreys (1952) and Jeffreys & Jeffreys (1956) are concerned only with the effect on equations of motion of rotating the frame of reference or axes. Consequently, their definition of what they call a *Cartesian tensor* refers only to the orthogonal group and not, in general to arbitrary linear or non-linear transformation of coordinates. Their notation differs from that used here, most noticeably through the absence of superscripts.

Other useful references, again with a bias towards applications to physics, include McConnell (1931) and Lawden (1968). Thomas (1965, Section 6) emphasises the importance of transformation groups in the definition of a tensor.

For more recent work, again connected mainly with mathematical physics, see Richtmyer (1981).

In the theory of differential geometry, which is concerned with describing the local behaviour of curves and surfaces in space, notions of curvature and torsion are required that are independent of the choice of coordinate system on the surface. This requirement leads naturally to the notion of a tensor under the group of arbitrary invertible parameterizations of the surface. Gaussian and Riemannian curvature as well as mean curvature are invariants derived from such tensors. This work has a long history going back to Levi-Cività, Ricci, Riemann and Gauss's celebrated *theorem egregium*. Details can be found in the books by Eisenhart (1926), Weatherburn (1950), Sokolnikoff (1951) and Stoker (1969). For a more recent treatment of Riemannian geometry, see Willmore (1982) or Spivak (1970).

For a discussion of the geometry of generalized inverses, see Kruskal (1975) and the references therein.

The notion of a *spinor* is connected with rotations in Euclidean space and has applications in the theory of special relativity. Inevitably, there are strong similarities with quaternions, which are also useful for studying rotations. Tensors arise naturally in the study of such objects. See, for example, Cartan (1981).

Despite the widespread use in statistics of multiply-indexed arrays, for example, in the study of factorial, fractional factorial and other designs, explicit use of tensor methods is rare, at least up until the past few years. For an exception, see Takemura (1983). The reasons for this neglect are unclear: matrix notation abounds and is extraordinarily convenient provided that we do not venture far beyond linear models and second-moment assumptions. In any case, the defects and shortcomings of matrix notation become clearly apparent as soon as we depart from linear models or need to study moments beyond the second. For example, many of the quantities that arise in later chapters of this book cannot be expressed using matrix notation. This is the realm of the tensor, and it is our aim in the remainder of this book to demonstrate to the reader that great simplification can result from judicious choice of notation.

1.9   Further results and exercises 1

1.1   Derive the transformation laws (1.7) for log likelihood derivatives.

1.2   Show that if ω^{ijk} is a contravariant tensor and ω_{ijk} is a covariant tensor, then $\omega^{ijk}\omega_{ijk}$ is an invariant.

1.3 Show directly, using the notation in (1.7), that $U_r U_s I^{rs}$ is invariant under the group of invertible transformations acting on the parameter space.

1.4 Let $i_{rs} = -E\{U_{rs}; \theta\}$ and let i^{rs} be the matrix inverse. Under which group of transformations is $U_r U_s i^{rs}$ an invariant?

1.5 Show, using the notation in (1.7), that

$$V_{ij} = U_{ij} - \kappa_{ij,k} i^{kl} U_l$$

is a covariant tensor, where $\kappa_{rs,t}$ is the covariance of U_{rs} and U_t.

1.6 Let a_j^i be the elements of a square matrix, not necessarily symmetrical, and let its inverse, b_i^j satisfy $a_j^i b_k^j = \delta_k^i = a_k^j b_j^i$. Show that the derivatives satisfy

$$\frac{\partial b_i^j}{\partial a_s^r} = -b_r^j b_i^s$$

$$\frac{\partial a_i^j}{\partial b_s^r} = -a_r^j a_i^s$$

1.7 Show that the spectral decomposition of the the symmetric matrix $\mathbf{A}$

$$\mathbf{A} = \mathbf{Q} \Lambda \mathbf{Q}^T, \quad \mathbf{Q}\mathbf{Q}^T = \mathbf{I}, \quad \Lambda = \text{diag}\{\lambda_1, ..., \lambda_p\}$$

is unique up to permutations of the columns of $\mathbf{Q}$ and the elements of Λ if the eigenvalues of $\mathbf{A}$ are distinct.

1.8 Show that there exists a linear transformation $Y^r = a_i^r X^i$ from X to Y such that the quadratic form $\omega_{ij} X^i X^j$ may be written as $\epsilon_{rs} Y^r Y^s$ where $\epsilon_{ii} = \pm 1$ and $\epsilon_{ij} = 0$ if $i \neq j$.

1.9 Consider the decomposition $\omega_{ij} = \tau_i^r \tau_j^s \epsilon_{rs}$ of the symmetric covariant tensor ω_{ij}, where the notation is that used in (1.14). Define

$$\omega_{ij}^+ = \tau_i^r \tau_j^s |\epsilon_{rs}|.$$

Show that ω_{ij}^+ is a covariant tensor and that the scalar

$$s = \omega^{ij} \omega_{ij}^+$$

is independent of the choice of τ_i^r and also of the choice of generalized inverse ω^{ij}. Show that s is the signature of ω_{ij}.

1.10 In the notation of the previous exercise, let

$$\omega^{ij} = \gamma_r^i \gamma_s^j \epsilon^{rs},$$

where $\epsilon^{rs} = \epsilon_{rs}$ and $\gamma_r^i \tau_j^r = \delta_j^i$. Show that ω^{ij} is the Moore-Penrose inverse of ω_{ij}.

1.11 Show that the identity matrix is a generalized inverse of any projection matrix. Show also that a projection matrix, not necessarily symmetric, is its own generalized inverse satisfying (1.9) and (1.11). Under what conditions is a projection matrix self-inverse in the Moore-Penrose sense?

1.12 Let ω^{ij}, with inverse ω_{ij}, be the components of a $p \times p$ symmetric matrix of rank p. Show that

$$\gamma^{rs,ij} = \omega^{ri}\omega^{sj} + \omega^{rj}\omega^{si},$$

regarded as a $p^2 \times p^2$ matrix with rows indexed by (r, s) and columns by (i, j), is symmetric with rank $p(p+1)/2$. Show also that $\omega_{ri}\omega_{sj}/2$ is a generalized inverse. Find the Moore-Penrose generalized inverse.

1.13 Consider the linear mapping from R^p to itself given by

$$\bar{X}^r = \omega_i^r X^i$$

where ω_i^r is nonsingular. Show that, under simultaneous change of coordinates

$$Y^r = a_i^r X^i, \quad \bar{Y}^r = a_i^r \bar{X}^i,$$

ω_i^r transforms as a mixed tensor. By comparing the volume of a set, B say, in the X coordinate system with the volume of the transformed set, $\bar{B}$, interpret the determinant of ω_i^r as an invariant. Give similar interpretations of the remaining $p - 1$ invariants, e.g. in terms of surface area and so on.

1.14 Let $\pi_1, ..., \pi_k$ be positive numbers adding to unity and define the multinomial covariance matrix

$$\omega_{ij} = \begin{cases} \pi_i(1 - \pi_i) & i = j \\ -\pi_i\pi_j & i \neq j \end{cases}.$$

Show that ω_{ij} has rank $k - 1$ and that

$$\omega^{ij} = \begin{cases} 1/\pi_i & i = j \\ 0 & \text{otherwise} \end{cases}$$

is a generalized inverse of rank k. Find the Moore-Penrose generalized inverse.

1.15 Let the $p \times q$ matrix $\mathbf{A}$, with components a^r_i, be considered as defining a linear transformation from the domain, R^q, to the range in R^p. Interpret the *singular values* of $\mathbf{A}$ as invariants under independent orthogonal transformation of the domain and range spaces. For the definition of singular values and their application in numerical linear algebra, see Chambers (1977, Section 5.e).

1.16 Let $\mathbf{A}$, $\mathbf{A}^{-1}$ and $\mathbf{X}$ be symmetric matrices with components a_{ij}, a^{ij} and x_{ij} respectively. Show that the Taylor expansion for the log determinant of $\mathbf{A} + \mathbf{X}$ about the origin may be written

$$\log \det(\mathbf{A} + \mathbf{X}) = \log \det(\mathbf{A}) + x_{ij}a^{ij} - x_{ij}x_{kl}a^{ik}a^{jl}/2$$
$$+ x_{ij}x_{kl}x_{mn}a^{ik}a^{jm}a^{ln}/3 + \dots$$

Describe the form of the general term in this expansion. Compare with Exercise 1.6 and generalize to asymmetric matrices.

1.17 Justify the claim that Kronecker's delta, δ^j_i, is a tensor.

1.18 Show that δ_{ij} is a tensor under the orthogonal group but not under any larger group.

1.19 Show that δ_{ijk}, δ_{ijkl},... are tensors under the *symmetric* group but not under any larger group.

[The symmetric group, which is most conveniently represented by the set of permutation matrices, arises naturally in the study of sample moments based on simple random samples, where the order in which the observations are recorded is assumed to be irrelevant. See Chapter 4.]

1.20 Show that the symmetric group is a subgroup of the orthogonal group, which, in turn, is a subgroup of the general linear group.

1.21 *Hooke's Law*: In the mechanics of deformable solids, the components of the *stress tensor*, p_{ij}, measure force per unit area in the following sense. Let $\mathbf{e}_i$ be the unit vector in the ith coordinate direction and let $\bar{\mathbf{e}}_i$ be the orthogonal plane. Then p_{ii} is the force per unit area normal to the plane, also called the *normal stress*, and p_{ij}, $j \neq i$ are the *shear stresses* acting in the plane. The components of the *strain tensor*, q_{ij}, which are dimensionless, measure percentage deformation or percentage change in length. Both arrays are symmetric tensors under the orthogonal group.

In the case of *elastic* deformation of an *isotropic* material, Hooke's law in its most general form states that the relationship between stress and strain is linear. Thus,

$$p_{rs} = b_{rs}^{ij} q_{ij},$$

where b_{rs}^{ij} is an *isotropic* fourth-order tensor given by

$$b_{rs}^{ij} = \lambda \delta^{ij} \delta_{rs} + 2\mu \delta_r^i \delta_s^j,$$

for constants λ, μ that are characteristic of the material.

Show that the inverse relationship giving the strains in terms of the stresses may be written in the form

$$q_{ij} = \left(\lambda' \delta^{rs} \delta_{ij} + 2\mu' \delta_i^r \delta_j^s \right) p_{rs},$$

where the new constants are given by

$$\mu' = \frac{1}{4\mu}, \qquad \lambda' + 2\mu' = \frac{\lambda + \mu}{\mu(3\lambda + 2\mu)} = E^{-1}.$$

In the terminology used in Mechanics, E is known as *Young's modulus* or *modulus of elasticity*, μ is called the *rigidity* or *shear modulus* and $\sigma = \lambda/\{2(\lambda + \mu)\}$ is *Poisson's ratio*. Note that $E = 2(1 + \sigma)\mu$, implying that two independent constants entirely determine the three-dimensional elastic properties of the material. (Murnaghan, 1951, Chapters 3,4; Jeffreys & Jeffreys 1956, Section 3.10; Drucker, 1967, Chapter 12).

1.22 The array ϵ_{ijk} of order $3 \times 3 \times 3$ defined by

$$\epsilon_{123} = \epsilon_{231} = \epsilon_{312} = 1$$
$$\epsilon_{213} = \epsilon_{132} = \epsilon_{321} = -1$$
$$\epsilon_{ijk} = 0 \quad \text{otherwise,}$$

is known as the *alternating tensor* (Ames & Murnaghan, 1929, p. 440). For any 3×3 matrix a_r^i, show that

$$\epsilon_{ijk} a_r^i a_s^j a_t^k = \epsilon_{rst} \det(\mathbf{A}).$$

Hence show that ϵ_{ijk} is an isotropic tensor under O^+, the orthogonal group with positive determinant (Jeffreys & Jeffreys, 1956, Sections 2.07, 3.03).

Write down the generalization of the alternating tensor appropriate for a $p \times p \times p$ array.

CHAPTER 2

Elementary theory of cumulants

2.1 Introduction

This chapter deals with the elementary theory of cumulants in the multivariate case as well as the univariate case. Little prior knowledge is assumed other than some familiarity with moments and the notion of mathematical expectation, at least in the univariate case. In what follows, all integrals and infinite sums are assumed to be convergent unless otherwise stated. If the random variable X has a density function $f_X(x)$ defined over $-\infty < x < \infty$, then the expectation of the function $g(X)$ is just

$$E\{g(X)\} = \int_{-\infty}^{\infty} g(x) f_X(x) dx. \qquad (2.1)$$

If the distribution of X is discrete, the integral is replaced by a sum over the discrete values. More generally, to take care of both the discrete case and the continuous case simultaneously, we may replace $f_X(x)dx$ in the above integral by $dF_X(x)$, where F_X is a probability measure. Such differences, however, need not concern us here. All that is required is a knowledge of some elementary properties of the integral and the expectation operator. In the particular case where $g(X)$ is the rth power of X, (2.1) is called the rth *moment* of X.

In the multivariate case, essentially the same definitions may be used except that the integral over R^1 is replaced by an integral over R^p. Not only do we have to consider the moments of each component of X but also the cross moments such as $E(X^1 X^2)$ and $E(X^1 X^1 X^2)$. As always, in the multivariate case, superscripts denote components and not powers. The moments of a given component of X give information regarding the marginal distribution of that component. The cross moments are required to give information concerning the joint distribution of the components.

Cumulants are normally introduced as functions of the moments. It is entirely natural to inquire at the outset why it is preferable to work with cumulants rather than moments since the two are entirely equivalent. A single totally convincing answer to this query is difficult to find and, in a sense, Chapters 2 to 6 provide several answers. Simplicity seems to be the main criterion as the following brief list shows.

 (i) Most statistical calculations using cumulants are simpler than the corresponding calculations using moments.
(ii) For independent random variables, the cumulants of a sum are the sums of the cumulants.
(iii) For independent random variables, the cross cumulants or mixed cumulants are zero.
(iv) Edgeworth series used for approximations to distributions are most conveniently expressed using cumulants.
 (v) Where approximate normality is involved, higher-order cumulants can usually be neglected but not higher-order moments.

2.2   Generating functions

2.2.1   *Definitions*

As always, we begin with the random variable X whose components are $X^1, ..., X^p$. All arrays bearing superscripts refer to these components. Unless otherwise specified, the moments of X about the origin are assumed finite and are denoted by

$$\kappa^i = E(X^i), \quad \kappa^{ij} = E(X^i X^j),$$
$$\kappa^{ijk} = E(X^i X^j X^k) \tag{2.2}$$

and so on. Moments about the mean, also called central moments, are rarely considered explicitly here, except as a special case of the above with $\kappa^i = 0$. The indices need not take on distinct values and there may, of course, be more than p indices, implying repetitions. Thus, for example, κ^{11} is the mean square of X^1 about the origin and κ^{222} is the mean cube of X^2, the second component of X. In this context, superscripts must not be confused with powers and powers should, where possible, be avoided. In this book, powers are avoided for the most part, the principal exceptions arising in Sections 2.5 and 2.6 where the connection with other notations is described and interpretations are given.

Consider now the infinite series

$$M_X(\xi) = 1 + \xi_i \kappa^i + \xi_i \xi_j \kappa^{ij}/2! + \xi_i \xi_j \xi_k \kappa^{ijk}/3!$$
$$+ \xi_i \xi_j \xi_k \xi_l \kappa^{ijkl}/4! + ..., \tag{2.3}$$

which we assume to be convergent for all $|\xi|$ sufficiently small. The sum may be written

$$M_X(\xi) = E\{\exp(\xi_i X^i)\}$$

and the moments are just the partial derivatives of $M_X(\xi)$ evaluated at $\xi = 0$.

The cumulants are most easily defined via their generating function,

$$K_X(\xi) = \log M_X(\xi),$$

which has an expansion

$$K_X(\xi) = \xi_i \kappa^i + \xi_i \xi_j \kappa^{i,j}/2! + \xi_i \xi_j \xi_k \kappa^{i,j,k}/3!$$
$$+ \xi_i \xi_j \xi_k \xi_l \kappa^{i,j,k,l}/4! + ... \tag{2.4}$$

This expansion implicitly defines all the cumulants, here denoted by κ^i, $\kappa^{i,j}$, $\kappa^{i,j,k}$ and so on, in terms of the corresponding moments. The major departure from standard statistical notation is that we have chosen to use the same letter for both moments and cumulants. Both are indexed by a set rather than by a vector, the only distinction arising from the commas, which are considered as separators for the cumulant indices. Thus the cumulants are indexed by a set of indices fully partitioned and the moments by the same set unpartitioned. One curious aspect of this convention is that the notation does not distinguish between moments with one index and cumulants with one index. This is convenient because the first moments and first cumulants are identical.

The infinite series expansion (2.3) for $M_X(\xi)$ may be divergent for all real $|\xi| > 0$ either because some of the higher-order moments are infinite or because the moments, though finite, increase sufficiently rapidly to force divergence (Exercise 2.2). In such cases, it is entirely legitimate to work with the finite series expansions up to any specified number of terms. This device can be justified by taking ξ to be purely imaginary, in which case, the integral $E\{\exp(\xi_i X^i)\}$ is convergent and its Taylor approximation for small imaginary ξ is just the truncated expansion. Similarly for $K_X(\xi)$. Thus cumulants of any order are well defined if the corresponding moment and all lower-order marginal moments are finite: see (2.9) and Exercise 2.1. One difficulty that arises in using moment or cumulant calculations to prove results of a general nature is that the infinite set of moments is, in general, not sufficient to determine the joint distribution uniquely. Feller (1971, Section VII.3) gives a pair of non-identical univariate density functions having identical moments of all orders. Non-uniqueness occurs only when the function $M_X(\xi)$ is not analytic at the origin. Thus, for a large class of problems, non-uniqueness can be avoided by including the condition that the series expansion for $M_X(\xi)$ be convergent for $|\xi| < \delta$ where $\delta > 0$ (Moran, 1968, Section 6.4; Billingsley, 1985, Exercise 30.5). In the univariate case, other conditions limiting the rate of increase of the even moments are given by Feller (1971, Sections VII.3, VII.6 and XV.4).

2.2.2 Some examples

Multivariate normal distribution: The multivariate normal density with mean vector κ^i and covariance matrix $\kappa^{i,j}$ may be written

$$(2\pi)^{-p/2}|\kappa^{i,j}|^{-1/2}\exp\{-\tfrac{1}{2}(x^i - \kappa^i)(x^j - \kappa^j)\kappa_{i,j}\},$$

where $\kappa_{i,j}$ is the matrix inverse of $\kappa^{i,j}$ and x ranges over R^p. The moment generating function may be found by completing the square in the exponent and using the fact that the density integrates to 1. This gives

$$M_X(\xi) = \exp\{\xi_i\kappa^i + \tfrac{1}{2}\xi_i\xi_j\kappa^{i,j}\}$$

and

$$K_X(\xi) = \xi_i\kappa^i + \tfrac{1}{2}\xi_i\xi_j\kappa^{i,j}.$$

In other words, for the normal distribution, all cumulants of order three or more are zero and, as we might expect, the second cumulant is just the covariance array.

Multinomial distribution: For a second multivariate example, we take the multinomial distribution on k categories with index m

and parameter $\pi = \pi_1, ..., \pi_k$. The joint distribution or probability function may be written

$$\operatorname{pr}(X^1 = x^1, ..., X^k = x^k) = \binom{m}{x^1, ..., x^k} \prod_{j=1}^{k} \pi_j^{x^j}$$

where $0 \le x^j \le m$ and $\sum_j x^j = m$. The moment generating function may be found directly using the multinomial theorem, giving

$$M_X(\xi) = \sum_x \binom{m}{x^1, ..., x^k} \prod_{j=1}^{j=k} \exp(\xi_j x^j) \pi_j^{x^j} = \{\sum \pi_j \exp(\xi_j)\}^m.$$

The cumulant generating function is

$$K_X(\xi) = m \log\{\sum \pi_j \exp(\xi_j)\}.$$

Thus all cumulants are finite and have the form $m \times$ (function of π). The first four are given in Exercise 2.16.

Student's distribution: Our third example involves a univariate distribution whose moments of order three and higher are infinite. The t distribution on three degrees of freedom has density function

$$f_X(x) = \frac{2}{\pi\sqrt{3}(1 + x^2/3)^2} \qquad -\infty < x < \infty.$$

The moment generating function, $M_X(\xi) = E\{\exp(\xi X)\}$, diverges for all real $|\xi| > 0$ so that the function $M_X(\xi)$ is not defined or does not exist for real ξ. However, if we write $\xi = i\varsigma$ where ς is real, we find

$$M_X(i\varsigma) = \int_{-\infty}^{\infty} \frac{2 \exp(i\varsigma x)}{\pi\sqrt{3}(1 + x^2/3)^2} \, dx.$$

The integrand has poles of order 2 at $x = \pm i\sqrt{3}$ but is analytic elsewhere in the complex plane. If $\varsigma > 0$, the integral may be evaluated by deforming the contour into the positive complex half-plane, leaving the residue at $x = +i\sqrt{3}$. If $\varsigma < 0$, it is necessary to deform in the other direction, leaving the residue at $x = -i\sqrt{3}$. This procedure gives

$$M_X(i\varsigma) = \exp(-\sqrt{3}|\varsigma|)\{1 + \sqrt{3}|\varsigma|\}$$

$$K_X(i\varsigma) = -\sqrt{3}|\varsigma| + \log(1 + \sqrt{3}|\varsigma|)$$

$$= \begin{cases} -3\varsigma^2/2 + \sqrt{3}\varsigma^3 - \dots & \text{if } \varsigma > 0 \\ -3\varsigma^2/2 - \sqrt{3}\varsigma^3 - \dots & \text{if } \varsigma < 0 \end{cases}$$

Thus $K_X(i\varsigma)$ has a unique Taylor expansion only as far as the quadratic term. It follows that $E(X) = 0$, $\text{var}(X) = 3$ and that the higher-order cumulants are not defined.

2.3 Cumulants and moments

To establish the relationships connecting moments with cumulants, we write $M_X(\xi) = \exp\{K_X(\xi)\}$ and expand to find

$$
\begin{aligned}
1 + \xi_i\kappa^i &+ \xi_i\xi_j(\kappa^{i,j}/2! + \kappa^i\kappa^j/2!) \\
&+ \xi_i\xi_j\xi_k\kappa^{i,j,k}/3! + \xi_i\xi_j\xi_k\xi_l\kappa^{i,j,k,l}/4! + \dots \\
&+ \xi_i\xi_j\xi_k\kappa^i\kappa^{j,k}/2! + \xi_i\xi_j\xi_k\xi_l\{\kappa^i\kappa^{j,k,l}/6 + \kappa^{i,j}\kappa^{k,l}/8\} + \dots \\
&+ \xi_i\xi_j\xi_k\kappa^i\kappa^j\kappa^k/3! + \xi_i\xi_j\xi_k\xi_l\kappa^i\kappa^j\kappa^{k,l}/4 + \dots \\
&+ \xi_i\xi_j\xi_k\xi_l\kappa^i\kappa^j\kappa^k\kappa^l/4! + \dots \\
&+ \dots
\end{aligned}
\tag{2.5}
$$

After combining terms and using symmetry, we find the following expressions for moments in terms of cumulants:

$$
\begin{aligned}
\kappa^{ij} &= \kappa^{i,j} + \kappa^i\kappa^j \\
\kappa^{ijk} &= \kappa^{i,j,k} + (\kappa^i\kappa^{j,k} + \kappa^j\kappa^{i,k} + \kappa^k\kappa^{i,j}) + \kappa^i\kappa^j\kappa^k \\
&= \kappa^{i,j,k} + \kappa^i\kappa^{j,k}[3] + \kappa^i\kappa^j\kappa^k \\
\kappa^{ijkl} &= \kappa^{i,j,k,l} + \kappa^i\kappa^{j,k,l}[4] + \kappa^{i,j}\kappa^{k,l}[3] + \kappa^i\kappa^j\kappa^{k,l}[6] \\
&\quad + \kappa^i\kappa^j\kappa^k\kappa^l,
\end{aligned}
\tag{2.6}
$$

where, for example,

$$\kappa^{i,j}\kappa^{k,l}[3] = \kappa^{i,j}\kappa^{k,l} + \kappa^{i,k}\kappa^{j,l} + \kappa^{i,l}\kappa^{j,k}$$

is the sum over the three 2^2 partitions of four indices. The bracket notation is simply a convenience to avoid listing explicitly all 15 partitions of four indices in the last equation (2.6). Only the five distinct types, each corresponding to a partition of the *number* 4, together with the number of partitions of each type, need be listed. The following is a complete list of the 15 partitions of four items, one column for each of the five types.

$$
\begin{array}{ccccc}
ijkl & i|jkl & ij|kl & i|j|kl & i|j|k|l \\
 & j|ikl & ik|jl & i|k|jl & \\
 & k|ijl & il|jk & i|l|jk & \\
 & l|ijk & & j|k|il & \\
 & & & j|l|ik & \\
 & & & k|l|ij & \\
\end{array}
$$

In the case of fifth-order cumulants, there are 52 partitions of seven different types and our notation makes the listing of such partitions feasible for sets containing not more than eight or nine items. Such lists are given in Tables 1,2 of the Appendix, and these may be used to find the expressions for moments in terms of cumulants. We find, for example, from the partitions of five items, that

$$
\begin{aligned}
\kappa^{ijklm} = {} & \kappa^{i,j,k,l,m} + \kappa^{i}\kappa^{j,k,l,m}[5] + \kappa^{i,j}\kappa^{k,l,m}[10] \\
& + \kappa^{i}\kappa^{j}\kappa^{k,l,m}[10] + \kappa^{i}\kappa^{j,k}\kappa^{l,m}[15] \\
& + \kappa^{i}\kappa^{j}\kappa^{k}\kappa^{l,m}[10] + \kappa^{i}\kappa^{j}\kappa^{k}\kappa^{l}\kappa^{m}.
\end{aligned}
$$

If $\kappa^{i} = 0$, all partitions having a unit part (a block containing only one element) can be ignored. The formulae then simplify to

$$
\begin{aligned}
\kappa^{ij} &= \kappa^{i,j}, \quad \kappa^{ijk} = \kappa^{i,j,k} \\
\kappa^{ijkl} &= \kappa^{i,j,k,l} + \kappa^{i,j}\kappa^{k,l}[3] \\
\kappa^{ijklm} &= \kappa^{i,j,k,l,m} + \kappa^{i,j}\kappa^{k,l,m}[10] \\
\kappa^{ijklmn} &= \kappa^{i,j,k,l,m,n} + \kappa^{i,j}\kappa^{k,l,m,n}[15] + \kappa^{i,j,k}\kappa^{l,m,n}[10] \\
& \quad + \kappa^{i,j}\kappa^{k,l}\kappa^{m,n}[15].
\end{aligned}
$$

These are the formulae for the central moments in terms of cumulants.

The reverse formulae giving cumulants in terms of moments may be found either by formal inversion of (2.6) or by expansion of $\log M_X(\xi)$ and combining terms. The expressions obtained for the first four cumulants are

$$\kappa^{i,j} = \kappa^{ij} - \kappa^i \kappa^j$$
$$\kappa^{i,j,k} = \kappa^{ijk} - \kappa^i \kappa^{jk}[3] + 2\kappa^i \kappa^j \kappa^k \qquad (2.7)$$
$$\kappa^{i,j,k,l} = \kappa^{ijkl} - \kappa^i \kappa^{jkl}[4] - \kappa^{ij} \kappa^{kl}[3] + 2\kappa^i \kappa^j \kappa^{kl}[6] - 6\kappa^i \kappa^j \kappa^k \kappa^l.$$

Again, the sum is over all partitions of the indices but this time, the coefficient $(-1)^{\nu-1}(\nu-1)!$ appears, where ν is the number of blocks of the partition. The higher-order formulae follow the same pattern and the list of partitions in Tables 1,2 of the Appendix may be used for cumulants up to order eight.

More generally, the relationships between moments and cumulants may be written as follows. Let $\Upsilon = \{v_1, ..., v_\nu\}$ be a partition of a set of p indices into ν non-empty blocks. (Υ and v are the upper and lower cases of the Greek letter 'upsilon'). For example, if $p = 4$ and $\nu = 2$ we might have $\Upsilon = \{(i,j),(k,l)\}$, $\{(i,k),(j,l)\}$ or $\{(i),(j,k,l)\}$ or any one of the four other possible partitions into 2 blocks. The partition comprising just a single block is denoted by Υ_1 and the partition comprising p unit blocks is denoted by Υ_p. There is no standard statistical notation for these partitions and it is sometimes convenient to use the alternatives, 1 and 0, borrowed from lattice theory: see Section 3.6. The moment involving all p indices is written $\kappa(\Upsilon_1)$ and the corresponding cumulant $\kappa(\Upsilon_p)$. In the above example, $\Upsilon_1 = \{(i,j,k,l)\}$ is the partition into one block, $\kappa(\Upsilon_1) = \kappa^{ijkl}$ is the corresponding moment, $\Upsilon_4 = \{(i),(j),(k),(l)\}$ is the partition into 4 blocks and $\kappa(\Upsilon_4) = \kappa^{i,j,k,l}$ is the corresponding cumulant. Cumulants involving only those indices in block v_j are written $\kappa(v_j)$ and the corresponding moment is written $\mu(v_j)$.

By examining the general term in expansion (2.5) for $M_X(\xi)$, it is not difficult to see that the expression for the moment $\kappa(\Upsilon_1)$ in terms of cumulants may be written

$$\kappa(\Upsilon_1) = \sum_\Upsilon \kappa(v_1)...\kappa(v_\nu) \qquad (2.8)$$

where the sum extends over all partitions of the indices. Equivalently, the above may be expressed as a double sum, first over ν and then over all partitions of the indices into ν blocks.

The corresponding expression for the cumulant $\kappa(\Upsilon_p)$ in terms of the moments is

$$\kappa(\Upsilon_p) = \sum_{\Upsilon}(-1)^{\nu-1}(\nu-1)!\mu(v_1)...\mu(v_\nu). \qquad (2.9)$$

Note that the block sizes do not enter into either of the above expressions.

In fact, we may take (2.9) as an alternative definition of cumulant, more directly applicable than the definition relying on generating functions. The advantage of (2.9) as a definition is that it makes explicit the claim made at the end of Section 2.2, that cumulants of order r are well defined when the corresponding rth-order moment and all lower-order marginal moments are finite. Note that, apart from the univariate case, $\kappa^{ijk} < \infty$ does not imply that $\kappa^{i,j,k} < \infty$: see Exercise 2.1.

2.4   Linear and affine transformations

The objective in this section is to examine how the moment arrays κ^{ij}, κ^{ijk}, ... and the cumulant arrays $\kappa^{i,j}$, $\kappa^{i,j,k}$, ... change when we make a simple transformation from the original variables $X^1, ..., X^p$ to new variables $Y = Y^1, ..., Y^q$. If Y is a linear function of X, we may write

$$Y^r = a_i^r X^i,$$

where a_i^r is an array of constants. It is not difficult to see that the moments of Y are

$$a_i^r \kappa^i, \quad a_i^r a_j^s \kappa^{ij}, \quad a_i^r a_j^s a_k^t \kappa^{ijk}, ...$$

while the cumulants are

$$a_i^r \kappa^i, \quad a_i^r a_j^s \kappa^{i,j}, \quad a_i^r a_j^s a_k^t \kappa^{i,j,k}, ... \qquad (2.10)$$

In other words, under linear transformation, both moments and cumulants transform like contravariant tensors. Note however, that the matrix a_i^r need not have full rank.

Affine transformations involve a change of origin according to the equation

$$Y^r = a^r + a_i^r X^i.$$

The cumulants of Y, derived at the end of this section, are

$$a^r + a_i^r \kappa^i, \quad a_i^r a_j^s \kappa^{i,j}, \quad a_i^r a_j^s a_k^t \kappa^{i,j,k}, \quad a_i^r a_j^s a_k^t a_l^u \kappa^{i,j,k,l} \quad (2.11)$$

and so on. The change of origin affects only the mean vector or first cumulant. For this reason, cumulants are sometimes called semi-invariants. On the other hand, the moments of Y are

$$a^r + a_i^r \kappa^i,$$
$$a^r a^s + a^r a_i^s \kappa^i[2] + a_i^r a_j^s \kappa^{ij},$$
$$a^r a^s a^t + a^r a^s a_i^t \kappa^i[3] + a^r a_i^s a_j^t \kappa^{ij}[3] + a_i^r a_j^s a_k^t \kappa^{ijk}$$

and so on, where $a^r a_i^s \kappa^i[2] = a^r a_i^s \kappa^i + a^s a_i^r \kappa^i$. Thus, unlike the cumulants, the moments do not transform in a pleasant way under affine transformation of coordinates.

The transformation law for cumulants is similar to the transformation law of Cartesian tensors, (Jeffreys, 1952), the only difference being the dependence of the first cumulant on the choice of origin. To avoid any ambiguity of terminology, we use the term *cumulant tensor* to describe any array of quantities that transforms according to (2.11) under affine transformation of X. In Section 3.3, we develop rules for the transformation of cumulant tensors under non-linear transformation of X. These rules are quite different from tensor transformation laws that arise in differential geometry or in theoretical physics.

To prove (2.11) we use the method of generating functions, giving

$$M_Y(\xi) = E[\exp\{\xi_r(a^r + a_i^r X^i)\}]$$
$$= \exp(\xi_r a^r) M_X(\xi_r a_i^r).$$

In other words,

$$K_Y(\xi) = \xi_r a^r + K_X(\xi_r a_i^r)$$

from which expressions (2.10) and (2.11) follow directly.

2.5   Univariate cumulants and power notation

Much of the literature on cumulants concentrates on the univariate case, $p = 1$ and uses the condensed *power* notation κ_r for the rth cumulant of X^1, written in this section as X without indices. In this and in the following section, we move quite freely from power notation to index notation and back: this should cause no confusion on account of the different positions of indices in the two notations. Following Kendall & Stuart (1977, Chapter 3) we write

$$\mu'_r = E(X^r) \quad \text{and} \quad \mu_r = E(X - \mu'_1)^r$$

where the superscript here denotes a power. Expressions (2.6) giving moments in terms of cumulants become

$$\mu'_2 = \kappa_2 + \kappa_1^2$$
$$\mu'_3 = \kappa_3 + 3\kappa_1\kappa_2 + \kappa_1^3$$
$$\mu'_4 = \kappa_4 + 4\kappa_1\kappa_3 + 3\kappa_2^2 + 6\kappa_1^2\kappa_2 + \kappa_1^4$$

where superscripts again denote powers. The reverse formulae (2.7) become

$$\kappa_2 = \mu'_2 - (\mu'_1)^2$$
$$\kappa_3 = \mu'_3 - 3\mu'_1\mu'_2 + 2(\mu'_1)^3$$
$$\kappa_4 = \mu'_4 - 4\mu'_1\mu'_3 - 3(\mu'_2)^2 + 12(\mu'_1)^2\mu'_2 - 6(\mu'_1)^4.$$

In this notation, the permutation factors, previously kept in [.], and the arithmetic factors $(-1)^{\nu-1}(\nu-1)!$ become combined, with the result that the essential simplicity of the formulae (2.7) disappears. For this reason, unless we are dealing with a single random variable or a set of independent and identically distributed random variables, it is usually best to use index notation, possibly reverting to power notation at the last step of the calculations.

An extended version of power notation is sometimes used for bivariate cumulants corresponding to the random variables X and Y. For example, if $\mu'_{rs} = E(X^rY^s)$, the corresponding cumulant may be written κ_{rs}. Note that in this notation, $\kappa_{rs} \neq \kappa_{sr}$. In other words, with power notation, the cumulants are indexed by a vector whereas, with index notation, the cumulants are indexed by a set or, more generally, by the partitions of a set.

To establish the relationships between bivariate moments and bivariate cumulants in this notation, we simply convert the terms in (2.7) into power notation. For $r = 2$, $s = 1$ this gives

$$\kappa_{21} = \mu'_{21} - 2\mu'_{10}\mu'_{11} - \mu'_{01}\mu'_{20} + 2(\mu'_{10})^2\mu'_{01}$$

and, for $r = 2$, $s = 2$,

$$\kappa_{22} = \mu'_{22} - 2\mu'_{10}\mu'_{12} - 2\mu'_{01}\mu'_{21} - \mu'_{20}\mu'_{02} - 2(\mu'_{11})^2$$
$$+ 2(\mu'_{10})^2\mu'_{02} + 2(\mu'_{01})^2\mu'_{20} + 8\mu'_{01}\mu'_{10}\mu'_{11} - 6(\mu'_{10})^2(\mu'_{01})^2.$$

Additional, more impressive formulae of this type may be found in David, Kendall & Barton (1966, Tables 2.1.1 and 2.1.2). The formula for κ_{44}, for example, occupies 24 lines.

Simpler formulae are available in terms of central moments, but it is clear that the above notation conceals the simplicity of the formulae. For example, if μ'_{01} and μ'_{10} are both equal to zero, the formulae for κ_{44}, κ_{35}, κ_{26} and so on, can be found from the list of partitions of $1,2,...,8$ in Table 2 of the Appendix. For this reason, power notation is best avoided at least in formal manipulations. It is, however, very useful and convenient in the univariate case and is also useful more generally for interpretation.

2.6 Interpretation of cumulants

Although the definition of cumulants given in Sections 2.2 and 2.3 covered both univariate and mixed cumulants, their interpretations are best considered separately. Roughly speaking, mixed cumulants have an interpretation in terms of dependence or independence: univariate cumulants have a simple interpretation in terms of the shape of the marginal distribution. Of course, mixed cumulants could be interpreted also in terms of the shape of the joint distribution but such interpretations are not given here. We deal first with univariate cumulants using the notation of Section 2.5 and work our way up to mixed cumulants involving four distinct variables.

The first cumulant of X, denoted by κ_1, is the mean value and the second cumulant, κ_2, is the variance. In rigid body mechanics, κ_1 is the x-coordinate of the centre of gravity and κ_2 is the moment of inertia about the axis $x = \kappa_1$ of a uniform laminar body of unit mass in the (x, y) plane, bounded by $0 \leq y \leq f(x)$ where $f(x)$ is the density of X.

The third cumulant of X is a measure of asymmetry in the sense that $\kappa_3 = E(X - \kappa_1)^3$ is zero if X is symmetrically distributed. Of course $\kappa_3 = 0$ does not, on its own, imply symmetry: to guarantee symmetry, we require all odd cumulants to vanish and the distribution to be determined by its moments. For an example of an asymmetrical distribution whose odd cumulants are zero, see Kendall & Stuart (1977, Exercise 3.26), which is based on the note by Churchill (1946). The usual measure of skewness is the standardized third cumulant, $\rho_3 = \kappa_3/\kappa_2^{3/2}$, which is unaffected by affine transformations $X \to a + bX$ with $b > 0$. If $b < 0$, $\rho_3 \to -\rho_3$. Third and higher-order standardized cumulants given by $\rho_r = \kappa_r/\kappa_2^{r/2}$, can be interpreted as summary measures of departure from normality in the sense that if X is normal, all cumulants of order three or more are zero. This aspect is developed in greater detail in Chapter 5 where Edgeworth expansions are introduced.

Suppose now we have two random variables X^1 and X^2. With index notation, the mixed cumulants are denoted by $\kappa^{1,2}$, $\kappa^{1,1,2}$, $\kappa^{1,2,2}$, $\kappa^{1,1,1,2}$, ... The corresponding quantities in power notation are κ_{11}, κ_{21}, κ_{12}, κ_{31} and so on. To the extent that third and higher-order cumulants can be neglected, we find from the bivariate normal approximation that

$$E(X^2|X^1 = x^1) \simeq \kappa_{01} + (\kappa_{11}/\kappa_{20})(x^1 - \kappa_{10})$$
$$E(X^1|X^2 = x^2) \simeq \kappa_{10} + (\kappa_{11}/\kappa_{02})(x^2 - \kappa_{01})$$

(2.12)

so that $\kappa_{11} > 0$ implies positive dependence in the sense of increasing conditional expectations. Refined versions of the above, taking third- and fourth-order cumulants into account, are given in Chapter 5.

The simplest interpretations of bivariate cumulants are given in terms of independence. If X^1 and X^2 are independent, then all mixed cumulants involving X^1 and X^2 alone are zero. Thus, $\kappa^{1,2} = \kappa^{1,1,2} = \kappa^{1,2,2} = ... = 0$ or, more concisely using power notation, $\kappa_{rs} = 0$ for all $r, s \geq 1$. Provided that the moments determine the joint distribution, the converse is also true, namely that if $\kappa_{rs} = 0$ for all $r, s \geq 1$, then X^1 and X^2 are independent. The suggestion here is that if $\kappa_{rs} = 0$ for $r, s = 1, ..., t$, say, then X^1 and X^2 are approximately independent in some sense. However it is difficult to make this claim rigorous except in the asymptotic sense of Chapter 5 where all higher-order cumulants are negligible.

Consider now the case of three random variables whose mixed cumulants may be denoted by κ_{111}, κ_{211}, κ_{121}, κ_{112}, κ_{311}, κ_{221}, and so on by an obvious extension of the notation of Section 2.5. It is not difficult to see that if X^1 is independent of (X^2, X^3), or if X^2 is independent of (X^1, X^3), or if X^3 is independent of (X^1, X^2) then $\kappa_{111} = 0$ and, in fact, more generally, $\kappa_{rst} = 0$ for all $r, s, t \geq 1$. These independence relationships are most succinctly expressed using generalized power notation and it is for this reason that we switch here from one notation to the other.

More generally, if we have any number of random variables that can be partitioned into two independent blocks, then all mixed cumulants involving indices from both blocks are zero. Note that if X^1 and X^2 are independent, it does *not* follow from the above that, say, $\kappa^{1,2,3} = 0$. For example, if $X^3 = X^1 X^2$, then it follows from (3.2) that $\kappa^{1,2,3} = \kappa_{20}\kappa_{02}$ and this is strictly positive unless X^1 or X^2 is degenerate. For a more interesting example, see the bivariate exponential recurrence process $\{X_j, Y_j\}$, $j = \dots -1, 0, 1, \dots$ described in Exercise 2.35, in which each X_j is independent of the marginal process $\{Y_j\}$ and conversely, Y_j is independent of the marginal process $\{X_j\}$, but the two processes are dependent.

To see how the converse works, we note that $\kappa_{rs0} = 0$ for all $r, s \geq 1$ implies independence of X^1 and X^2, but only if the joint distribution is determined by its moments. Similarly, $\kappa_{r0s} = 0$ implies independence of X^1 and X^3. This alone does not imply that X^1 is independent of the pair (X^2, X^3): for this we require, in addition to the above, that $\kappa_{rst} = 0$ for all $r, s, t \geq 1$.

2.7 The central limit theorem

2.7.1 *Sums of independent random variables*

Suppose that $X_1, \dots, X_n$ are n independent vector-valued random variables where X_r has components $X_r^1, \dots, X_r^p$. We do not assume that the observations are identically distributed and so the cumulants of X_r are denoted by κ_r^i, $\kappa_r^{i,j}$, $\kappa_r^{i,j,k}$ and so on. One of the most important properties of cumulants is that the joint cumulants of $X_\bullet = X_1 + \dots + X_n$ are just the sums of the corresponding cumulants of the individual variables. Thus we may write $\kappa_\bullet^i$, $\kappa_\bullet^{i,j}$, $\kappa_\bullet^{i,j,k}$ for the joint cumulants of $X_\bullet$ where, for example, $\kappa_\bullet^{i,j} = \sum_{r=1}^n \kappa_r^{i,j}$.

This property is not shared by moments and there is, therefore, some risk of confusion if we were to write $\kappa_\bullet^{ij} = E(X_\bullet^i X_\bullet^j)$ because

$$\kappa_\bullet^{ij} = \kappa_\bullet^{i,j} + \kappa_\bullet^i \kappa_\bullet^j \neq \sum_r \kappa_r^{ij}.$$

For this reason, it is best to avoid the notation $\kappa_\bullet^{ij}$.

To demonstrate the additive property of cumulants we use the method of generating functions and write

$$M_{X_\bullet}(\xi) = E[\exp\{\xi_i(X_1^i + \ldots + X_n^i)\}].$$

By independence we have,

$$M_{X_\bullet}(\xi) = M_{X_1}(\xi) \ldots M_{X_n}(\xi)$$

and thus

$$K_{X_\bullet}(\xi) = K_{X_1}(\xi) + \ldots + K_{X_n}(\xi).$$

The required result follows on extraction of the appropriate coefficients of ξ.

In the particular case where the observations are identically distributed, we have that $\kappa_\bullet^i = n\kappa^i$, $\kappa_\bullet^{i,j} = n\kappa^{i,j}$ and so on. If the observations are not identically distributed, it is sometimes convenient to define average cumulants by writing

$$\kappa_\bullet^i = n\bar{\kappa}^i, \quad \kappa_\bullet^{i,j} = n\bar{\kappa}^{i,j}, \quad \kappa_\bullet^{i,j,k} = n\bar{\kappa}^{i,j,k}$$

and so on where, for example, $\bar{\kappa}^{i,j}$ is the average covariance of components i and j.

2.7.2 Standardized sums

With the notation of the previous section, we write

$$Y^i = n^{-1/2}\{X_\bullet^i - n\bar{\kappa}^i\}.$$

The cumulants of $X_\bullet$ are $n\bar{\kappa}^i, n\bar{\kappa}^{i,j}, n\bar{\kappa}^{i,j,k}$ and so on. It follows from Section 2.4 that the cumulants of Y are

$$0, \quad \bar{\kappa}^{i,j}, \quad n^{-1/2}\bar{\kappa}^{i,j,k}, \quad n^{-1}\bar{\kappa}^{i,j,k,l}, \ldots$$

the successive factors decreasing in powers of $n^{-1/2}$. Of course, the average cumulants are themselves implicitly functions of n and without further, admittedly mild, assumptions, there is no guarantee that $n^{-1/2}\bar{\kappa}^{i,j,k}$ or $n^{-1}\bar{\kappa}^{i,j,k,l}$ will be negligible for large n. We avoid such difficulties in the most direct way, simply by assuming that $\bar{\kappa}^{i,j}, \bar{\kappa}^{i,j,k}, \ldots$ have finite limits as $n \to \infty$ and that the limiting covariance matrix, $\bar{\kappa}^{i,j}$ is positive definite. In many problems, these assumptions are entirely reasonable but they do require checking. See Exercise 2.10 for a simple instance of failure of these assumptions.

The cumulant generating function of Y is

$$K_Y(\xi) = \xi_i\xi_j\bar{\kappa}^{i,j}/2! + n^{-1/2}\xi_i\xi_j\xi_k\bar{\kappa}^{i,j,k}/3! + \ldots \qquad (2.13)$$

Under the assumptions just given, and for complex ξ, the remainder in this series after r terms is $O(n^{-r/2})$. Now, $\xi_i\xi_j\bar{\kappa}^{i,j}/2$ is the cumulant generating function of a normal random variable with zero mean and covariance matrix $\bar{\kappa}^{i,j}$. Since convergence of the cumulant generating function implies convergence in distribution, subject to continuity of the limiting function at the origin (Moran, 1968, Section 6.2), we have just proved a simple version of the central limit theorem for independent but non-identically distributed random variables. In fact, it is not necessary here to use the generating function directly. Convergence of the moments implies convergence in distribution provided that the limiting moments uniquely determine the distribution, as they do in this case. For a more accurate approximation to the density of Y, we may invert (2.13) formally, leading to an asymptotic expansion in powers of $n^{-1/2}$. This expansion is known as the Edgeworth series after F.Y. Edgeworth (1845–1926). Note however, that although the error in (2.13) after two terms is $O(n^{-1})$, the error in probability calculations based on integrating the formal inverse of (2.13) need not be $O(n^{-1})$. In discrete problems, the error is typically $O(n^{-1/2})$.

Stronger forms of the central limit theorem that apply under conditions substantially weaker than those assumed here, are available in the literature. In particular, versions are available in which finiteness of the higher-order cumulants is not a requirement. Such theorems, on occasion, have statistical applications but they sometimes suffer from the disadvantage that the error term for finite n may be large and difficult to quantify, even in an asymptotic sense. Often the error is $o(1)$ as opposed to $O(n^{-1/2})$ under the kind of assumptions made here. Other forms of the central limit theorem are available in which certain specific types of dependence are permitted. For example, in applications related to time series, it is often reasonable to assume that observations sufficiently separated in time must be nearly independent. With additional mild assumptions, this ensures that the asymptotic cumulants of derived statistics are of the required order of magnitude in n and the central limit result follows.

2.8 Derived scalars

Suppose we are interested in the distribution of the statistic

$$T^2 = (X^i - \kappa_0^i)(X^j - \kappa_0^i)\kappa_{i,j}$$

where $\kappa_{i,j}$ is the matrix inverse or generalized inverse of $\kappa^{i,j}$. The Mahalanobis statistic, T^2, is a natural choice that arises if we are testing the hypothesis $H_0 : \kappa^i = \kappa_0^i$, where the higher-order cumulants are assumed known. One reason for considering this particular statistic is that it is invariant under affine transformation of X. Its distribution must therefore depend on scalars derived from the cumulants of X that are invariant under affine nonsingular transformations

$$X^r \to a^r + a_i^r X^i$$

where a_i^r is a $p \times p$ matrix of rank p. These scalars are the multivariate generalizations of $\rho_3^2, \rho_4, \rho_6, \rho_5^2$ and so on, in the univariate case: see Section 2.6.

To obtain the multivariate generalization of $\rho_3^2 = \kappa_3^2/\kappa_2^3$, we first write $\kappa^{i,j,k}\kappa^{l,m,n}$ as the generalization of κ_3^2. Division by κ_2 generalizes to multiplication by $\kappa_{r,s}$. Thus we require a scalar derived from

$$\kappa^{i,j,k}\kappa^{l,m,n}\kappa_{r,s}\kappa_{t,u}\kappa_{v,w}$$

by contraction, i.e. by equating pairs of indices and summing. This operation can be done in just two distinct ways giving two non-negative scalars

$$\bar{\rho}_{13}^2 = \rho_{13}^2/p = \kappa^{i,j,k}\kappa^{l,m,n}\kappa_{i,j}\kappa_{k,l}\kappa_{m,n}/p, \qquad (2.14)$$

$$\bar{\rho}_{23}^2 = \rho_{23}^2/p = \kappa^{i,j,k}\kappa^{l,m,n}\kappa_{i,l}\kappa_{j,m}\kappa_{k,n}/p. \qquad (2.15)$$

Similarly, to generalize $\rho_4 = \kappa_4/\kappa_2^2$, we obtain just the single expression

$$\bar{\rho}_4 = \rho_4/p = \kappa^{i,j,k,l}\kappa_{i,j}\kappa_{k,l}/p. \tag{2.16}$$

In the univariate case, (2.14) and (2.15) reduce to the same quantity and $\bar{\rho}_4$ satisfies the familiar inequality $\bar{\rho}_4 \geq \bar{\rho}_3^2 - 2$. The multivariate generalization of this inequality applies most directly to $\bar{\rho}_{13}^2$, giving

$$\bar{\rho}_4 \geq \bar{\rho}_{13}^2 - 2.$$

Equality is achieved if and only if the joint distribution of the Xs is concentrated on some conic in p-space in which the coefficients of the quadratic term are $\kappa_{i,j}$. The support of the distribution may be degenerate at a finite number of points but it is assumed here that it is not contained in any lower dimensional subspace. Otherwise, the covariance matrix would be rank deficient and p would be replaced by the rank of the subspace. In the univariate case, equality is achieved if and only if the distribution is concentrated on two points: see Exercise 2.12.

The corresponding inequality for $\bar{\rho}_{23}^2$ is

$$\bar{\rho}_4 \geq \bar{\rho}_{23}^2 - p - 1.$$

See Exercise 2.14. This limit is attained if and only if the joint distribution is concentrated on $p + 1$ points not contained in any linear subspace of R^p. The inequality for $\bar{\rho}_{23}^2$ is obtained by taking the trace of the residual covariance matrix of the products after linear regression on the linear terms. The trace vanishes only if this matrix is identically zero. Thus, achievement of the bound for $\bar{\rho}_{23}^2$ implies achievement of the bound for $\bar{\rho}_{13}^2$ and also that the higher-order cumulants are determined by those up to order four. See Section 3.8.

The simplest example that illustrates the difference between the two skewness scalars is the multinomial distribution with index m and parameter vector $\pi_1, ..., \pi_k$. The joint cumulants are given in Exercise 2.16 and the covariance matrix $\kappa^{i,j} = m\{\pi_i\delta_{ij} - \pi_i\pi_j\}$, has rank $p = k - 1$. The simplest generalized inverse is $\kappa_{i,j} = \{m\pi_i\}^{-1}$ for $i = j$ and zero otherwise. Substitution of the expressions in Exercise 2.16 for the third and fourth cumulants into (2.14) – (2.16) gives

$$m(k-1)\bar{\rho}_{13}^2 = \sum_j \pi_j^{-1} - k^2$$

$$m(k-1)\bar{\rho}_{23}^2 = \sum_j (1-\pi_j)(1-2\pi_j)/\pi_j$$

$$m(k-1)\bar{\rho}_4 = \sum_j \pi_j^{-1} - k^2 - 2(k-1).$$

Thus $\bar{\rho}_{13}^2$ is zero for the uniform multinomial distribution even though $\kappa^{i,j,k}$ is not identically zero. On the other hand, $\bar{\rho}_{23}^2$ is zero only if $\kappa^{i,j,k}$ is identically zero and, in the case of the multinomial distribution, this cannot occur unless $k = 2$ and $\pi = \frac{1}{2}$. For additional interpretations of the differences between these two scalars: see Exercise 2.15.

The above invariants are the three scalars most commonly encountered in theoretical work such as the expansion of the log likelihood ratio statistic or computing the variance of T^2. However, they are not the only invariant functions of the first four cumulants. A trivial example is $\kappa^{i,j}\kappa_{i,j} = p$, or more generally, the rank of $\kappa^{i,j}$. Also, if we were to generalize $\rho_4^2 = \kappa_4^2/\kappa_2^4$ by considering quadratic expressions in $\kappa^{i,j,k,l}$, there are two possibilities in addition to $(\bar{\rho}_4)^2$. These are

$$\bar{\rho}_{14}^2 = \kappa^{i,j,k,l}\kappa^{r,s,t,u}\kappa_{i,j}\kappa_{k,r}\kappa_{l,s}\kappa_{t,u}/p,$$

$$\bar{\rho}_{24}^2 = \kappa^{i,j,k,l}\kappa^{r,s,t,u}\kappa_{i,r}\kappa_{j,s}\kappa_{k,t}\kappa_{l,u}/p.$$

In addition to these, there are integer invariants of a qualitatively different kind, obtained by extending the notions of rank and signature to multi-way arrays. Kruskal (1977) defines the rank of a three-way asymmetrical array in a way that is consistent with the standard definition for matrices. In addition, the four-way array, $\kappa^{i,j,k,l}$ can be thought of as a symmetric $p^2 \times p^2$ matrix whose rank and signature are invariants. See Section 1.5.2 and Exercise 1.9. There may also be other invariants unconnected with the notions of rank or signature but none have appeared in the literature. However, it seems unnatural to consider $\kappa^{i,j,k,l}$ as a two-way array and, not surprisingly, the integer invariants just mentioned do not arise in the usual statistical calculations. For completeness, however, it would be good to know the complete list of all invariants that can be formed from, say, the first four cumulants.

To see that (2.14)–(2.16) are indeed invariant under affine transformation, we note that

$$\kappa^{i,j,k} \to a_i^r a_j^s a_k^t \kappa^{i,j,k} \quad \text{and} \quad \kappa_{i,j} \to b_r^i b_s^j \kappa_{i,j},$$

where b_r^i is the matrix inverse of a_i^r. Direct substitution followed by cancellation reveals the invariance. This invariance property of scalars derived by contraction of tensors is an elementary consequence of the tensor transformation property. Provided that we work exclusively with tensors, it is not necessary to check that scalars derived in this way are invariant.

2.9 Conditional cumulants

Suppose we are given the conditional joint cumulants of the random variables $X^1, ..., X^p$ conditional on some event A. How do we combine the conditional cumulants to obtain the unconditional joint cumulants? In the case of moments, the answer is easy because

$$E(X^1 X^2 ...) = E_A E(X^1 X^2 ...|A). \qquad (2.17)$$

In other words, the unconditional moments are just the average of the conditional moments. However, it is not difficult to show, for example, that the covariance of X^i and X^j satisfies

$$\begin{aligned} \kappa^{i,j} =& E_A\{\text{cov}(X^i, X^j | A)\} \\ &+ \text{cov}_A\{E(X^i|A), E(X^j|A)\}. \end{aligned} \qquad (2.18)$$

To see how these expressions generalize to cumulants of arbitrary order, we denote the conditional cumulants by λ^i, $\lambda^{i,j}$, $\lambda^{i,j,k}$ and we use the identity connecting the moment generating functions

$$M_X(\xi) = E_A M_{X|A}(\xi).$$

Expansion of this identity and comparison of coefficients gives

$$\kappa^i = E_A\{\lambda^i\}$$
$$\kappa^{i,j} + \kappa^i \kappa^j = E_A\{\lambda^{i,j} + \lambda^i \lambda^j\}$$
$$\kappa^{i,j,k} + \kappa^i \kappa^{j,k}[3] + \kappa^i \kappa^j \kappa^k = E_A\{\lambda^{i,j,k} + \lambda^i \lambda^{j,k}[3] + \lambda^i \lambda^j \lambda^k\}.$$

Expression (2.18) for the unconditional covariance follows from the second expression above. On using this result in the third expression, we find

$$\kappa^{i,j,k} = E(\lambda^{i,j,k}) + \kappa_2(\lambda^i, \lambda^{j,k})[3] + \kappa_3(\lambda^i, \lambda^j, \lambda^k).$$

The generalization is easy to see though, for notational reasons, a little awkward to prove, namely

$$\kappa(\Upsilon_p) = \sum_\Upsilon \kappa_\nu \{\lambda(v_1), ..., \lambda(v_\nu)\}$$

with summation over all partitions of the p indices. In this expression, $\lambda(v_j)$ is the conditional mixed cumulant of the random variables whose indices are in v_j and $\kappa_\nu\{\lambda(v_1), ..., \lambda(v_\nu)\}$ is the νth order cumulant of the ν random variables listed as arguments. For details of a proof see the papers by Brillinger (1969) or Speed (1983).

In many circumstances, it is required to compute conditional cumulants from joint unconditional cumulants, the converse of the result just described. A little reflection soon shows that the converse problem is considerably more difficult and the best that can be expected are approximate conditional cumulants. Expansions of this type are given in Chapter 5.

However, if $M_{X,Y}(\xi, \varsigma) = E\{\exp(\xi_i X^i + \varsigma_r Y^r)\}$ is the joint moment generating function of X, Y, we can at least write down an expression for the conditional moment generating function $M_{X|Y}(\xi)$. Since

$$M_{X,Y}(\xi, \varsigma) = \int \exp(\varsigma_r Y^r) M_{X|Y}(\xi) f_Y(y) dy,$$

we may invert the integral transform to find

$$k\, M_{X|Y}(\xi) f_Y(y) = \int_{c-i\infty}^{c+i\infty} M_{X,Y}(\xi, \varsigma) \exp(-\varsigma_r y^r) d\varsigma.$$

Division by $f_Y(y)$ gives the conditional moment generating function in the form

$$M_{X|Y}(\xi) = \frac{\int M_{X,Y}(\xi, \varsigma) \exp(-\varsigma_r y^r) d\varsigma}{\int M_{X,Y}(0, \varsigma) \exp(-\varsigma_r y^r) d\varsigma}. \tag{2.19}$$

This expression, due to Bartlett (1938), can be used for generating expansions or approximations for conditional moments: it is rarely used directly for exact calculations. See, however, Moran (1968, Section 6.14).

2.10 Bibliographic notes

Cumulants were first defined by Thiele in about 1889. Thiele's work is mostly in Danish and the most accessible English translation is of his book *Theory of Observations* reprinted in *The Annals of Mathematical Statistics* (1931), pp.165-308. In Chapter 6 of that book, Thiele defines the univariate cumulants and calls them *half-invariants* on account of their simple transformation properties. He also derives a version of the central limit theorem by showing that the higher-order cumulants of a standardized linear combination of random variables converge rapidly to zero provided that 'the coefficient of any single term is not so great ... that it throws all the other terms into the shade', a delightful statement closely approximating the spirit of the Lindeberg-Feller condition.

For an excellent readable account of the central limit theorem and the Lindeberg condition, see LeCam (1986).

Fisher (1929), in an astonishing tour de force, rediscovered cumulants, recognized their superiority over moments, developed the corresponding sample cumulants and cross-cumulants, gave the formulae for the cumulants and cross-cumulants of the sample cumulants and formulated combinatorial rules for computing the cumulants of such statistics. Both Thiele and Fisher used what we call 'power notation' for cumulants and cross-cumulants and their achievements are the more remarkable for that reason.

Formulae giving univariate moments in terms of cumulants and vice versa are listed in Kendall & Stuart (1977, Chapter 3): multivariate versions of these formulae are given in Chapter 13. See also David, Kendall & Barton (1966, Tables 2.1.1 and 2.1.2) and David & Barton (1962, Chapter 9). Similar formulae are given by Brillinger (1975, Chapter 2).

The derived scalars $\bar{\rho}_4$ and $\bar{\rho}_{23}^2$ were given by Mardia (1970) as summary measures of multivariate kurtosis and skewness. The additional skewness scalar, $\bar{\rho}_{13}^2$ is given by McCullagh & Cox (1986) who show how it arises in calculations involving likelihood ratio tests. See also Davis (1980) who shows how the scalar $\psi_p = p(\bar{\rho}_{13}^2 -$

$\bar{\rho}_{23}^2$) arises in calculations concerning the effect of non-normality on the distribution of Wilks's Λ.

2.11 Further results and exercises 2

2.1 Let X have density function given by

$$f_X(x) = \begin{cases} 2/x^3 & x \geq 1 \\ 0 & \text{otherwise.} \end{cases}$$

Show that the mean of X is finite but that all higher-order moments are infinite. Find an expression for the density of $Y = 1/X$ and show that all moments and cumulants are finite. Let $\mu'_{rs} = E(X^r Y^s)$ be the joint moment of order $r + s$ (using the power notation of Section 2.5). Show that $\mu'_{21} = 2$ but that the corresponding cumulant, κ_{21}, is infinite.

2.2 Let X be a standard normal random variable and set $Y = \exp(X)$. Show that the rth moment of Y about the origin is $\mu'_r = \exp(r^2/2)$. Hence find expressions for the first four cumulants of Y. Show that the series expansions for $M_Y(\xi)$ and $K_Y(\xi)$ about $\xi = 0$ are divergent for all real $\xi > 0$ even though all cumulants are finite (Heyde, 1963).

2.3 Let X be a scalar random variable. Prove by induction on r that the derivatives of $M_X(\xi)$ and $K_X(\xi)$, if they exist, satisfy

$$M_X^{(r)}(\xi) = \sum_{j=1}^{r} \binom{r-1}{j-1} M_X^{(r-j)}(\xi) K_X^{(j)}(\xi) \qquad r \geq 1.$$

Hence show that

$$\mu'_r = \kappa_r + \sum_{j=1}^{r-1} \binom{r-1}{j-1} \kappa_j \mu'_{r-j}$$

and, for $r \geq 4$, that

$$\mu_r = \kappa_r + \sum_{j=2}^{r-2} \binom{r-1}{j-1} \kappa_j \mu_{r-j}$$

(Thiele, 1897, eqn. 22; Morris, 1982).

2.4 If $\mu(\Upsilon)$ and $\kappa(\Upsilon)$ denote the ordinary moment and the ordinary cumulant corresponding to the indices in $\Upsilon = \{i_1, ..., i_p\}$, show that

$$\mu(\Upsilon) = \kappa(\Upsilon) + \sum_{\{v_1, v_2\}} \kappa(v_1)\mu(v_2)$$

where $\{v_1, v_2\}$ is a partition of Υ into two non-empty blocks and the sum extends over all partitions such that $i_1 \in v_1$. What purpose does the condition $i_1 \in v_1$ serve? Show that this result generalizes the univariate identity in Exercise 2.3.

2.5 Show that the central and non-central moments satisfy

$$\mu'_r = \sum_{j=0}^{r} \binom{r}{j} \mu_{r-j} (\mu'_1)^j$$

$$\mu_r = \sum_{j=0}^{r} \binom{r}{j} \mu'_{r-j} (-\mu'_1)^j.$$

(Kendall and Stuart, 1977, p. 58).

2.6 The density function of Student's distribution on ν degrees of freedom is

$$\frac{(1 + t^2/\nu)^{-(\nu+1)/2}}{\nu^{1/2} B(\frac{1}{2}, \nu/2)} \qquad -\infty < t < \infty,$$

where $B(.,.)$ is the beta function. Show that the odd moments that exist are zero and that the even moments are

$$\mu_{2r} = \frac{1.3....(2r-1)\nu^r}{(\nu-2)...(\nu-2r)} \qquad 2r < \nu.$$

Hence show that

$$\rho_4 = 6/(\nu - 4) \quad \text{for} \quad \nu > 4$$
$$\rho_6 = 240/\{(\nu-4)(\nu-6)\} \quad \text{for} \quad \nu > 6.$$

2.7 Prove directly using (2.7) that if X^i is independent of the pair (X^j, X^k), then $\kappa^{i,j,k} = 0$.

2.8 Show that if X^1 is independent of (X^2, X^3), then the cumulant κ_{rst} of order $r + s + t$ with $r, s, t \geq 1$ is equal to zero.

2.9   Derive expressions (2.7) for cumulants in terms of moments (i) directly from (2.6) and (ii) by expansion of $\log M_X(\xi)$.

2.10   Let $h_r(x)$ be the standardized Hermite polynomial of degree r satisfying $\int h_r(x)h_s(x)\phi(x)dx = \delta_{rs}$ where $\phi(x)$ is the standard normal density. If $X_i = h_i(Z)$ where Z is a standard normal variable, show that $X_1, \ldots$ are uncorrelated but not independent. Show also that the second cumulant of $n^{1/2}\bar{X}$ is exactly one but that the third cumulant does not converge to zero. Construct a similar example in which the first three cumulants converge to those of the standard normal density, but where the central limit theorem does not apply.

2.11   Verify that $\bar{\rho}_4 = p^{-1}\kappa^{i,j,k,l}\kappa_{i,j}\kappa_{k,l}$ is invariant under affine non-singular transformation of X.

2.12   By considering the expression

$$\operatorname{var}\{\kappa_{i,j}X^iX^j - c_iX^i\}$$

with $c_i = \kappa_{i,r}\kappa_{s,t}\kappa^{r,s,t}$, show that $\bar{\rho}_4 \geq \bar{\rho}_{13}^2 - 2$, where $\bar{\rho}_4$ and $\bar{\rho}_{13}^2$ are defined by (2.14) and (2.16). In addition, by examining the expression

$$\int (a + a_ix^i + a_{ij}x^ix^j)^2 f_X(x)\, dx,$$

show that $\bar{\rho}_4 = \bar{\rho}_{13}^2 - 2$ if and only if the joint distribution of X is concentrated on a particular class of conic.

2.13   Show that $\bar{\rho}_{23}^2 \geq 0$ with equality only if $\kappa^{i,j,k} = 0$ identically.

2.14   Show that $\kappa^{ij,kl} - \kappa^{i,j,r}\kappa^{k,l,s}\kappa_{r,s}$, regarded as a $p^2 \times p^2$ symmetric matrix, is non-negative definite. By examining the trace of this matrix in the case where $\kappa^i = 0$, $\kappa^{i,j} = \delta^{ij}$, show that

$$\bar{\rho}_4 \geq \bar{\rho}_{23}^2 - p - 1,$$

with equality if and only if the joint distribution is concentrated on $p+1$ points not contained in any linear subspace of R^p. Deduce that $\bar{\rho}_4 = \bar{\rho}_{23}^2 - p - 1$ implies $\bar{\rho}_4 = \bar{\rho}_{13}^2 - 2$ and that $\bar{\rho}_4 > \bar{\rho}_{13}^2 - 2$ implies $\bar{\rho}_4 > \bar{\rho}_{23}^2 - p - 1$.

2.15   Show that $\bar{\rho}_{13}^2 = 0$ if and only if every linear combination a_iX^i is uncorrelated with the quadratic form $\kappa_{i,j}X^iX^j$. (Take $\kappa^i = 0$.) Show also that $\bar{\rho}_{23}^2 = 0$ if and only if every linear combination a_iX^i is uncorrelated with every quadratic form $a_{ij}X^iX^j$.

2.16 Show that the multinomial distribution with index m and probability vector $\pi_1, ..., \pi_k$ has cumulant generating function

$$m\{k(\theta + \xi) - k(\theta)\},$$

where $k(\theta) = \log[\Sigma \exp(\theta_j)]$ and

$$\pi_i = \exp(\theta_i)/\sum \exp(\theta_j).$$

Hence show that the first four cumulants are

$$\kappa^i = m\pi_i$$
$$\kappa^{i,j} = m\{\pi_i \delta_{ij} - \pi_i \pi_j\}$$
$$\kappa^{i,j,k} = m\{\pi_i \delta_{ijk} - \pi_i \pi_j \delta_{ik}[3] + 2\pi_i \pi_j \pi_k\}$$
$$\kappa^{i,j,k,l} = m\{\pi_i \delta_{ijkl} - \pi_i \pi_j (\delta_{ik}\delta_{jl}[3] + \delta_{jkl}[4]) + 2\pi_i \pi_j \pi_k \delta_{il}[6]$$
$$- 6\pi_i \pi_j \pi_k \pi_l\},$$

where, for example, $\delta_{ijk} = 1$ if $i = j = k$ and zero otherwise, and no summation is implied where indices are repeated at the same level.

2.17 Evaluate explicitly the fourth cumulants of the multinomial distribution for the five distinct index patterns.

2.18 Show, for the multinomial distribution with index $m = 1$, that the moments are $\kappa^i = \pi_i$, $\kappa^{ij} = \pi_i \delta_{ij}$, $\kappa^{ijk} = \pi_i \delta_{ijk}$ and so on, where no summation is implied. Hence give an alternative derivation of the first four cumulants in Exercise 2.16.

2.19 For the multinomial distribution with $p = \text{rank}(\kappa^{i,j}) = k - 1$, show that

$$(k - 1)\bar{\rho}_{13}^2 = m^{-1}\{\sum \pi_j^{-1} - k^2\},$$
$$(k - 1)\bar{\rho}_{23}^2 = m^{-1}\{\sum \pi_j^{-1} - 3k + 2\}$$
$$(k - 1)\bar{\rho}_4 = m^{-1}\{\sum \pi_j^{-1} - k^2 - 2(k - 1)\},$$

and hence that

$$\bar{\rho}_4 = \bar{\rho}_{13}^2 - 2/m = \bar{\rho}_{23}^2 - k/m,$$

showing that the inequalities in Exercises 2.12 and 2.14 are sharp for $m = 1$. Show also that the minimum value of $\bar{\rho}_{23}^2$ for the multinomial distribution is $(k - 2)/m$.

2.20 Hölder's inequality for a pair of random variables X and Y is

$$E|XY| \leq \{E|X|^p\}^{1/p}\{E|Y|^q\}^{1/q}$$

where $p^{-1} + q^{-1} = 1$. Deduce from the above that

$$\{E|X_1 X_2 ... X_r|\}^r \leq E|X_1|^r ... E|X_r|^r$$

for random variables $X_1, ..., X_r$. Hence prove that if the diagonal elements of cumulant tensors are finite then all other elements are finite.

2.21 Using (2.6) and (2.7), express $\kappa^{i,jkl} = \text{cov}(X^i, X^j X^k X^l)$ in terms of ordinary moments and hence, in terms of ordinary cumulants.

2.22 Let $a = a^1, ..., a^p$ and $b = b^1, ..., b^p$, where $a^j \leq b^j$, be the coordinates of two points in R^p and denote by (a, b) the Cartesian product in R^p of the intervals (a^j, b^j). Let $f(x) = f(x^1, ..., x^p)$ be a p-dimensional joint density function and define

$$F(a, b) = \int_{x \in (a, b)} f(x)\, dx$$

where $dx = dx^1 ... dx^p$. Express $F(a, b)$ in terms of the cumulative distribution function $F(x) \equiv F(-\infty, x)$ evaluated at points with coordinates $x^j = a^j$ or b^j. Comment briefly on the similarities with and differences from (2.9).

2.23 Let a^{ij} be the elements of a square matrix, not necessarily symmetrical, and let its inverse, a_{ij}, satisfy $a^{ij}a_{kj} = \delta^i_k = a^{ji}a_{jk}$, being careful regarding the positions of the indices. Show that the derivatives satisfy

$$\partial a_{ij}/\partial a^{rs} = -a_{is}a_{rj}$$
$$\partial a^{ij}/\partial a_{rs} = -a^{is}a^{rj}.$$

2.24 Show that if a_i are the components of a vector of coefficients satisfying $a_i a_j \kappa^{i,j} = 0$, then

$$a_i \kappa^{i,j} = 0, \quad a_i \kappa^{i,j,k} = 0, \quad a_i \kappa^{i,j,k,l} = 0$$

and so on. Hence deduce that the choice of generalized inverse has no effect on the scalars derived in Section 2.8.

2.25 Let $X_1, ..., X_n$ be independent and identically distributed scalar random variables having cumulants $\kappa_1, \kappa_2, \kappa_3, ...$. Show that

$$2\kappa_2 = E(X_1 - X_2)^2$$
$$3\kappa_3 = E(X_1 + \omega X_2 + \omega^2 X_3)^3 \quad (\omega = e^{2\pi i/3})$$
$$4\kappa_4 = E(X_1 + \omega X_2 + \omega^2 X_3 + \omega^3 X_4)^4 \quad (\omega = e^{2\pi i/4})$$

where $i^2 = -1$. Hence, by writing $\omega = e^{2\pi i/r}$ and

$$r\kappa_r = \lim_{n \to \infty} n^{-1}[X_1 + \omega X_2 + \omega^2 X_3 + ... + \omega^{nr-1} X_{nr}]^r,$$

give an interpretation of cumulants as coefficients in the Fourier transform of the randomly ordered sequence $X_1, X_2, ...$. Express κ_r as a *symmetric* function of $X_1, X_2, ...$, (Good, 1975, 1977).

2.26 Show that if $\Upsilon = \{v_1, ..., v_\nu\}$ is a partition of the indices $j_1, ..., j_n$ and $\omega = e^{2\pi i/n}$ is a primitive nth root of unity, then

$$\sum_j \omega^{j_1 + ... + j_n} \delta(v_1)...\delta(v_\nu) = \begin{cases} 0 \text{ if } \Upsilon < 1 \\ n \text{ if } \Upsilon = 1 \end{cases}$$

where the sum extends over all positive integer vectors having components in the range $(1, n)$.

2.27 Let $X_1, ..., X_n$ be independent and identically distributed p-dimensional random vectors having cumulants $\kappa^r, \kappa^{r,s}, \kappa^{r,s,t}, ...$. Define the random vector $Z_{(n)}$ by

$$Z_{(n)}^r = \sum_{j=1}^n X_j^r \exp(2\pi i j/n)$$

where $i^2 = -1$. Using the result in the previous exercise or otherwise, show that the nth-order moments of $Z_{(n)}$ are the same as the nth-order cumulants of X, i.e.

$$E(Z_{(n)}^{r_1}...Z_{(n)}^{r_n}) = \kappa^{r_1,...,r_n},$$

(Good, 1975, 1977). Hence give an interpretation of mixed cumulants as Fourier coefficients along the lines of the interpretation in Exercise 2.25.

2.28 Let $X_1, ..., X_n$ be independent χ_1^2 random variables. Show that the joint cumulant generating function is $-\frac{1}{2} \sum_i \log(1 - 2\xi_i)$. Show also that the joint cumulant generating function of $Y_1 = \sum X_j$ and $Y_2 = \sum \lambda_j X_j$ is $-\frac{1}{2} \sum_i \log(1 - 2\xi_1 - 2\lambda_i \xi_2)$. Hence show that the joint cumulants of (Y_1, Y_2) are given by

$$\kappa_{rs} = 2^{r+s-1}(r + s - 1)! \sum \lambda_i^s$$

using the notation of Section 2.5.

2.29 Show that if the ratio $R = Y/X$ is independent of X, then the moments of the ratio are the ratio of the moments, i.e.

$$\mu_r'(R) = \mu_r'(Y)/\mu_r'(X).$$

2.30 In the notation of Exercise 2.28, let $R = Y_2/Y_1$. Show that the first four cumulants of this ratio are

$$\kappa_1(R) = \kappa_1(\lambda)$$
$$\kappa_2(R) = 2\kappa_2(\lambda)/(n + 2)$$
$$\kappa_3(R) = 8\kappa_3(\lambda)/\{(n + 2)(n + 4)\}$$
$$\kappa_4(R) = 48\kappa_4(\lambda)/\{(n + 2)(n + 4)(n + 6)\}$$
$$+ 48n\kappa_2^2(\lambda)/\{(n + 2)^2(n + 4)(n + 6)\}$$

and explain what is meant by the notation $\kappa_r(\lambda)$.

2.31 Using the notation of Exercises 2.28 and 2.30, show that if the eigenvalues decrease exponentially fast, say $\lambda_j = \lambda^j$, with $|\lambda| < 1$, then nR has a non-degenerate limiting distribution for large n, with cumulants

$$\kappa_r(nR) \simeq 2^{r-1}(r - 1)!\lambda^r/(1 - \lambda^r).$$

Show that this result is false if λ is allowed to depend on n, say $\lambda_j = 1 - 1/j$.

2.32 Using (2.18), show that if X and Y are independent real-valued random variables, then the variance of the product is

$$\text{var}(XY) = \kappa_{10}^2 \kappa_{02} + \kappa_{01}^2 \kappa_{20} + \kappa_{20}\kappa_{02}.$$

2.33 Let X_1 and X_2 be independent and identically distributed p-dimensional random variables with zero mean and identity covariance matrix. The spherical polar representation of X is written (R, θ) where θ has $p - 1$ components and $R = |X|$ is the Euclidean norm of X. Show that

$$p\bar\rho_{13}^2 = E(R_1^3 R_2^3 \cos\theta_{12})$$
$$p\bar\rho_{23}^2 = E(R_1^3 R_2^3 \cos^3\theta_{12})$$

where $R_1 R_2 \cos\theta_{12} = X_1^i X_2^j \delta_{ij}$, so that θ_{12} is the random angle between X_1 and X_2. Hence give a geometrical interpretation of these two scalars in the special case where $p = 2$. Show also that

$$4p\bar\rho_{23}^2 - 3p\bar\rho_{13}^2 = E\{R_1^3 R_2^3 \cos(3\theta_{12})\}$$

and that this quantity is non-negative if $p \le 2$.

2.34 Let X be a scalar random variable and write

$$M_X^{(n)}(\xi) = 1 + \mu_1'\xi + \mu_2'\xi^2/2! + \ldots + \mu_n'\xi^n/n!$$

for the truncated moment generating function. The zeros of this function, $a_1^{-1}, \ldots, a_n^{-1}$, not necessarily real, are defined by

$$M_X^{(n)}(\xi) = (1 - a_1\xi)(1 - a_2\xi)\ldots(1 - a_n\xi).$$

Show that the symmetric functions of the as

$$\langle rs\ldots u \rangle = \sum_{i<j<\ldots<l} a_i^r a_j^s \ldots a_k^u$$

are semi-invariants of X (unaffected by the transformation $X \to X + c$) if and only if the powers $r, s, \ldots, u$ that appear in the symmetric function are at least 2. Show also that the cumulants are given by the particular symmetric functions

$$\kappa_r = -(r - 1)! \sum_i a_i^r = -(r - 1)! < r > \qquad (r < n).$$

Express the semi-invariant $\langle 22 \rangle$ in terms of κ_2 and κ_4. (MacMahon, 1884, 1886; Cayley, 1885).

2.35   Let $\{\epsilon_j\}$, $\{\epsilon_j'\}$ and $\{Z_j\}$, $j = \ldots - 1, 0, 1, \ldots$ be three doubly infinite, mutually independent sequences of independent unit exponential random variables. The bivariate sequence $\{X_j, Y_j\}$, $j = \ldots - 1, 0, 1, \ldots$, defined by

$$X_{j+1} = \begin{cases} X_j - Z_j & \text{if } X_j > Z_j \\ \epsilon_{j+1} & \text{otherwise} \end{cases}$$

$$Y_{j+1} = \begin{cases} Y_j - Z_j & \text{if } Y_j > Z_j \\ \epsilon_{j+1}' & \text{otherwise} \end{cases}$$

is known as a bivariate *exponential recurrence* process. Show that

 (i) X_j and Y_j are unit exponential random variables.
 (ii) $\text{cov}(X_j, X_{j+h}) = 2^{-h}$ where $h \geq 0$.
 (iii) X_i is independent of Y_j.
 (iv) X_i is independent of the sequence $\{Y_j\}$.

Hence deduce that all third-order mixed cumulants involving both Xs and Ys are zero. Show also that

$$\text{cum}(X_j, X_{j+1}, Y_j, Y_{j+1}) = 1/12$$

(McCullagh, 1984c).

2.36   In the two-dimensional case, show that the homogeneous cubic form $\kappa^{i,j,k} w_i w_j w_k$ can be written, using power notation, in the form

$$Q_3(w) = \kappa_{30} w_1^3 + \kappa_{03} w_2^3 + 3\kappa_{21} w_1^2 w_2 + 3\kappa_{12} w_1 w_2^2.$$

By transforming to polar coordinates, show that

$$Q_3(w) = r^3 \{\tau_1 \cos(\theta - \epsilon_1) + \tau_3 \cos(3\theta - 3\epsilon_3)\},$$

where

$$16\tau_1^2 = 9(\kappa_{30} + \kappa_{12})^2 + 9(\kappa_{03} + \kappa_{21})^2$$
$$16\tau_3^2 = (\kappa_{30} - 3\kappa_{12})^2 + (\kappa_{03} - 3\kappa_{21})^2.$$

Find similar expressions for ϵ_1 and ϵ_3 in terms of the κs.

2.37   By taking X to be a two-dimensional standardized random variable with zero mean and identity covariance matrix, interpret $Q_3(w)$, defined in the previous exercise, as a directional standardized skewness. [Take w to be a unit vector.] Show that, in the polar representation, $\epsilon_3 - \epsilon_1$ is invariant under rotation of X, but changes sign under reflection. Find an expression for this semi-invariant in terms of κ_{30}, κ_{03}, κ_{21} and κ_{12}. Discuss the statistical implications of the following conditions:

(i) $4\rho_{23}^2 - 3\rho_{13}^2 = 0$;
(ii) $\rho_{13}^2 = 0$;
(iii) $\epsilon_3 - \epsilon_1 = 0$.

2.38 In the two-dimensional case, show that the homogeneous quartic form $\kappa^{i,j,k,l} w_i w_j w_k w_l$ can be written, using power notation, in the form

$$Q_4(w) = \kappa_{40} w_1^4 + \kappa_{04} w_2^4 + 4\kappa_{31} w_1^3 w_2 + 4\kappa_{13} w_1 w_2^3 + 6\kappa_{22} w_1^2 w_2^2.$$

By transforming to polar coordinates, show that

$$Q_4(w) = r^4 \{\tau_0 + \tau_2 \cos(2\theta - 2\epsilon_2) + \tau_4 \cos(4\theta - 4\epsilon_4)\}.$$

Show that $8\tau_0 = 3\kappa_{40} + 3\kappa_{04} + 6\kappa_{22}$ is invariant under orthogonal transformation of X. Find similar expressions for τ_2, τ_4, ϵ_2 and ϵ_4 in terms of the κs.

2.39 By taking X to be a two-dimensional standardized random variable with zero mean and identity covariance matrix, interpret $Q_4(w)$, defined in the previous exercise, as a directional standardized kurtosis. Taking ϵ_1 as defined in Exercises 2.36 and 2.37, show, using the polar representation, that that $\epsilon_2 - \epsilon_1$ and $\epsilon_4 - \epsilon_1$ are both invariant under rotation of X, but change sign under reflection. Find expressions for these semi-invariants in terms of the κs. Interpret τ_0 as the mean directional kurtosis and express this as a function of $\bar{\rho}_4$. Discuss the statistical implications of the following conditions:

(i) $\tau_0 = 0$;
(ii) $\tau_2 \doteq 0$;
(iii) $\tau_4 = 0$;
(iv) $\epsilon_4 - \epsilon_2 = 0$.

2.40 In the one-dimensional case, the functions of the cumulants of X that are invariant under affine transformation are

$$\frac{\kappa_{2r}}{\kappa_2^r}, \quad \frac{\kappa_{2r+1}^2}{\kappa_2^{2r+1}}, \quad r = 1, 2, \ldots$$

using power notation. All other invariants can be expressed as functions of this sequence. By extending the results described in the previous four exercises, describe the corresponding complete list of invariants and semi-invariants in the bivariate case.

2.41   Discuss the difficulties encountered in extending the above argument beyond the bivariate case.

2.42   *Spherically symmetric random variables*: A random variable X is said to be spherically symmetric if its distribution is unaffected by orthogonal transformation. Equivalently, in spherical polar coordinates, the radial vector is distributed independently of the angular displacement. Show that the odd cumulants of such a random variable are zero and that the even cumulants must have the form

$$\kappa^{i,j} = \tau_2 \delta^{ij}, \quad \kappa^{i,j,k,l} = \tau_4 \delta^{ij}\delta^{kl}[3], \quad \kappa^{i,j,k,l,m,n} = \tau_6 \delta^{ij}\delta^{kl}\delta^{mn}[15]$$

and so on, for some set of coefficients $\tau_2, \tau_4, \ldots$. Show that the standardized cumulants are

$$\bar{\rho}_4 = \tau_4(p+2)/\tau_2^2, \quad \bar{\rho}_6 = \tau_6(p+2)(p+4)/\tau_2^3$$

and hence that $\tau_4 \geq -2\tau_2^2/(p+2)$.

2.43   For the previous exercise, show that

$$\tau_6 \geq -\tau_2\tau_4 - \tau_2^3 + \frac{3(\tau_4 + \tau_2^2)^2}{(p+4)\tau_2}.$$

CHAPTER 3

Generalized cumulants

3.1 Introduction and definitions

In Chapter 2 we examined in some detail how cumulant tensors transform under affine transformation of X. Cumulants of order two and higher transform like Cartesian tensors, but the first-order cumulant does not. In this chapter, we show how cumulant tensors transform under non-linear or non-affine transformation of X. The algorithm that we describe relies heavily on the use of index notation and is easy to implement with the assistance of suitable tables. Applications are numerous. For example, the maximum likelihood estimator and the maximized likelihood ratio statistic can be expressed as functions of the log likelihood derivatives at the true but unknown parameter point, θ. The distribution of these derivatives at the true θ is known as a function of θ. With the methods developed here, we may compute moments or cumulants of any derived statistic, typically as an asymptotic approximation, to any required order of approximation.

In general, it is a good deal more convenient to work with polynomial functions rather than, say, exponential or logarithmic functions of X. The first step in most calculations is therefore to expand the function of interest as a polynomial in X and to truncate at an appropriate point. The essential ingredient when working with polynomial functions is to develop a notation capable of coping with generalized cumulants of the type

$$\kappa^{i,jk} = \operatorname{cov}(X^i, X^j X^k).$$

It seems obvious and entirely natural to denote this quantity by $\kappa^{i,jk}$, thereby indexing the set of generalized cumulants by partitions of the indices. In other words, to each partition there corresponds a unique cumulant and to each cumulant there corresponds a unique partition. For example,

$$\kappa^{i,jkl} = \mathrm{cov}(X^i, X^j X^k X^l)$$
$$\kappa^{ij,kl} = \mathrm{cov}(X^i X^j, X^k X^l) \qquad (3.1)$$
$$\kappa^{i,j,kl} = \mathrm{cum}(X^i, X^j, X^k X^l).$$

Thus $\kappa^{i,j,kl}$, the third-order cumulant of the three variables X^i, X^j and the product $X^k X^l$, is said to be of order $\alpha = 3$ and degree $\beta = 4$. The order is the number of blocks of the partition and the degree is the number of indices. *Ordinary* cumulants with $\alpha = \beta$ and *ordinary* moments with $\alpha = 1$ are special cases of generalized cumulants. The order of the indices in (3.1) is immaterial provided only that the partition is preserved. In this way, the notion of symmetry under index permutation is carried over to generalized cumulants.

3.2   The fundamental identity for generalized cumulants

Just as moments can be expressed as combinations of ordinary cumulants according to (2.6), so too generalized cumulants can be expressed in a similar way. First, we give the expressions for the four generalized cumulants listed above and then the general formula is described. The following four formulae may be derived from first principles using (2.6) and (2.7).

$$\kappa^{i,jk} = \kappa^{ijk} - \kappa^i \kappa^{jk}$$
$$= \kappa^{i,j,k} + \kappa^j \kappa^{i,k} + \kappa^k \kappa^{i,j}$$
$$\kappa^{i,jkl} = \kappa^{ijkl} - \kappa^i \kappa^{jkl}$$
$$= \kappa^{i,j,k,l} + \kappa^j \kappa^{i,k,l}[3] + \kappa^{i,j} \kappa^{k,l}[3] + \kappa^{i,j} \kappa^k \kappa^l[3]$$
$$\kappa^{ij,kl} = \kappa^{ijkl} - \kappa^{ij} \kappa^{kl}$$
$$= \kappa^{i,j,k,l} + \kappa^i \kappa^{j,k,l}[2] + \kappa^k \kappa^{i,j,l}[2] + \kappa^{i,k} \kappa^{j,l}[2]$$
$$\qquad + \kappa^i \kappa^k \kappa^{j,l}[4]$$
$$\kappa^{i,j,kl} = \kappa^{ijkl} - \kappa^i \kappa^{jkl} - \kappa^j \kappa^{ikl} - \kappa^{ij} \kappa^{kl} + 2\kappa^i \kappa^j \kappa^{kl}$$
$$= \kappa^{i,j,k,l} + \kappa^k \kappa^{i,j,l}[2] + \kappa^{i,k} \kappa^{j,l}[2]. \qquad (3.2)$$

Again, the bracket notation has been employed, but now the interpretation of groups of terms depends on the context. Thus, for example, in the final expression above, $\kappa^{i,k} \kappa^{j,l}[2] = \kappa^{i,k} \kappa^{j,l} + \kappa^{i,l} \kappa^{j,k}$

must be interpreted in the context of the partition on the left, namely $i|j|kl$. The omitted partition of the same type is $ij|kl$, corresponding to the cumulant product $\kappa^{i,j}\kappa^{k,l}$. Occasionally, this shorthand notation leads to ambiguity and if so, it becomes necessary to list the individual partitions explicitly. However, the reader quickly becomes overwhelmed by the sheer number of terms that complete lists involve. For this reason we make every effort to avoid explicit complete lists.

An alternative notation, useful in order to avoid the kinds of ambiguity alluded to above, is to write $\kappa^{i,k}\kappa^{j,l}[2]_{ij}$ for the final term in (3.2). However, this notation conflicts with the summation convention and, less seriously, $\kappa^{i,k}\kappa^{j,l}[2]_{ij}$ is the same as $\kappa^{i,k}\kappa^{j,l}[2]_{kl}$. For these reasons, the unadorned bracket notation will be employed here only where there is no risk of ambiguity.

From the above examples, it is possible to discern, at least qualitatively, the rule that is being applied in expressing generalized cumulants in terms of ordinary cumulants. An arbitrary cumulant of order α involving β random variables may be written as $\kappa(\Upsilon^*)$ where $\Upsilon^* = \{v_1^*,...,v_\alpha^*\}$ is a partition of β indices into α non-empty blocks. Rather conveniently, every partition that appears on the right in (3.2) has coefficient $+1$ and, in fact, the general expression may be written

$$\kappa(\Upsilon^*) = \sum_{\Upsilon \vee \Upsilon^* = 1} \kappa(v_1)...\kappa(v_\nu), \qquad (3.3)$$

where the sum is over all $\Upsilon = \{v_1,...,v_\nu\}$ such that Υ and Υ^* are not both sub-partitions of any partition other than the full set, $\Upsilon_1 = \{(1, 2, ..., \beta)\}$. We refer to this set as the set of partitions *complementary* to Υ^*. The notation and terminology used here are borrowed from lattice theory where $\Upsilon \vee \Upsilon^*$ is to be read as the *least upper bound* of Υ and Υ^*. A proof of this result is given in Section 3.6, using properties of the lattice of set partitions. In practice, the following graph-theoretical description of the condition $\Upsilon \vee \Upsilon^* = 1$ seems preferable and is easier to visualize.

Any partition, say $\Upsilon^* = \{v_1^*,...,v_\alpha^*\}$, can be represented as a graph on β vertices. The edges of the graph consist of all pairs (i,j) that are in the same block of Υ^*. Thus the graph of Υ^* is the union of α disconnected complete graphs and we use the notation Υ^* interchangeably for the graph and for the partition.

Since Υ and Υ^* are two graphs sharing the same vertices, we may define the edge sum graph $\Upsilon \oplus \Upsilon^*$, whose edges are the union of the edges of Υ and Υ^*. The condition that $\Upsilon \oplus \Upsilon^*$ be connected is identical to the condition $\Upsilon \vee \Upsilon^* = 1$ in (3.3). For this reason, we use the terms *connecting partition* and *complementary partition* interchangeably. In fact, this graph-theoretical device provides a simple way of determining the least upper bound of two or more partitions: the blocks of $\Upsilon \vee \Upsilon^*$ are just the connected components of the graph $\Upsilon \oplus \Upsilon^*$. The connections in this case need not be direct. In other words, the blocks of $\Upsilon \vee \Upsilon^*$ do not, in general, correspond to cliques of $\Upsilon \oplus \Upsilon^*$.

Consider, for example, the (4,6) cumulant $\kappa^{12,34,5,6}$ with $\Upsilon^* = 12|34|5|6$. Each block of the partition $\Upsilon = 13|24|56$ joins two blocks of Υ^*, but Υ and Υ^* are both sub-partitions of $1234|56$ and so the condition in (3.3) is not satisfied. In the graph-theoretical representation, the first four vertices of $\Upsilon \oplus \Upsilon^*$ are connected and also the last two, but the vertices fall into two disconnected sets. On the other hand, $\Upsilon = 1|23|456$ satisfies the condition $\Upsilon \vee \Upsilon^* = 1$, required in (3.3).

Expression (3.3) gives generalized cumulants in terms of ordinary cumulants. The corresponding expression for generalized cumulants in terms of ordinary moments is

$$\kappa(\Upsilon^*) = \sum_{\Upsilon \geq \Upsilon^*} (-1)^{\nu-1}(\nu - 1)!\, \mu(v_1)...\mu(v_\nu) \qquad (3.4)$$

where $\mu(v_j)$ is an ordinary moment and the sum is over all partitions Υ such that Υ^* is a sub-partition of Υ. This expression follows from the development in Section 2.4 and can be regarded as effectively equivalent to (2.9). See also Section 3.6.

It is not difficult to see that (2.8) and (2.9) are special cases of (3.3) and (3.4). If we take $\Upsilon^* = \Upsilon_1$, the unpartitioned set, then every partition Υ satisfies the condition required in (3.3), giving (2.8). On the other hand, if we take $\Upsilon^* = \Upsilon_p$, the fully partitioned set, then Υ^* is a sub-partition of every partition. Thus every partition contributes to (3.4) in this case, giving (2.9).

3.3   Cumulants of homogeneous polynomials

For definiteness, consider two homogeneous polynomials

$$P_2 = a_{ij}X^iX^j \quad \text{and} \quad P_3 = a_{ijk}X^iX^jX^k$$

of degree 2 and 3 respectively. In many ways, it is best to think of P_2 and P_3, not as quadratic and cubic forms, but as linear forms in pairs of variables and triples of variables respectively. From this point of view, we can see immediately that

$$E(P_2) = a_{ij}\kappa^{ij} = a_{ij}\{\kappa^{i,j} + \kappa^i\kappa^j\}$$

$$E(P_3) = a_{ijk}\kappa^{ijk} = a_{ijk}\{\kappa^{i,j,k} + \kappa^i\kappa^{j,k}[3] + \kappa^i\kappa^j\kappa^k\}$$

$$\begin{aligned}
\text{var}(P_2) &= a_{ij}a_{kl}\kappa^{ij,kl} \\
&= a_{ij}a_{kl}\{\kappa^{i,j,k,l} + \kappa^i\kappa^{j,k,l}[2] + \kappa^k\kappa^{i,j,l}[2] \\
&\quad + \kappa^{i,k}\kappa^{j,l}[2] + \kappa^i\kappa^k\kappa^{j,l}[4]\} \\
&= a_{ij}a_{kl}\{\kappa^{i,j,k,l} + 4\kappa^i\kappa^{j,k,l} + 2\kappa^{i,k}\kappa^{j,l} + 4\kappa^i\kappa^k\kappa^{j,l}\}
\end{aligned}$$

$$\begin{aligned}
\text{cov}(P_2, P_3) &= a_{ij}a_{klm}\kappa^{ij,klm} \\
&= a_{ij}a_{klm}\{\kappa^{i,j,k,l,m} + \kappa^i\kappa^{j,k,l,m}[2] + \kappa^k\kappa^{i,j,l,m}[3] \\
&\quad + \kappa^{i,k}\kappa^{j,l,m}[6] + \kappa^{k,l}\kappa^{i,j,m}[3] + \kappa^i\kappa^k\kappa^{j,l,m}[6] + \kappa^k\kappa^l\kappa^{i,j,m}[3] \\
&\quad + \kappa^{i,k}\kappa^{j,l}\kappa^m[6] + \kappa^{i,k}\kappa^{l,m}\kappa^j[6] + \kappa^{i,k}\kappa^j\kappa^l\kappa^m[6]\}
\end{aligned}$$

For the final expression above where $\Upsilon^* = ij|klm$, the list of 42 complementary partitions can be found in Table 1 of the Appendix, where Υ^* is coded numerically as 123|45. Since the arrays of coefficients a_{ij} and a_{ijk} are symmetrical, the permutation factors in [.] can be changed into ordinary arithmetic factors. Thus the 42 complementary partitions contribute only 10 distinct terms. These cannot be condensed further except in special cases. In the expression for $\text{var}(P_2)$, on the other hand, the two classes of partitions $i|jkl[2]$ and $k|ijl[2]$ make equal contributions and further condensation is then possible as shown above.

In specific applications, it is often the case that the arrays a_{ij} and a_{ijk} have some special structure that can be exploited in order to simplify expressions such as those listed above. Alternatively, it may be that the cumulants have simple structure characteristic of independence, exchangeability, or identical distributions. In such cases, the joint cumulants listed above can be condensed further using power notation.

By way of illustration, we suppose that a_{ij} is a residual covariance matrix of rank r, most commonly written as $\mathbf{I} - \mathbf{X}(\mathbf{X}^T\mathbf{X})^{-1}\mathbf{X}^T$ in the notation of linear models where $\mathbf{X}$ is a model matrix of known constants. We suppose in addition that the random variables $X^i - \kappa^i$ are independent and identically distributed. It follows from $a_{ij}\kappa^j = 0$ that

$$E(P_2) = a_{ij}\kappa^{i,j} = \kappa_2 \sum a_{ii} = r\kappa_2$$

$$\mathrm{var}(P_2) = a_{ij}a_{kl}\{\kappa^{i,j,k,l} + 2\kappa^{i,k}\kappa^{j,l}\}$$

$$= \kappa_4 \sum a_{ii}^2 + 2r\kappa_2^2$$

In this example, P_2 is just the residual sum of squares on r degrees of freedom after linear regression on $\mathbf{X}$, where the theoretical errors are independent and identically distributed but not necessarily normal.

3.4   Polynomial transformations

It is not difficult now to develop the transformation law for cumulants under arbitrary non-linear transformation to new variables Y. The formulae developed in Section 3.2 refer to the particular polynomial transformation

$$Y^1 = \prod_{j \in v_1^*} X^j, \quad Y^2 = \prod_{j \in v_2^*} X^j, \ \ldots, Y^\alpha = \prod_{j \in v_\alpha^*} X^j.$$

To the extent that any continuous function can be approximated with arbitrary accuracy by means of a polynomial, there is little loss of generality in considering polynomial transformations

$$Y^r = a^r + a_i^r X^i + a_{ij}^r X^i X^j + a_{ijk}^r X^i X^j X^k + \ldots. \tag{3.5}$$

It is necessary at this stage to insist that the infinite expansion (3.5) be convergent, not necessarily for all X, but at least for all X for which the probability is appreciable.

To state the transformation law of cumulants in a concise way, it is helpful to abbreviate (3.5) by using 'matrix' notation as follows:

$$Y^r = (A_0^r + A_1^r + A_2^r + A_3^r + ...)X \qquad (3.6)$$

where, for example, $A_2^r X$ is understood to represent a vector whose components are quadratic or bilinear forms in X. We will further abbreviate (3.6) by introducing the operators $P^r = A_0^r + A_1^r + A_2^r + ...$ and writing

$$Y^r = P^r X. \qquad (3.7)$$

The cumulant generating function of $Y = Y^1, ..., Y^q$ may now be written purely formally as

$$K_Y(\xi) = \exp(\xi_r P^r) \kappa_X \qquad (3.8)$$

where $\exp(\xi_r P^r)$ is an operator acting on the cumulants of X as follows.

$$K_Y(\xi) = \{1 + \xi_r P^r + \xi_r \xi_s P^r P^s / 2! \\ + \xi_r \xi_s \xi_t P^r P^s P^t / 3! + ...\} \kappa_X. \qquad (3.9)$$

We define $1\kappa_X = 0$ and

$$P^r \kappa_X = a^r + a_i^r \kappa^i + a_{ij}^r \kappa^{ij} + a_{ijk}^r \kappa^{ijk} + ...$$

Compound operators $P^r P^s$ and $P^r P^s P^t$ acting on κ_X produce generalized cumulants of order $\alpha = 2$ and $\alpha = 3$ respectively. For example, $A_1^r A_2^s \kappa_X = a_i^r a_{jk}^s \kappa^{i,jk}$, which is not to be confused with $A_2^r A_1^s \kappa_X = a_{ij}^r a_k^s \kappa^{ij,k}$. In like manner, third-order compound operators produce terms such as

$$A_1^r A_1^s A_1^t \kappa_X = a_i^r a_j^s a_k^t \kappa^{i,j,k}$$
$$A_1^r A_2^s A_1^t \kappa_X = a_i^r a_{jk}^s a_l^t \kappa^{i,jk,l}.$$

Compound operators involving A_0 produce terms such as

$$A_0^r A_1^s \kappa_X = a^r a_i^s \kappa^{,i} = 0$$
$$A_1^r A_0^s A_2^t \kappa_X = a_i^r a^s a_{kl}^t \kappa^{i,,kl} = 0$$

and these are zero because they are mixed cumulants involving, in effect, one variable that is a constant.

Expansion (3.9) for the first four cumulants gives

$$K_Y(\xi) = \xi_r\{a^r + a_i^r\kappa^i + a_{ij}^r\kappa^{ij} + a_{ijk}^r\kappa^{ijk} + ...\}$$

$$+ \xi_r\xi_s\{a_i^r a_j^s\kappa^{i,j} + a_i^r a_{jk}^s\kappa^{i,jk}[2] + a_{ij}^r a_{kl}^s\kappa^{ij,kl} + ...\}/2!$$

$$+ \xi_r\xi_s\xi_t\{a_i^r a_j^s a_k^t\kappa^{i,j,k} + a_i^r a_j^s a_{kl}^t\kappa^{i,j,kl}[3] + a_i^r a_{jk}^s a_{lm}^t\kappa^{i,jk,lm}[3] + ...\}/3!$$

$$+ \xi_r\xi_s\xi_t\xi_u\{a_i^r a_j^s a_k^t a_l^u\kappa^{i,j,k,l} + a_i^r a_j^s a_k^t a_{lm}^u\kappa^{i,j,k,lm}[4] + ...\}/4!$$

$$+ ...\tag{3.10}$$

The leading terms in the above expansions are the same as (2.11), the law governing affine transformation.

The proof of (3.8) is entirely elementary and follows from the definition of generalized cumulants, together with the results of Section 2.4. A similar, though less useful formal expression can be developed for the moment generating function $M_X(\xi)$, which may be written

$$M_Y(\xi) = \exp(\xi_r P^r) * \kappa_X \tag{3.11}$$

where $1 * \kappa_X = 1$, $P^r * \kappa_X = P^r\kappa_X$ as before, and commas are omitted in the application of compound operators giving

$$A_0^r A_1^s * \kappa_X = a^r a_i^s\kappa^i$$

$$A_1^r A_1^s * \kappa_X = a_i^r a_j^s\kappa^{ij}$$

$$A_1^r A_0^s A_2^t * \kappa_X = a_i^r a^s a_{jk}^t\kappa^{ijk}$$

and so on. Once again, the proof follows directly from the definition of moments.

In practice, in order to make use of (3.10), it is usually necessary to re-express all generalized cumulants in terms of ordinary cumulants. This exercise involves numerous applications of (3.3) and the formulae become considerably longer as a result.

3.5   Classifying complementary partitions

In order to use the fundamental identity (3.3) for the generalized cumulant $\kappa(\Upsilon^*)$ we need to list all partitions complementary to Υ^*. If the number of elements of Υ^* is more than, say six, the number of complementary partitions can be very large indeed. If the listing is done by hand, it is difficult to be sure that no complementary partitions have been omitted. On the other hand, programming a computer to produce the required list is not an easy task. Further, it is helpful to group the partitions into a small number of equivalence classes in such a way that all members of a given class make the same contribution to the terms that occur in (3.10).

Suppose, for example, that we require the covariance of the two scalars

$$a_{ij}X^iX^j \quad \text{and} \quad b_{ij}X^iX^j$$

both of which are quadratic in X. We find

$$\text{cov}\{a_{ij}X^iX^j, \ b_{ij}X^iX^j\} = a_{ij}b_{kl}\kappa^{ij,kl}.$$

To simplify the following expressions, we take $\kappa^i = 0$, giving

$$a_{ij}b_{kl}\{\kappa^{i,j,k,l} + \kappa^{i,k}\kappa^{j,l} + \kappa^{i,l}\kappa^{j,k}\}.$$

Since, by assumption, $a_{ij} = a_{ji}$ and $b_{ij} = b_{ji}$, this expression reduces to

$$a_{ij}b_{kl}\{\kappa^{i,j,k,l} + 2\kappa^{i,k}\kappa^{j,l}\}.$$

Thus the two partitions $ik|jl$ and $il|jk$ make identical contributions to the covariance and therefore they belong to the same equivalence class. This classification explains the grouping of terms in (3.2).

The classification of complementary partitions is best described in terms of the intersection matrix $M = \Upsilon^* \cap \Upsilon$, where m_{ij} is the number of elements in $v_i^* \cap v_j$. This matrix is defined only up to row and column permutations. Two partitions, Υ_1 and Υ_2, whose intersection matrices are M_1 and M_2, are regarded as equivalent if $M_1 = M_2$ after suitably permuting the blocks of Υ_1 and Υ_2 or the columns of M_1 and M_2. It is essential in this comparison that the ith rows of M_1 and M_2 refer to the same block of Υ^*.

To take a simple example, suppose that $\Upsilon^* = 12|34|5$, $\Upsilon_1 = 135|24$, $\Upsilon_2 = 123|45$ and $\Upsilon_3 = 134|25$ with intersection matrices

$$M_1 = \begin{matrix} 1 & 1 \\ 1 & 1 \\ 1 & 0 \end{matrix} \qquad M_2 = \begin{matrix} 2 & 0 \\ 1 & 1 \\ 0 & 1 \end{matrix} \qquad M_3 = \begin{matrix} 1 & 1 \\ 2 & 0 \\ 0 & 1 \end{matrix}$$

These matrices are all distinct. However, the partitions $145|23$, $235|14$ and $245|13$ have M_1 as intersection matrix, the partition $124|35$ has M_2 as intersection matrix and $234|15$ has M_3 as intersection matrix. Thus these eight complementary partitions are written as

$$135|24[4] \quad \cup \quad 123|45[2] \quad \cup \quad 134|25[2].$$

It is not difficult to see that if Υ_1 and Υ_2 are equivalent partitions in the sense just described, they must have the same number of blocks and also identical block sizes. Further, when permuting columns, we need only consider blocks of equal size: by convention, the blocks are usually arranged in decreasing size.

Tables 1 and 2 in the Appendix give lists of complementary partitions classified according to the above scheme. A typical element of each equivalence class is given and the number of elements in that class follows in [.].

3.6   Elementary lattice theory

3.6.1   *Generalities*

For our purposes, a lattice may be defined as a finite partially ordered set $\mathcal{L}$ having the additional property that for every pair of elements $a, b \in \mathcal{L}$ there is defined a unique greatest lower bound $c = a \wedge b$ and a unique least upper bound $d = a \vee b$ where $c, d \in \mathcal{L}$. These additional properties should not be taken lightly and, in fact, some commonly occurring partially ordered sets are not lattices because not every pair has a unique least upper bound or greatest lower bound. One such example is described in Exercise 3.34.

Lattices of various types arise rather frequently in statistics and probability theory. For example, in statistics, the class of all factorial models, which can be described using the operators

$+$ and $*$ on factors A, B, C,..., forms a lattice known as the *free distributive lattice*. Typical elements of this lattice are $a = A+B*C$ and $b = A*B+B*C+C*A$, each corresponding to a factorial model. In the literature on discrete data, these models are also called *hierarchical*, but this usage conflicts with standard terminology in the analysis of variance where *hierarchical* refers to the presence of several variance components. In this particular example, $a < b$ because a is a sub-model of b, and the partial order has a useful statistical interpretation.

In probability theory or in set theory where $A_1, A_2, ...$ are subsets of Ω, it is sometimes useful to consider the lattice with elements Ω, A_i, $A_i \cap A_j$, $A_i \cap A_j \cap A_k$ and so on. We say that $a < b$ if $a \subset b$; the lattice so formed is known as the *binary lattice*. The celebrated inclusion-exclusion principle for calculating $\text{pr}(A_1 \cup A_2 \cup ...)$ is a particular instance of (3.12) below (Rota, 1964). Exercise 2.22 provides a third example relevant both to statistics and to probability theory.

In this Section, however, we are concerned only with the lattice of set partitions where $a < b$ if a is a sub-partition of b. The least upper bound, $a \vee b$ was described in Section 3.2 as the partition whose blocks are the connected vertices of the graph $a \oplus b$: the greatest lower bound $a \wedge b$ is the partition whose blocks are the non-empty intersections of the blocks of a and b. Figure 3.1 gives the Hasse diagrams of the partition lattices of sets of up to four items. Each partition in the ith row of one of these diagrams contains i blocks. Partitions in the same row are unrelated in the partial order: partitions in different rows may or may not be related in the partial order. Notice that $123|4$ has three immediate descendants whereas $13|24$ has only two.

Suppose now that $\mathcal{L}$ is an arbitrary lattice and that $f(.)$ is a real-valued function defined on $\mathcal{L}$. We define the new function, $F(.)$ on $\mathcal{L}$ by

$$F(b) = \sum_{a \leq b} f(a),$$

analogous to integrating over the interval $[0, b]$. The formal inverse operation analogous to differentiation may be written

$$f(b) = \sum_{a \leq b} m(a, b) F(a) \tag{3.12}$$

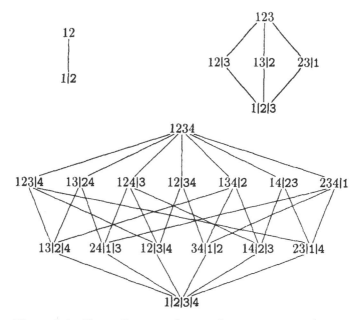

Figure 3.1: *Hasse diagrams for smaller partition lattices.*

where $m(a, b)$ is known as the Möbius function for the lattice. See Exercise 2.22 for a justification of this analogy with differential and integral calculus. The Möbius function may be obtained inductively by $m(a, a) = 1$,

$$m(a, c) = - \sum_{a \leq b < c} m(a, b),$$

and $m(a, b) = 0$ otherwise. It follows, for fixed a and c, that the sum over the lattice interval $[a, c]$,

$$\sum_{a \leq b \leq c} m(a, b) = \delta(a, c)$$

is zero unless $a = c$. In this sense, the Möbius function can be thought of as a series of contrasts or difference operators on all lattice intervals, $[a, c]$ where $a < c$.

It is not difficult to see that the function $m(a, b)$ so defined satisfies (3.12). If we think of $F(.)$ and $f(.)$ as column vectors indexed by the lattice elements, then we may write $F(b) = \sum_a h(a, b) f(a)$ where $h(a, b) = 1$ if $a \leq b$ and zero otherwise. To verify (3.12), we need only observe that $m(a, b)$ is the matrix inverse of $h(a, b)$. This follows immediately from the definition of the Möbius function, which may be written

$$\sum_b m(a, b) h(b, c) = \delta(a, c) = \sum_b m(b, c) h(a, b).$$

The actual value of the Möbius function depends on the structure of $\mathcal{L}$, but for the partition lattice, it is convenient to remember that

$$m(\Upsilon, 1) = (-1)^{\nu-1}(\nu - 1)! \qquad (3.13)$$

where ν is the number of blocks of the partition Υ, and 1 is an abbreviation for Υ_1, the greatest element of the lattice. More generally, $m(\Upsilon, \Upsilon^*)$ is the product of terms like (3.13). For example, $m(1|2|3|4, 12|34) = (-1)^2$ while $m(1|2|3|4, 123|4) = 2 \times 1$.

3.6.2 Cumulants and the partition lattice

In Section 2.3 we examined the series expansion of $\log M_X(\xi)$ and showed that the moment $\kappa(\Upsilon_1)$ could be expressed as

$$\kappa(\Upsilon_1) = \sum_{0 \leq \Upsilon \leq 1} \kappa(v_1)...\kappa(v_\nu) \qquad (3.14)$$

where 0 is an abbreviation for the least element, Υ_p. On multiplying together expressions such as (3.14), one expression for each of the blocks of Υ^*, we find

$$\mu(v_1^*)...\mu(v_\alpha^*) = \sum_{0 \leq \Upsilon \leq \Upsilon^*} \kappa(v_1)...\kappa(v_\nu). \qquad (3.15)$$

In matrix notation, for the special case of four variables, we may write the above equation as shown in Table 3.1.

The matrix equation is entirely equivalent to the set of identities (2.6), which give second-, third- and fourth-order moments in

Table 3.1 *Moment products expressed in terms of cumulants*

Moment product	c1	c2	c3	c4	c5	c6	c7	c8	c9	c10	c11	c12	c13	c14	c15	Cumulant product
κ^{rstu}	1	1	1	1	1	1	1	1	1	1	1	1	1	1	1	$\kappa^{r,s,t,u}$
$\kappa^{rst}\kappa^u$		1							1	1		1			1	$\kappa^{r,s,t}\kappa^u$
$\kappa^{rsu}\kappa^t$			1						1		1		1		1	$\kappa^{r,s,u}\kappa^t$
$\kappa^{rtu}\kappa^s$				1						1	1			1	1	$\kappa^{r,t,u}\kappa^s$
$\kappa^{stu}\kappa^r$					1							1	1	1	1	$\kappa^{s,t,u}\kappa^r$
$\kappa^{rs}\kappa^{tu}$						1			1					1	1	$\kappa^{r,s}\kappa^{t,u}$
$\kappa^{rt}\kappa^{su}$							1			1			1		1	$\kappa^{r,t}\kappa^{s,u}$
$\kappa^{ru}\kappa^{st}$ $=$								1			1	1			1	$\kappa^{r,u}\kappa^{s,t}$
$\kappa^{rs}\kappa^t\kappa^u$									1						1	$\kappa^{r,s}\kappa^t\kappa^u$
$\kappa^{rt}\kappa^s\kappa^u$										1					1	$\kappa^{r,t}\kappa^s\kappa^u$
$\kappa^{ru}\kappa^s\kappa^t$											1				1	$\kappa^{r,u}\kappa^s\kappa^t$
$\kappa^{st}\kappa^r\kappa^u$												1			1	$\kappa^{s,t}\kappa^r\kappa^u$
$\kappa^{su}\kappa^r\kappa^t$													1		1	$\kappa^{s,u}\kappa^r\kappa^t$
$\kappa^{tu}\kappa^r\kappa^s$														1	1	$\kappa^{t,u}\kappa^r\kappa^s$
$\kappa^r\kappa^s\kappa^t\kappa^u$															1	$\kappa^r\kappa^s\kappa^t\kappa^u$

terms of cumulants. To include fifth-order quantities, it is necessary to use a 52×52 matrix. Equation (3.15) has the advantage over the matrix equation in that it recognizes explicitly the lattice structure and also in that the number of variables need not be stated explicitly.

Application of the inversion formula (3.12) gives

$$\kappa(v_1^*)...\kappa(v_\alpha^*) = \sum_{0 \leq \Upsilon \leq \Upsilon^*} m(\Upsilon, \Upsilon^*)\mu(v_1)...\mu(v_\nu).$$

The matrix with elements $m(\Upsilon_i, \Upsilon_j)$ is also upper triangular and is given explicitly in Section 4.5.3 for the case of four variables. If $\alpha = 1$, we may write

$$\kappa(\Upsilon_p) = \sum_{0 \leq \Upsilon \leq 1} (-1)^{\nu-1}(\nu - 1)!\,\mu(v_1)...\mu(v_\nu).$$

More generally if each index in the above expression refers to a product of variables, we may deduce that

$$\kappa(\Upsilon^*) = \sum_{\Upsilon \geq \Upsilon^*} (-1)^{\nu-1}(\nu - 1)!\,\mu(v_1)...\mu(v_\nu), \qquad (3.16)$$

the expression for a generalized cumulant in terms of ordinary moments. Finally, substitution of (3.15) gives

$$\kappa(\Upsilon^*) = \sum_{\Upsilon \geq \Upsilon^*} (-1)^{\nu-1}(\nu-1)! \sum_{\Pi \leq \Upsilon} \kappa(\pi_1)...\kappa(\pi_\sigma) \qquad (3.17)$$

where σ is the number of blocks of Π. It is really quite straightforward to see that the only partitions making a contribution are those for which $\Pi \vee \Upsilon^* = 1$ and that these occur with unit coefficient. See Exercises 3.16 and 3.17 for a justification of this claim. Thus

$$\kappa(\Upsilon^*) = \sum_{\Upsilon \vee \Upsilon^* = 1} \kappa(v_1)...\kappa(v_\nu)$$

and the fundamental result (3.3) is proved.

In matrix notation, for the special case of four variables, we may write this equation as shown in Table 3.2.

Table 3.2 *Generalized cumulants expressed in terms of ordinary cumulants*

κ^{rstu}	1	1	1	1	1	1	1	1	1	1	1	1	1	1		$\kappa^{r,s,t,u}$
$\kappa^{rsu,t}$	1	1		1	1	1	1		1		1		1			$\kappa^{r,s,u}\kappa^t$
$\kappa^{rtu,s}$	1	1	1		1	1	1	1			1	1				$\kappa^{r,t,u}\kappa^s$
$\kappa^{stu,r}$	1	1	1	1		1	1	1	1	1	1					$\kappa^{s,t,u}\kappa^r$
$\kappa^{rs,tu}$	1	1	1	1	1		1	1		1	1	1	1			$\kappa^{r,s}\kappa^{t,u}$
$\kappa^{rt,su}$	1	1	1	1	1	1		1	1		1	1		1		$\kappa^{r,t}\kappa^{s,u}$
$\kappa^{ru,st}$ =	1	1	1	1	1	1	1		1	1		1	1			$\kappa^{r,u}\kappa^{s,t}$
$\kappa^{rs,t,u}$	1			1	1		1	1								$\kappa^{r,s}\kappa^t\kappa^u$
$\kappa^{rt,s,u}$	1		1		1	1		1								$\kappa^{r,t}\kappa^s\kappa^u$
$\kappa^{ru,s,t}$	1	1			1	1	1									$\kappa^{r,u}\kappa^s\kappa^t$
$\kappa^{st,r,u}$	1		1	1		1	1									$\kappa^{s,t}\kappa^r\kappa^u$
$\kappa^{su,r,t}$	1	1		1		1		1								$\kappa^{s,u}\kappa^r\kappa^t$
$\kappa^{tu,r,s}$	1	1	1				1	1								$\kappa^{t,u}\kappa^r\kappa^s$
$\kappa^{r,s,t,u}$	1															$\kappa^r\kappa^s\kappa^t\kappa^u$

When we come to deal with the theory of symmetric functions in Chapter 4, we will find that there are three convenient linear bases for symmetric functions of given degree. One basis, associated with generalized cumulants, is the set of generalized k-statistics; the second, associated with cumulant products is the set of polykays; the third, associated with moment products, is called symmetric means. The expression derived above, when translated, gives generalized k-statistics in terms of polykays. We now derive the inverse expression for cumulant products in terms of generalized cumulants or, equivalently, polykays in terms of generalized k-statistics. First invert (3.16) giving

$$m(\Upsilon^*, 1)\mu(v_1^*)...\mu(v_\alpha^*) = \sum_{\Upsilon \geq \Upsilon^*} m(\Upsilon^*, \Upsilon)\kappa(\Upsilon)$$

see e.g. Aigner, 1980, p.152. Now use the formula below (3.15) for cumulant products in terms of moment products giving

$$\kappa(v_1^*)...\kappa(v_\alpha^*) = \sum_{\Upsilon} \sigma(\Upsilon^*, \Upsilon)\kappa(\Upsilon) \qquad (3.18)$$

where $\sigma(b, c)$ is a symmetric array with elements

$$\sigma(b, c) = \sum_a \frac{m(a, b)m(a, c)}{m(a, 1)}$$

and summation may be restricted to $a \leq b \wedge c$.

Explicitly, if $p = 3$, we may write the five partitions in order as $a_1 = 123$, $a_2 = 1|23$, $a_3 = 2|13$, $a_4 = 3|12$ and $a_5 = 1|2|3$. The matrix giving generalized cumulants in terms of cumulant products in (3.3) is

$$\begin{pmatrix} 1 & 1 & 1 & 1 & 1 \\ 1 & 0 & 1 & 1 & 0 \\ 1 & 1 & 0 & 1 & 0 \\ 1 & 1 & 1 & 0 & 0 \\ 1 & 0 & 0 & 0 & 0 \end{pmatrix}$$

where a unit entry in position (i, j) corresponds to the criterion $a_i \vee a_j = 1$. The matrix inverse corresponding to (3.18) above is

$$\frac{1}{2} \begin{pmatrix} 0 & 0 & 0 & 0 & 2 \\ 0 & -1 & 1 & 1 & -1 \\ 0 & 1 & -1 & 1 & -1 \\ 0 & 1 & 1 & -1 & -1 \\ 2 & -1 & -1 & -1 & 1 \end{pmatrix}$$

Table 3.3 *Matrix giving cumulant products in terms of moment products*

	0	0	0	0	0	0	0	0	0	0	0	0	0	0	6
	0	-1	-1	-1	-1	1	1	1	-1	-1	2	-1	2	2	-2
	0	-1	-1	-1	-1	1	1	1	-1	2	-1	2	-1	2	-2
	0	-1	-1	-1	-1	1	1	1	2	-1	-1	2	2	-1	-2
	0	-1	-1	-1	-1	1	1	1	2	2	2	-1	-1	-1	-2
	0	1	1	1	1	-1	-1	-1	-2	1	1	1	1	-2	-1
1	0	1	1	1	1	-1	-1	-1	1	-2	1	1	-2	1	-1
−	0	1	1	1	1	-1	-1	-1	1	1	-2	-2	1	1	-1
6	0	-1	-1	2	2	-2	1	1	2	-1	-1	-1	-1	-1	1
	0	-1	2	-1	2	1	-2	1	-1	2	-1	-1	-1	-1	1
	0	2	-1	-1	2	1	1	-2	-1	-1	2	-1	-1	-1	1
	0	-1	2	2	-1	1	1	-2	-1	-1	-1	2	-1	-1	1
	0	2	-1	2	-1	1	-2	1	-1	-1	-1	-1	2	-1	1
	0	2	2	-1	-1	-2	1	1	-1	-1	-1	-1	-1	2	1
	6	-2	-2	-2	-2	-1	-1	-1	1	1	1	1	1	1	-1

Application of (3.18), using the second row of the above matrix, gives

$$\kappa^i \kappa^{j,k} = \{\kappa^{j,ik} + \kappa^{k,ij} - \kappa^{i,jk} - \kappa^{i,j,k}\}/2$$

and this particular expression can be verified directly.

For $p = 4$, the matrix $\sigma(.,.)$ in (3.18) is shown in Table 3.3.

The most important of these identities, or at least the ones that occur most frequently in the remainder of this book, are (3.15) and its inverse and the expression involving connecting partitions for generalized cumulants in terms of cumulant products. The inverse of the latter expression seems not to arise often. It is given here in order to emphasize that the relationship is invertible.

Finally, we note from (3.15) that any polynomial in the moments, homogeneous in the sense that every term is of degree 1 in each of p variables, can be expressed as a similarly homogeneous polynomial in the cumulants, and vice-versa for cumulants in terms of moments. In addition, from (3.16) we see that every generalized cumulant of degree p is uniquely expressible as a homogeneous polynomial in the moments. Inversion of (3.16) shows that every homogeneous polynomial in the moments is expressible as a *linear* function of generalized cumulants, each of degree p. Similarly, every homogeneous polynomial in the cumulants is expressible as a linear function of generalized cumulants. It follows that any polynomial in the cumulants or in the moments, homogeneous or not, is expressible as a *linear* function of generalized cumulants. Furthermore, this linear representation is unique because the generalized cumulants are linearly independent (Section 3.8). In fact, we have previously made extensive use of this important property of generalized cumulants. Expansion (3.10) for the cumulant generating function of the polynomial (3.5) is a *linear* function of generalized cumulants.

3.7 Some examples involving linear models

The notation used in this section is adapted so as to conform to the conventions for linear models. Thus $y = y^1, ..., y^n$ is the observed value of the random vector $Y = Y^1, ..., Y^n$ whose cumulants are κ^i, $\kappa^{i,j}$, $\kappa^{i,j,k}$ and so on. The common case where the observations are independent is especially important but, for the moment, the cumulant arrays are taken as arbitrary with no special structure. Now, Y and y lie in R^n but the usual linear model specifies that the mean vector with components κ^i lies in the p-dimensional subspace, S_p of R^n spanned by the vectors $x_1, ..., x_p$ with components x_r^i. Thus we may write

$$\kappa^i = x_r^i \beta^r$$

where β is a p-dimensional vector of parameters to be estimated.

It is important at this stage, to ensure that the notation distinguish between vectors in R^n, such as y^i and κ^i, and vectors in R^p such as β^r. To do so, we use the letters $r, s, t, ...$ in reference to R^p and $i, j, k, ...$ to refer to R^n.

Assume now that $\kappa^{i,j}$ is known and that its matrix inverse is $\kappa_{i,j}$. The cumulants of $Y_i = \kappa_{i,j} Y^j$ may be written κ_i, $\kappa_{i,j}$, $\kappa_{i,j,k}$, and so on, and y_i is a quantity that can be computed. The weighted least squares estimate, b^r of β^r satisfies

$$(x_r^i x_s^j \kappa_{i,j}) b^s = x_r^i \kappa_{i,j} Y^j = x_r^i Y_i. \tag{3.19}$$

Since the matrix $x_r^i x_s^j \kappa_{i,j}$ arises rather frequently, we denote it by $\beta_{r,s}$ and its inverse by $\beta^{r,s}$. The reason for this unusual choice of notation is that $\beta^r, \beta^{r,s}$ and $\beta_{r,s}$ play exactly the same roles in R^p as κ^i, $\kappa^{i,j}$ and $\kappa_{i,j}$ in R^n. Thus (3.19) becomes

$$\beta_{r,s} b^s = b_r = x_r^i Y_i$$

so that b^r and b_r play exactly the same role in R^p as Y^i and Y_i in R^n. The cumulants of b_r are

$$x^i_r \kappa_{i,j} \kappa^j = x^i_r \kappa_{i,j} x^j_s \beta^s = \beta_{r,s} \beta^s = \beta_r,$$

$$x^i_r x^j_s \kappa_{i,j} = \beta_{r,s}, \qquad x^i_r x^j_s x^k_t \kappa_{i,j,k} = \beta_{r,s,t},$$

$$x^i_r x^j_s x^k_t x^l_u \kappa_{i,j,k,l} = \beta_{r,s,t,u}$$

and so on. The cumulants of the least squares estimate b^r are obtained by raising indices giving β^r, $\beta^{r,s}$,

$$\beta^{r,s,t} = \beta^{r,u} \beta^{s,v} \beta^{t,w} \beta_{u,v,w}$$

and so on.
When we use the term tensor in reference to $\beta^{r,s}$, $\beta^{r,s,t}$ and so on, we refer not to transformation of the response vector Y but to changes in the coordinate system in S^p, i.e. to linear or non-linear transformation of β. The least squares equation (3.19) may be derived by differentiating

$$S^2(\beta) = (Y^i - x^i_r \beta^r)(Y^j - x^j_s \beta^s)\kappa_{i,j}$$

with respect to β and equating the derivative to zero. If we were to make a linear transformation to new parameters $\theta = \theta^1, ..., \theta^p$, the estimating equations corresponding to (3.19) would be

$$a^r_t(x^i_r x^j_s \kappa_{i,j})b^s = a^r_t x^i_r \kappa_{i,j} Y^i = a^r_t b_r$$

where $a^r_t = \partial \beta^r / \partial \theta^t$ is assumed to have full rank. Thus $b_r \to a^r_t b_r$ implying that b_r and all its cumulants are covariant. By a similar argument, b^r and the corresponding cumulants with superscripts are contravariant, but only under linear transformation of β. Under nonlinear transformation, the estimate b^r transforms in a more complicated way. Details are given in Chapter 7.

The residual sum of squares, $S^2 = S^2(b)$ may be written explicitly as the quadratic form

$$S^2(b) = Y^i Y^j (\kappa_{i,j} - \lambda_{i,j}),$$

where $\lambda^{i,j} = x^i_r x^j_s \beta^{r,s}$ and $\lambda_{i,j} = \kappa_{i,k} \kappa_{k,l} \lambda^{k,l}$ is one choice of generalized inverse. Note also that

$$\lambda_{i,j} x^j_r = \kappa_{i,k} \lambda^{k,l} \kappa_{l,j} x^j_r = \kappa_{i,j} x^j_r$$

and

$$\kappa^{i,j}\lambda_{i,j} = p.$$

These identities are more easily demonstrated using matrix notation. It follows that the cumulants of S and b are given by

$$E(S^2) = \kappa^{i,j}(\kappa_{i,j} - \lambda_{i,j}) = n - p$$

$$\text{var}(S^2) = (\kappa_{i,j} - \lambda_{i,j})(\kappa_{k,l} - \lambda_{k,l})\kappa^{ij,kl}$$

$$= (\kappa_{i,j} - \lambda_{i,j})(\kappa_{k,l} - \lambda_{k,l})\{\kappa^{i,j,k,l} + \kappa^{i,k}\kappa^{j,l}[2]\}$$

$$= 2(n - p) + \kappa^{i,j,k,l}(\kappa_{i,j} - \lambda_{i,j})(\kappa_{k,l} - \lambda_{k,l})$$

$$\text{cov}(b^r, S^2) = \beta^{r,s}x_s^i\kappa_{i,j}(\kappa_{k,l} - \lambda_{k,l})\kappa^{j,kl}.$$

In the important special case where the observations are independent, we may write, in an obvious notation,

$$\text{var}(S^2) = 2(n - p) + \sum_i \rho_4^i(1 - h_{ii})^2$$

where $\mathbf{H} = \mathbf{X}(\mathbf{X}^T\mathbf{W}\mathbf{X})^{-1}\mathbf{X}^T\mathbf{W}$ is the projection matrix producing fitted values and

$$\text{cov}(\mathbf{b}, S^2) = (\mathbf{X}^T\mathbf{W}\mathbf{X})^{-1}\mathbf{X}^T\mathbf{W}\mathbf{C}_3$$

where $c_3^i = \kappa_3^i(1 - h_{ii})/(\kappa_2^i)$. Alternatively, we may write

$$\text{cov}(\hat{\mu}, S^2) = \mathbf{H}\mathbf{C}_3$$

where $\hat{\mu} = \mathbf{X}\mathbf{b}$ is the vector of fitted values.

3.8   Cumulant spaces

We examine here the possibility of constructing distributions in R^p having specified cumulants or moments up to a given finite order, n. Such constructions are simpler in the univariate case and the discussion is easier in terms of moments than in terms of cumulants. Thus, we consider the question of whether or not there exists a distribution on the real line having moments μ_1', μ_2', ...,μ_n', the higher-order moments being left unspecified. If such a distribution exists, it follows that

$$\int (a_0 + a_1 x + a_2 x^2 + ...)^2 f_X(x) dx \geq 0 \qquad (3.20)$$

for any polynomial, with equality only if the density is concentrated at the roots of the polynomial. The implication is that for each $r = 1, 2, ..., [n/2]$, the matrix of order $r + 1 \times r + 1$

$$M_r' = \begin{pmatrix} 1 & \mu_1' & \mu_2' & \cdots & \mu_r' \\ \mu_1' & \mu_2' & \mu_3' & \cdots & \mu_{r+1}' \\ \mu_2' & \mu_3' & \mu_4' & \cdots & \mu_{r+2}' \\ \cdot & \cdot & \cdot & & \cdot \\ \cdot & \cdot & \cdot & & \cdot \\ \cdot & \cdot & \cdot & & \cdot \\ \mu_r' & \mu_{r+1}' & \mu_{r+2}' & \cdots & \mu_{2r}' \end{pmatrix}$$

must be non-negative definite. Equivalently, we may eliminate the first row and column and work with the matrix whose (r, s) element is $\mu_{r,s} = \mathrm{cov}(X^r, X^s)$, where indices here denote powers. The first two inequalities in this sequence are $\kappa_2 \geq 0$ and $\rho_4 \geq \rho_3^2 - 2$.

The above sequence of conditions is necessary but not sufficient to establish that a distribution having the required moments exists. By way of illustration, if we take $\mu_1' = \mu_2' = 1$, it follows that $\kappa_2 = 0$ giving $f_X(x) = \delta(x, 1)$. In other words, all higher-order moments are determined by these particular values of μ_1' and μ_2'. Similarly, if $\mu_1' = 0$, $\mu_2' = 1$, $\mu_3' = 0$, $\mu_4' = 1$, giving $\rho_4 = -2$, we must have

$$f_X(x) = \tfrac{1}{2}\delta(x, -1) + \tfrac{1}{2}\delta(x, 1)$$

because M_2' has a zero eigenvalue with eigenvector $(-1, 0, 1)$, implying that $f_X(x)$ must be concentrated at the roots of the polynomial

$x^2 - 1$. Since $\mu_1' = 0$, the weights must be $\frac{1}{2}$ at $x = \pm 1$. Again, all higher-order moments and cumulants are determined by this particular sequence of four moments. It follows that there is no distribution whose moments are $0, 1, 0, 1, 0, 2, \ldots$ even though the corresponding M_3' is positive semi-definite.

More generally, if for some $k \geq 1$, the matrices $M_1', \ldots, M_{k-1}'$ are positive definite and $|M_k'| = 0$, then $f_X(x)$ is concentrated on exactly k points. The k points are determined by the eigenvector of M_k' whose eigenvalue is zero. The probability mass at each point is determined by the first $k - 1$ moments. Since the value of any function at k distinct points can be expressed as a linear combination of the first $k - 1$ polynomials at those points, it follows that

$$\text{rank}(M_r') = \min(k, r). \qquad (3.21)$$

Equivalently, the above rank condition may be deduced from the fact that the integral (3.20) is a sum over k points. Thus M_r' is a sum of k matrices each of rank one. The positive definiteness condition together with (3.21) is sufficient to ensure the existence of $f_X(x)$.

In the case of multivariate distributions, similar criteria may be used to determine whether any particular sequence of arrays is a sequence of moments from some distribution. The elements of the multivariate version of M_r' are arrays suitably arranged. For example, the $(2, 2)$ component of M_r' is a square array of second moments whereas the $(1, 3)$ component is the same array arranged as a row vector. The existence of a density is guaranteed if, for each $r = 1, 2, \ldots$, M_r' is non-negative definite and the rank is maximal, namely

$$\text{rank}(M_r') = \binom{p + r}{r}.$$

See Exercise 3.31. In the case of rank degeneracy, additional conditions along the lines of (3.21) are required to ensure consistency.

The moment space, $\mathcal{M}_n$, is the set of all vectors $(\mu_1', \ldots, \mu_n')$ having n components that can be realized as moments of some distribution. Since the moments of the degenerate distribution $\delta(x, t)$ are t^r, any vector of the form $(t, t^2, \ldots, t^n)$ must lie in $\mathcal{M}_n$. In addition, since $\mathcal{M}_n$ is convex (Exercise 3.28), all convex combinations of such vectors lie in $\mathcal{M}_n$. Finally, $\mathcal{M}_n$ has dimension n because there are n linearly independent vectors of the above polynomial type. Hence, the moments are functionally independent in the sense that no non-trivial function of any finite set of moments is identically zero for all distributions. Discontinuous functions such as $1 - H\{\kappa_2\}$ that are identically zero on $\mathcal{M}_n$ are regarded as trivial. The argument just given applies also to the multivariate case where the basis vectors of $\mathcal{M}_n^{(p)}$ are $(t^i, t^j t^k, ...)$, superscripts now denoting components. It follows immediately that if $f_X(x)$ has a finite number of support points, say k, then $\mathrm{rank}(M_r') \leq k$ and that this limit is attained for sufficiently large r.

The cumulant space, $\mathcal{K}_n$, is defined in the obvious way as the set of all vectors that can be realized as cumulants of some distribution. The transformation from $\mathcal{M}_n$ to $\mathcal{K}_n$ is nonlinear and the convexity property of $\mathcal{M}_n$ is lost in the transformation. To see that $\mathcal{K}_n$ is not convex for $n \geq 4$, it is sufficient to observe that the set $\kappa_4 \geq -2\kappa_2^2$ is not convex in the (κ_2, κ_4) plane. However, it is true that if t_1 and t_2 are vectors in $\mathcal{K}_n$, then $t_1 + t_2$ is also in $\mathcal{K}_n$ (Exercise 3.29). As a consequence, if t is in $\mathcal{K}_n$ then λt is also in $\mathcal{K}_n$ for any positive integer $\lambda \geq 1$. This property invites one to suppose that λt lies in $\mathcal{K}_n$ for all $\lambda \geq 1$, not necessarily an integer, but Exercise 3.30 demonstrates that this is not so for $n \geq 6$. The claim is true by definition for infinitely divisible distributions and the counterexample is related to the fact that not all distributions are infinitely divisible. Finally, the transformation from $\mathcal{M}_n$ to $\mathcal{K}_n$ is one to one and continuous, implying that $\mathcal{K}_n$ has dimension n. As a consequence, there are no non-trivial functions of the cumulants that are zero for all distributions. A similar result with obvious modifications applies in the multivariate case.

Identity (3.3) shows that the generalized cumulants are functionally dependent. However, they are *linearly* independent as the following argument shows. Without loss of generality, we may consider an arbitrary linear combination of generalized cumulants, each of the same degree, with coefficients $c(\Upsilon)$. The linear combination, $\sum_{\Upsilon^*} c(\Upsilon^*) \kappa(\Upsilon^*)$ may be written as

$$\sum_{\Upsilon^*} c(\Upsilon^*) \sum_{\Upsilon \geq \Upsilon^*} (-1)^{\nu-1}(\nu - 1)! \, \mu(v_1)...\mu(v_\nu)$$
$$= \sum_{\Upsilon} (-1)^{\nu-1}(\nu - 1)! \, \mu(v_1)...\mu(v_\nu) C(\Upsilon) \tag{3.22}$$

where $C(\Upsilon) = \sum_{\Upsilon^* \leq \Upsilon} c(\Upsilon^*)$ is an invertible linear function of the original coefficients. The implication is that $\sum c(\Upsilon)\kappa(\Upsilon) = 0$ with $c \neq 0$ implies a syzygy in the moments. From the previous discussion, this is known to be impossible because the moments are functionally independent. A simple extension of this argument covers the case where the linear combination involves generalized cumulants of unequal degree.

3.9   Bibliographic notes

There is some difficulty in tracing the origins of the fundamental identity (3.3). Certainly, it is not stated in Fisher's (1929) paper on k-statistics but Fisher must have known the result in some form in order to derive his rules for determining the joint cumulants of k-statistics. In fact, Fisher's procedure was based on the manipulation of differential operators (Exercise 3.11) and involved an expression for $M_X(\xi)$ essentially the same as (3.11) above. His subsequent calculations for joint cumulants were specific to the k-statistics for which many of the partitions satisfying $\Upsilon \vee \Upsilon^* = 1$ vanish on account of the orthogonality of the k-statistics. Rather surprisingly, despite the large number of papers on k-statistics that appeared during the following decades, the first explicit references to the identity (3.3) did not appear until the papers by James (1958), Leonov & Shiryaev (1959) and James & Mayne (1962). The statement of the result in these papers is not in terms of lattices or graphs. James (1958) uses the term *dissectable* for intersection matrices that do not satisfy the condition in (3.3). He gives a series of rules for determining the moments or cumulants of any homogeneous polynomial symmetric function although his primary interest is in k-statistics. He notes that for k-statistics, only the pattern of non-zero elements of the intersection matrix is relevant, but that in general, the numerical values are required: see Chapter 4. Leonov & Shiryaev (1959) use the term *indecomposability* defined essentially as a connectivity condition on the matrix $\Upsilon^* \cap \Upsilon$: see Exercises 3.4 and 3.5.

Rota's (1964) paper is the source of the lattice-theory notation and terminology. The notation and the derivation of (3.3) are taken from McCullagh (1984b). Alternative derivations and alternative statements of the result can be found in Speed (1983). The earliest statement of the result in a form equivalent to (3.3), though in a different notation, appears to be in Malyshev (1980) under the colourful title of 'vacuum cluster expansions'.

There is also a more specialized literature concerned with variances of products: see, for example, Barnett (1955), or Goodman (1960, 1962). For a thorough discussion of moment spaces, the reader is referred to Karlin & Studden (1966).

3.10   Further results and exercises 3

3.1   Let $X^1, ..., X^n$ be independent and identically distributed. By expressing $\bar{X}$ as a linear form and the sample variance, s^2 as a quadratic form, show that $\text{cov}(\bar{X}, s^2) = \kappa_3/n$. Hence show that $\text{corr}(\bar{X}, s^2) \to \rho_3/(2 + \rho_4)^{1/2}$ as $n \to \infty$. Show also that $\text{cov}(\bar{X}^2, s^2) = \kappa_4/n + 2\kappa_1\kappa_3/n$ and show that the limiting correlation is non-zero in general for non-normal variables.

3.2   Let $X = X^1, ..., X^n$ be independent and let the residuals $X^i - \kappa^i$ be identically distributed with cumulants $\kappa_2\delta^{ij}, \kappa_3\delta^{ijk}$ and so on. Find the mean and variance of the quadratic form $a_{ij}X^iX^j$ and express the result in matrix notation (Seber, 1977, p. 9).

3.3   Using the results derived in the previous exercise, find an expression for the variance of the residual sum of squares in the standard linear regression model.

3.4   Let M be the intersection matrix $\Upsilon^* \cap \Upsilon$. Columns j_1 and j_2 are said to *hook* if, for some i, $m_{ij_1} > 0$ and $m_{ij_2} > 0$. The set of columns is said to *communicate* if there exists a sequence $j_1, j_2, ..., j_\nu$ such that columns j_l and j_{l+1} hook. The matrix M is said to be *indecomposable* if its columns communicate. Show that M is indecomposable if and only if M^T is indecomposable.

3.5   Show, in the terminology of Exercise 3.4, that M is indecomposable if and only if $\Upsilon \vee \Upsilon^* = 1$. (Leonov & Shiryaev, 1959; Brillinger, 1975, Section 2.3.)

3.6   If Υ^* is a 2^k partition (a partition of $2k$ elements into k blocks of 2 elements each), show that the number of 2^k partitions Υ satisfying $\Upsilon \vee \Upsilon^* = 1$, is $2^{k-1}(k - 1)!$.

3.7 If $X = X^1, ..., X^p$ are jointly normal with zero mean and co-variance matrix $\kappa^{i,j}$ of full rank, show that the rth order cumulant of $Y = \kappa_{i,j} X^i X^j$ is $2^{r-1} p(r-1)!$. Hence show that the cumulant generating function of Y is $-\frac{1}{2} p \log(1 - 2\xi)$ and therefore that Y has the χ^2 distribution on p degrees of freedom.

3.8 Show that $\kappa^{r,s} \kappa^{t,u} \xi_{rt} \xi_{su} = \mathrm{tr}\{(\kappa\xi)^2\}$ where $\kappa\xi$ is the usual matrix product.

3.9 For any positive definite matrix $\mathbf{A}$, define the matrix $\mathbf{B} = \log(\mathbf{A})$ by

$$\mathbf{A} = \exp(\mathbf{B}) = \mathbf{I} + \mathbf{B} + \mathbf{B}^2/2! + ... + \mathbf{B}^r/r! + ...$$

By inverting this series, or otherwise, show that

$$\log|\mathbf{A}| = \mathrm{tr}\log(\mathbf{A}),$$

where $|\mathbf{A}|$ is the determinant of $\mathbf{A}$.

3.10 If $X = X^1, ..., X^p$ are jointly normal with zero mean, show that the joint cumulant generating function, $\log E \exp(\xi_{ij} Y^{ij})$ of $Y^{ij} = X^i X^j$ is $K_Y(\xi) = -\frac{1}{2} \log|\mathbf{I} - 2\xi\kappa|$. Hence derive the cumulant generating function for the Wishart distribution.

3.11 Let X be a scalar random variable with moments μ_r and cumulants κ_r in the notation of Section 2.5. Show that the rth moment of the polynomial

$$Y = P(X) = a_0 + a_1 X + a_2 X^2 + ...$$

is given formally by $P(d) M_X(\xi)|_{\xi=0}$ where $d = d/d\xi$. Hence show that the moment generating function of Y is

$$M_Y(\varsigma) = \exp\{\varsigma P(d)\} M_X(\xi)|_{\xi=0}$$

in which the operator is supposed to be expanded in powers before attacking the operand (Fisher, 1929, Section 10).

3.12 Show that any polynomial expression in $X^1, ..., X^p$, say

$$Q_4 = a_0 + a_i X^i + a_{ij} X^i X^j/2! + a_{ijk} X^i X^j X^k/3!$$
$$+ a_{ijkl} X^i X^j X^k X^l/4!,$$

can be expressed as a homogeneous polynomial in $X^0, X^1, ..., X^p$

$$Q_4 = \sum_{ijkl=0}^{p} b_{ijkl} X^i X^j X^k X^l / 4!,$$

where $X^0 = 1$. Give expressions for the coefficients b in terms of the as. Find $E(Q_4)$ and $\text{var}(Q_4)$ and express the results in terms of the as.

3.13 Consider the first-order autoregressive process $Y_0 = \epsilon_0 = 0$, $Y_j = \beta Y_{j-1} + \epsilon_j$, $j = 1, ..., n$, where $|\beta| < 1$ and ϵ_j are independent $N(0, 1)$ random variables. Show that the log likelihood function has first derivative $U = \partial l / \partial \beta = T_1 - \beta T_2$ where $T_1 = \Sigma Y_j Y_{j-1}$ and $T_2 = \Sigma Y_{j-1}^2$ with summation from 1 to n. By expressing U as a quadratic form in ϵ, show that the first three cumulants of U are $E(U) = 0$,

$$\kappa_2(U) = \sum_{1 \le i < j \le n} \beta^{2j - 2i - 2}$$
$$= (1 - \beta^2)^{-1} \{ n - (1 - \beta^{2n}) / (1 - \beta^2) \} = E(T_2)$$
$$\kappa_3(U) = 6 \sum_{1 \le i < j < k \le n} \beta^{2k - 2i - 3}$$
$$= \frac{6 \{ n\beta(1 - \beta^2) - 2\beta + n\beta^{2n-1}(1 - \beta^2) + 2\beta^{2n+1} \}}{(1 - \beta^2)^3}.$$

3.14 Let $Y = Y^1, ..., Y^p$ have zero mean and covariance matrix $\kappa^{i,j}$. Show that the 'total variance', $\sigma^2 = E(Y^i Y^j \delta_{ij})$, is invariant under orthonormal transformation of Y. For any given direction, ϵ, define

$$\sigma_\epsilon^2 = \text{var}(\epsilon_i Y^i) = \epsilon_i \epsilon_j \kappa^{i,j}$$
$$\tau_\epsilon^2 = \epsilon_i \epsilon_j \{ \sigma^2 \delta^{ij} - \kappa^{i,j} \} = \epsilon_i \epsilon_j I^{ij}.$$

Give an interpretation of σ_ϵ^2 and τ_ϵ^2 as regression and residual variances respectively. Show also that, in mechanics, τ_ϵ^2 is the moment of inertia of a rigid body of unit mass about the axis ϵ. Hence, interpret I^{ij} as the *inertia tensor* (Synge & Griffith, 1949, Section 11.3; Jeffreys & Jeffreys, 1956, Section 3.08).

3.15 *MacCullagh's formula*: Using the notation of the previous exercise, let $X^i = Y^i + \kappa^i$, where κ^i are the components of a vector of length $\rho > 0$ in the direction ϵ. Show that

$$E\left(\frac{1}{|X|}\right) = \frac{1}{\rho} + \frac{1}{2\rho^3}\left(-\sigma^2 + 3\sigma_\epsilon^2\right) + O(\rho^{-4})$$

$$= \frac{1}{\rho} + \frac{1}{2\rho^3}\left(2\sigma^2 - 3\tau_\epsilon^2\right) + O(\rho^{-4}),$$

(MacCullagh, 1855; Jeffreys & Jeffreys, 1956, Section 18.09). [The above expression gives the potential experienced at an external point (the origin) due to a unit mass or charge distributed as $f_X(x)$. The correction term is sometimes called the *gravitational quadrupole* (Kibble, 1985, Chapter 6).]

3.16 Show that the sum

$$\sum_{\Upsilon \geq \Upsilon^\bullet} (-1)^{\nu-1}(\nu - 1)! = \delta(\Upsilon^*, 1)$$

is zero unless $\Upsilon^* = 1$.

3.17 By reversing the order of summation in (3.17), show that the generalized cumulant may be written

$$\kappa(\Upsilon^*) = \sum_{\Pi} \kappa(\pi_1)...\kappa(\pi_\sigma) \sum_{\substack{\Upsilon \geq \Upsilon^* \\ \Upsilon \geq \Pi}} (-1)^{\nu-1}(\nu - 1)!.$$

Hence, using the result of the previous exercise, prove the fundamental identity (3.3).

3.18 Let X be a scalar random variable whose distribution is Poisson with mean 1. Show that the cumulants of X of all orders are equal to 1. Hence show that the rth moment is

$$\mu_r' = E(X^r) = B_r$$

where B_r, the rth Bell number, is the number of partitions of a set of r elements. Hence derive a generating function for the Bell numbers.

3.19 Let $\Upsilon^* = \{v_1^*, ..., v_\alpha^*\}$ be a partition of a set of p elements into α non-empty blocks of sizes $|v_1^*|, ..., |v_\alpha^*|$. By using (3.16) or otherwise, show that the number of partitions complementary to Υ^* is

$$\sum_{\substack{\Upsilon \geq \Upsilon^* \\ \Upsilon = \{v_1, ..., v_\nu\}}} (-1)^{\nu-1}(\nu-1)! \, B_{|v_1|}...B_{|v_\nu|}.$$

3.20 Interpret the expression in Exercise 3.19 as the cumulant of order α

$$\text{cum}\left(X^{|v_1^*|}, ..., X^{|v_\alpha^*|}\right)$$

where X is defined in Exercise 3.18 and the superscripts denote powers. Simplify this result in the special case where Υ^* is a 2^k partition and compare with Exercise 3.6.

3.21 Show that the number of sub-partitions of Υ^* is given by

$$\sum_{\Upsilon \leq \Upsilon^*} 1 = B_{|v_1^*|}...B_{|v_\alpha^*|}.$$

3.22 Using the result given in Exercise 3.19, show that the total number of ordered pairs of partitions (Υ_1, Υ_2) satisfying $\Upsilon_1 \vee \Upsilon_2 = 1$ is

$$C_p^{(2)} = \sum_\Upsilon (-1)^{\nu-1}(\nu-1)! \, B_{|v_1|}^2...B_{|v_\nu|}^2$$

where the partitions contain p elements and B_r^2 is the square of the rth Bell number. Deduce also that $C_p^{(2)}$ is the pth cumulant of $Y = X_1 X_2$ where the Xs are independent Poisson random variables with unit mean.

3.23 Show that the number of ordered pairs (Υ_1, Υ_2) satisfying $\Upsilon_1 \vee \Upsilon_2 = \Upsilon^*$ for some fixed partition Υ^*, is

$$\sum_{\Upsilon \leq \Upsilon^*} m(\Upsilon, \Upsilon^*) B_{|v_1|}^2...B_{|v_\nu|}^2,$$

where $m(\Upsilon, \Upsilon^*)$ is the Möbius function for the partition lattice, defined below (3.12). Hence prove that the total number of ordered triplets $(\Upsilon_1, \Upsilon_2, \Upsilon_3)$ satisfying $\Upsilon_1 \vee \Upsilon_2 \vee \Upsilon_3 = 1$ is

$$C_p^{(3)} = \sum_\Upsilon (-1)^{\nu-1}(\nu-1)! \, B_{|v_1|}^3...B_{|v_\nu|}^3.$$

Show also, in the notation of Exercise 3.22, that $C_p^{(3)}$ is the pth cumulant of the triple product $Y = X_1 X_2 X_3$.

3.24 Generalize the result of the previous exercise to ordered k-tuplets of partitions. Give a simple explanation for this result.

3.25 An alternative way of representing a partition by means of a graph is to use p labelled edges instead of labelled vertices. In this form, the graph of $\Upsilon^* = \{v_1, ..., v_\nu\}$ comprises p labelled edges emanating from α unlabelled vertices, giving a total of $\alpha+p$ vertices of which p are 'free'. If $\Upsilon = \{v_1, ..., v_\nu\}$ is another partition of the same indices represented as a graph in the same way, we define the graph $\Upsilon \otimes \Upsilon^*$ by connecting corresponding free vertices of the two graphs. This gives a graph with $\alpha + \nu$ unlabelled vertices and p labelled edges, parallel edges being permitted. Show that the graph $\Upsilon \otimes \Upsilon^*$ is connected if and only if $\Upsilon \vee \Upsilon^* = 1$.

3.26 In the notation of the previous exercise, prove that all cycles in the graph $\Upsilon \otimes \Upsilon^*$ have even length. (A cycle is a path beginning and ending at the same vertex.) Such a graph is said to be even. Show that all even connected graphs have a unique representation as $\Upsilon \otimes \Upsilon^*$. Hence prove that the number of connected even graphs having p labelled edges is $(C_p^{(2)} + 1)/2$ where $C_p^{(2)}$ is defined in Exercise 3.22 (Gilbert, 1956).

3.27 In the terminology of the previous two exercises, what does $C_p^{(3)}$ in Exercise 3.23 correspond to?

3.28 By considering the mixture density, $pf_1(x) + (1 - p)f_2(x)$, show that the moment space, M_n, is convex.

3.29 By considering the distribution of the sum of two independent random variables, show that the cumulant space, K_n is closed under vector addition.

3.30 Show that there exists a unique distribution whose odd cumulants are zero and whose even cumulants are $\kappa_2 = 1$, $\kappa_4 = -2$, $\kappa_6 = 16$, $\kappa_8 = -272$, $\kappa_{10} = 7936, ...$ Let $M_r'(\lambda)$ be the moment matrix described in Section 3.8, corresponding to the cumulant sequence, $\lambda\kappa_1, \ldots, \lambda\kappa_{2r}$. Show that for the particular cumulant sequence above, the determinant of $M_r'(\lambda)$ is

$$|M_r'| = 1! \, 2! \, ...r! \, \lambda^r (\lambda - 1)^{r-1} ... (\lambda - r + 1),$$

for $r = 1, 2, 3, 4$. Hence prove that there is no distribution whose cumulants are $\{\lambda \kappa_r\}$ for non-integer $\lambda < 3$. Find the unique distribution whose cumulants are $\{\lambda \kappa_r\}$ for $\lambda = 1, 2, 3$.

3.31 By counting the number of distinguishable ways of placing r identical objects in $p+1$ labelled boxes, show that, in p dimensions,

$$\text{rank}(M_r') \leq \binom{p+r}{r},$$

where M_r' is defined in Section 3.8.

3.32 Show that achievement of the limit $\bar{\rho}_4 = \bar{\rho}_{23}^2 - p - 1$ implies the following constraint on the third cumulants

$$\bar{\rho}_{23}^2 - \bar{\rho}_{13}^2 - p + 1 = 0.$$

3.33 Show that, if complex-valued random variables are permitted, there are no restrictions on the moment spaces or on the cumulant spaces such as those discussed in Section 3.8.

3.34 Describe the usual partial order on the set of partitions of the *number* k. Explain why this set, together with the usual partial order, does *not* form a lattice for $k \geq 5$. Verify by direct inspection that the structure is a lattice for each $k \leq 4$.

CHAPTER 4

Sample cumulants

4.1 Introduction

We now address problems of a more inferential flavour, namely problems concerning point estimation or, better, interval estimation of population cumulants based on observed data. Moment and cumulant estimators have a lengthy history in the statistical literature going back to the work of K. Pearson, Chuprov, Student, Fisher and others in the early part of this century. In recent years, the volume of work on such topics has declined, partly because of concerns about robustness, sensitivity to outliers, misrecorded or miscoded values and so on. Estimates of the higher-order cumulants are sensitive to such errors in the data and, for some purposes, this sensitivity may be considered undesirable. In addition, the variance of such an estimate depends on cumulants up to twice the order of the estimand and these are even more difficult to estimate accurately. If there is insufficient data it may not be possible to estimate such cumulants at all. We are confronted immediately and forcibly with what Mosteller & Tukey (1977, Chapter 1) call 'the misty staircase'. That is to say that, to assess the variability of a primary statistic, we compute a secondary statistic, typically more variable than the primary statistic. It is then possible to estimate the variability of the secondary statistic by computing a third statistic and so ad infinitum. Fortunately, we usually stop long before this extreme stage.

Section 4.2 is concerned with simple random samples from an infinite population. The emphasis is on the various symmetric functions, i.e. functions of the data that are unaffected by re-ordering the data values. The functions having the most pleasing statistical properties are the k-statistics and, to a lesser extent, the generalized k-statistics. The k-statistics are unbiased estimates of ordinary cumulants and generalized k-statistics are estimates of generalized cumulants, including moments as a special case.

Section 4.3 is concerned with simple random samples from a finite population. The idea here is to construct statistics whose expectation under simple random sampling is just the value of the statistic computed in the whole population. The generalized k-statistics have this property but it turns out to be more convenient to work with linear combinations called 'polykays'. These are unbiased estimates of population polykays or, in the infinite population case, unbiased estimates of cumulant products. Again, they are most conveniently indexed by set partitions.

The remaining sections are concerned with estimates of cumulants and cumulant products in the presence of identifiable systematic structure, typically in the mean value. A common example is to estimate the second- and higher-order cumulants when the data are divided into k groups differing only in mean value. A second example is to estimate the cumulants of the error distribution based on residuals after linear regression. The difficulty here is that neither the raw data nor the observed residuals are identically distributed, so there is no compelling reason for restricting attention to symmetric functions in the usual sense. Indeed, better estimates can be obtained by using functions that are not symmetric: see Exercise 4.22.

4.2   k-statistics

4.2.1   Definitions and notation

Let $Y_1, ..., Y_n$ be independent and identically distributed p-dimensional random variables where Y_i has components $Y_i^1, ..., Y_i^p$. The cumulants and generalized cumulants of Y_i are written κ^r, $\kappa^{r,s}$, $\kappa^{r,s,t}$, $\kappa^{r,st}$ and so on. No subscripts are necessary because of the assumption that the observations are identically distributed. For each generalized cumulant, κ, with appropriate superscripts, there is a unique polynomial symmetric function, denoted by k with matching superscripts, such that k is an unbiased estimate of κ. The lower-order k-statistics are very familiar, though perhaps by different names. Thus, for example, the simplest k-statistic

$$k^r = n^{-1} \sum_i Y_i^r = \bar{Y}^r \qquad (4.1)$$

is just the sample mean, an unbiased estimate of κ^r. Also,

$$k^{r,s} = \sum_i (Y_i^r - \bar{Y}^r)(Y_i^s - \bar{Y}^s)/(n-1)$$
$$= n^{-1}\phi^{ij}Y_i^r Y_j^s, \qquad (4.2)$$

where $\phi^{ii} = 1$ and $\phi^{ij} = -1/(n-1)$ for $i \neq j$, is just the usual sample covariance matrix. It is well known that $k^{r,s}$ is an unbiased estimate of $\kappa^{r,s}$.

We need not restrict attention to ordinary cumulants alone. A straightforward calculation shows that

$$k^{rs} = n^{-1}\sum_i Y_i^r Y_i^s \qquad (4.3)$$

is an unbiased estimate of κ^{rs} and that

$$k^{r,st} = n^{-1}\sum_{ij} \phi^{ij} Y_i^r Y_j^s Y_j^t \qquad (4.4)$$

is an unbiased estimate of $\kappa^{r,st}$.

The four statistics (4.1) to (4.4) are all examples of k-statistics. Following the terminology of Chapter 3, we refer to (4.1) and (4.2) as ordinary k-statistics and to (4.3) and (4.4) as generalized k-statistics. It is important at the outset to emphasize that while

$$\kappa^{rs} \equiv \kappa^{r,s} + \kappa^r \kappa^s,$$

the corresponding expression with κ replaced by k is false. In fact, we may deduce from (4.1) to (4.3) that

$$nk^{rs} = (n-1)k^{r,s} + nk^r k^s.$$

Equivalently, we may write

$$k^{rs} - k^{r,s} = n^{-1}\sum_{ij}(\delta^{ij} - \phi^{ij})Y_i^r Y_j^s$$
$$= \sum{}^{\#} Y_i^r Y_j^s / n^{(2)},$$

where $n^{(2)} = n(n-1)$, which is an unbiased estimate of the product $\kappa^r \kappa^s$, and is not the same as $k^r k^s$. The symbol, $\sum^{\#}$, which occurs frequently in the calculations that follow, denotes summation over unequal values of the indices, $i, j, \ldots$.

4.2.2   Some general formulae for k-statistics

It follows from the definition of moments that

$$k^r = n^{-1}\delta^i Y_i^r$$
$$k^{rs} = n^{-1}\delta^{ij} Y_i^r Y_j^s$$
$$k^{rst} = n^{-1}\delta^{ijk} Y_i^r Y_j^s Y_k^t$$

and so on, where $\delta^{ijk} = 1$ if $i = j = k$ and zero otherwise, are unbiased estimates of the moments κ^r, κ^{rs}, κ^{rst} and so on. To construct unbiased estimates of the ordinary cumulants, we write

$$k^{r,s} = n^{-1}\phi^{ij} Y_i^r Y_j^s$$
$$k^{r,s,t} = n^{-1}\phi^{ijk} Y_i^r Y_j^s Y_k^t \qquad (4.5)$$
$$k^{r,s,t,u} = n^{-1}\phi^{ijkl} Y_i^r Y_j^s Y_k^t Y_l^u$$

and so on, and aim to choose the coefficients ϕ to satisfy the criterion of unbiasedness. The k-statistics are required to be symmetric in two different senses. First, they are required to be symmetric functions in the sense that they are unaffected by permuting the n observations $Y_1, ..., Y_n$. As a consequence, for any permutation $\pi_1, ..., \pi_n$ of the first n integers, it follows that

$$\phi^{ijkl} = \phi^{\pi_i \pi_j \pi_k \pi_l}.$$

This means, for example, that $\phi^{iijj} = \phi^{1122}$ but it does not follow from the above criterion that ϕ^{1221} is the same as ϕ^{1122}. In fact, however, it turns out that the coefficients ϕ are symmetric under index permutation. This follows not from the requirement that the ks be symmetric functions, but from the requirement that the k-statistics, like the corresponding cumulants, be symmetric under index permutation. In Section 4.3.2, symmetric functions will be introduced for which the coefficients are not symmetric in this second sense. It follows that ϕ^{ijk} can take on at most three distinct values depending on whether $i = j = k$, $i = j \neq k$ or all three indices are distinct. Similarly, ϕ^{ijkl} can take on at most five distinct values, namely ϕ^{1111}, ϕ^{1112}, ϕ^{1122}, ϕ^{1123} and ϕ^{1234}, corresponding to the five partitions of the *number* 4.

On taking expectations in (4.5), we find that the following identities must be satisfied by the coefficients, ϕ.

$$\kappa^{r,s} = n^{-1}\phi^{ij}\left(\kappa^{r,s}\delta_{ij} + \kappa^r\kappa^s\delta_i\delta_j\right)$$

$$\kappa^{r,s,t} = n^{-1}\phi^{ijk}\left(\kappa^{r,s,t}\delta_{ijk} + \kappa^r\kappa^{s,t}\delta_i\delta_{jk}[3] + \kappa^r\kappa^s\kappa^t\delta_i\delta_j\delta_k\right)$$

$$\kappa^{r,s,t,u} = n^{-1}\phi^{ijkl}\left(\kappa^{r,s,t,u}\delta_{ijkl} + \kappa^r\kappa^{s,t,u}\delta_i\delta_{jkl}[4] + \kappa^{r,s}\kappa^{t,u}\delta_{ij}\delta_{kl}[3]\right.$$
$$\left. + \kappa^r\kappa^s\kappa^{t,u}\delta_i\delta_j\delta_{kl}[6] + \kappa^r\kappa^s\kappa^t\kappa^u\delta_i\delta_j\delta_k\delta_l\right).$$

Thus we must have

$$\begin{aligned}
\phi^{ij}\delta_{ij} &= n, & \phi^{ij}\delta_i &= 0, \\
\phi^{ijk}\delta_{ijk} &= n, & \phi^{ijk}\delta_{ij} &= 0, & \phi^{ijk}\delta_i &= 0, \\
\phi^{ijkl}\delta_{ijkl} &= n, & \phi^{ijkl}\delta_{ijk} &= 0, & \phi^{ijkl}\delta_{ij} &= 0, & \phi^{ijkl}\delta_i &= 0,
\end{aligned} \tag{4.6}$$

and so on. From these formulae, we find that, for i, j, k, l all distinct,

$$\begin{aligned}
\phi^{ii} = \phi^{iii} &= \phi^{iiii} = 1 \\
\phi^{ij} = \phi^{iij} = \phi^{iiij} &= \phi^{iijj} = -1/(n-1) \\
\phi^{ijk} = \phi^{iijk} &= 2/\{(n-1)(n-2)\} \\
\phi^{ijkl} &= -6/\{(n-1)(n-2)(n-3)\}.
\end{aligned}$$

One can then show by induction that the general expression for the coefficients ϕ is

$$(-1)^{\nu-1}\Big/\binom{n-1}{\nu-1} = (-1)^{\nu-1}(\nu-1)!/(n-1)^{(\nu-1)} \tag{4.7}$$

where $\nu \leq n$ is the number of distinct indices, and

$$(n-1)^{(\nu-1)} = (n-1)(n-2)...(n-\nu+1).$$

There are no unbiased estimates for cumulants of order greater than n.

Many of the pleasant statistical properties of k-statistics stem from the orthogonality of the ϕ-arrays and the δ-arrays as shown in (4.6). More generally, if v_1, v_2 are sets of indices, we may write

$\langle\phi(v_1), \delta(v_2)\rangle$ for the sum over those indices in $v_1 \cap v_2$. This notation gives

$$\langle\phi(v_1),\, \delta(v_2)\rangle = \begin{cases} 0, & \text{if } v_2 \subset v_1; \\ n, & \text{if } v_2 = v_1; \\ \delta(v_2 - v_1), & \text{if } v_1 \subset v_2, \end{cases} \qquad (4.8)$$

no simplification being possible otherwise. In the above expression, the symbol $\subset$ is to be interpreted as meaning 'proper subset of'. For an alternative derivation of expression (4.7) for ϕ, we may proceed as follows. Unbiased estimates of the moments are given by

$$k^{rs} = n^{-1}\sum Y_i^r Y_i^s, \quad k^{rst} = n^{-1}\sum Y_i^r Y_i^s Y_i^t$$

and so on. Unbiased estimates of products of moments are given by the so-called symmetric means

$$k^{(r)(s)} = \sum{}^{\#} Y_i^r Y_j^s/n^{(2)}, \qquad k^{(r)(st)} = \sum{}^{\#} Y_i^r Y_j^s Y_j^t/n^{(2)},$$

$$k^{(rs)(tu)} = \sum{}^{\#} Y_i^r Y_i^s Y_j^t Y_j^u/n^{(2)} \quad k^{(r)(stu)} = \sum{}^{\#} Y_i^r Y_j^s Y_j^t Y_j^u/n^{(2)}$$

$$k^{(r)(s)(t)} = \sum{}^{\#} Y_i^r Y_j^s Y_k^t/n^{(3)}, \qquad k^{(r)(st)(u)} = \sum{}^{\#} Y_i^r Y_j^s Y_j^t Y_k^u/n^{(3)}$$

and so on, with summation extending over unequal subscripts. It is a straightforward exercise to verify that $E(k^{(r)(st)(u)}) = \kappa^r \kappa^{st} \kappa^u$ and similarly for the remaining statistics listed above. All linear combinations of these statistics are unbiased for the corresponding parameter. Thus

$$k^{rst} - k^{(r)(st)}[3] + 2k^{(r)(s)(t)} \qquad (4.9)$$

is an unbiased estimate of $\kappa^{rst} - \kappa^r \kappa^{st}[3] + 2\kappa^r \kappa^s \kappa^t = \kappa^{r,s,t}$. By expressing (4.9) as a symmetric cubic polynomial with coefficients ϕ^{ijk}, it can be seen that the coefficients must satisfy (4.7), the numerator coming from the Möbius function and the denominator from the above sums over unequal subscripts.

A similar argument applies to higher-order *k*-statistics.

4.2.3 *Joint cumulants of ordinary k-statistics*

Before giving a general expression for the joint cumulants of *k*-statistics, it is best to examine a few simple cases. Consider first the covariance of k^r and k^s, which may be written

$$\text{cov}(k^r, k^s) = n^{-2}\phi^i \phi^j \kappa_{i,j}^{r,s}$$

where $\kappa_{i,j}^{r,s}$ is the covariance of Y_i^r and Y_j^s. This covariance is zero unless $i = j$, in which case we may write $\kappa_{i,j}^{r,s} = \kappa^{r,s}\delta_{ij}$ because the random variables are assumed to be identically distributed. Thus

$$\text{cov}(k^r, k^s) = n^{-2}\phi^i\phi^j\delta_{ij}\kappa^{r,s} = \kappa^{r,s}/n.$$

Similarly, for the covariance of $k^{r,s}$ and k^t, we may write

$$\text{cov}(k^{r,s}, k^t) = n^{-2}\phi^{ij}\phi^k\kappa_{ij,k}^{rs,t}.$$

On expansion of the generalized cumulant using (3.3), and after taking independence into account, we find

$$\text{cov}(k^{r,s}, k^t) = n^{-2}\phi^{ij}\phi^k\left(\kappa^{r,s,t}\delta_{ijk} + \kappa^s\kappa^{r,t}\delta_j\delta_{ik}[2]\right).$$

Application of (4.8) gives $\text{cov}(k^{r,s}, k^t) = n^{-1}\kappa^{r,s,t}$. Similarly, the covariance of two sample covariances is

$$
\begin{aligned}
\text{cov}(k^{r,s}, k^{t,u}) &= n^{-2}\phi^{ij}\phi^{kl}\kappa_{ij,kl}^{rs,tu} \\
&= n^{-2}\phi^{ij}\phi^{kl}\{\kappa^{r,s,t,u}\delta_{ijkl} + \kappa^r\kappa^{s,t,u}\delta_i\delta_{jkl}[4] \\
&\quad + \kappa^{r,t}\kappa^{s,u}\delta_{ik}\delta_{jl}[2] + \kappa^r\kappa^t\kappa^{s,u}\delta_i\delta_l\delta_{jk}[4]\} \\
&= n^{-1}\kappa^{r,s,t,u} + \kappa^{r,t}\kappa^{s,u}[2]\sum\nolimits_{ij}\phi^{ij}\phi^{ij}/n^2 \\
&= n^{-1}\kappa^{r,s,t,u} + \kappa^{r,t}\kappa^{s,u}[2]/(n-1).
\end{aligned}
$$

In the case of third and higher joint cumulants of the ks, it is convenient to introduce a new but obvious notation. For the joint cumulant of $k^{r,s}$, $k^{t,u}$, and $k^{v,w}$ we write

$$\kappa_k(r, s|t, u|v, w) = n^{-3}\phi^{ij}\phi^{kl}\phi^{mn}\kappa_{ij,kl,mn}^{rs,tu,vw}.$$

In the expansion for the 3,6 cumulant, all partitions having a unit part can be dropped because of (4.8). The third-order joint cumulant then reduces to

$$
\begin{aligned}
\kappa_k(r, s|t, u|v, w) = n^{-3}\phi^{ij}\phi^{kl}\phi^{mn}\big(&\kappa^{r,s,t,u,v,w}\delta_{ijklmn} \\
&+ \kappa^{r,s,t,v}\kappa^{u,w}\delta_{ijkm}\delta_{ln}[12] + \kappa^{r,s,t}\kappa^{u,v,w}\delta_{ijk}\delta_{lmn}[6] \\
&+ \kappa^{r,t,v}\kappa^{s,u,w}\delta_{ikm}\delta_{jln}[4] + \kappa^{r,t}\kappa^{s,v}\kappa^{u,w}\delta_{ik}\delta_{jm}\delta_{ln}[8]\big),
\end{aligned}
$$

where the third term has zero contribution on account of orthogonality. To simplify this expression further, we need to evaluate the coefficients of the cumulant products. For example, the coefficient of the final term above may be written as

$$n^{-3} \sum_{ijk} \phi^{ij} \phi^{ik} \phi^{jk}$$

and this sum, known as pattern function, has the value $(n-1)^{-2}$. See Table 4.1 under the pattern coded $12/13/23$. On evaluating the remaining coefficients, we find

$$\kappa_k(r,s|t,u|v,w) = \kappa^{r,s,t,u,v,w}/n^2 + \kappa^{r,s,t,v}\kappa^{u,w}[12]/\{n(n-1)\}$$
$$+ \kappa^{r,t,v}\kappa^{s,u,w}[4](n-2)/\{n(n-1)^2\} + \kappa^{r,t}\kappa^{s,v}\kappa^{u,w}[8]/(n-1)^2.$$

In the univariate case where k^r and $k^{r,s}$ are commonly written as $\bar{Y}$ and s^2, we may deduce from the cumulants listed above that

$$\mathrm{var}(\bar{Y}) = \kappa_2/n, \qquad \mathrm{cov}(\bar{Y}, s^2) = \kappa_3/n,$$
$$\mathrm{var}(s^2) = \kappa_4/n + 2\kappa_2^2/(n-1)$$

and the third cumulant of s^2 is

$$\kappa_3(s^2) = \kappa_6/n^2 + 12\kappa_4\kappa_2/\{n(n-1)\} + 4(n-2)\kappa_3^2/\{n(n-1)^2\}$$
$$+ 8\kappa_2^3/(n-1)^2.$$

More generally, the joint cumulant of several k-statistics can be represented by a partition, say Υ^*. On the right of the expression for $\kappa_k(\Upsilon^*)$ appear cumulant products corresponding to the partitions Υ complementary to Υ^*, multiplied by a coefficient that depends on n and on the intersection matrix $\Upsilon^* \cap \Upsilon$. This coefficient is zero for all partitions Υ having a unit block and also for certain other partitions that satisfy the first condition in (4.8), possibly after simplification by the third condition in (4.8). For example, if $\Upsilon^* = \{ij|kl|mn\}$ and $\Upsilon = \{ijk|lmn\}$, then the coefficient

$$\phi^{ij}\phi^{kl}\phi^{mn}\delta_{ijk}\delta_{klm} = \phi^{kl}\phi^{mn}\delta_k\delta_{lmn} = 0$$

is zero even though no block of Υ is a subset of a block of Υ^*. In general, the coefficient of the complementary partition $\Upsilon = \{v_1, ..., v_\nu\}$ may be written as

$$n^{-\alpha}\langle\phi(\Upsilon^*), \delta(\Upsilon)\rangle = \sum \phi(v_1^*)...\phi(v_\alpha^*)\,\delta(v_1)...\delta(v_\nu)$$

with summation over all indices. With this notation, the joint cumulant of several k-statistics may be written as

$$\kappa_k(\Upsilon^*) = \sum_{\Upsilon \vee \Upsilon^* = 1} n^{-\alpha} \langle \phi(\Upsilon^*), \delta(\Upsilon) \rangle \, \kappa(v_1)...\kappa(v_\nu). \qquad (4.10)$$

The main difficulty in using this formula lies in computing the coefficients $\langle \phi(\Upsilon^*), \delta(\Upsilon) \rangle$.

4.2.4   Pattern matrices and pattern functions

We now examine various ways of expressing and computing the coefficients $\langle \phi(\Upsilon^*), \delta(\Upsilon) \rangle$, also called *pattern functions*, that arise in (4.10). Evidently the coefficient is a function of n that depends on the intersection matrix $\Upsilon^* \cap \Upsilon$. In fact, since the value of ϕ depends only on the number of distinct indices and not on the number of repetitions of any index, it follows that $\langle \phi(\Upsilon^*), \delta(\Upsilon) \rangle$ must depend only on the pattern of non-zero values in $\Upsilon^* \cap \Upsilon$ and not on the actual intersection numbers. The so-called *pattern matrix* is determined only up to separate independent permutations of the rows and columns.

To take a few simple examples, suppose that $\Upsilon^* = ijk|lmn$, $\Upsilon_1 = ijl|kmn$ and $\Upsilon_2 = il|jkmn$. The intersection matrices are

$$\Upsilon^* \cap \Upsilon_1 = \begin{matrix} 2 & 1 \\ 1 & 2 \end{matrix} \quad \text{and} \quad \Upsilon^* \cap \Upsilon_2 = \begin{matrix} 1 & 2 \\ 1 & 2 \end{matrix},$$

which are different, but the pattern matrices are identical. The pattern functions $\langle \phi(\Upsilon^*), \delta(\Upsilon) \rangle$, written explicitly as sums of coefficients, are

$$\sum \phi^{iij}\phi^{ijj} \quad \text{and} \quad \sum \phi^{ijj}\phi^{ijj},$$

both of which reduce to $\sum (\phi^{ij})^2 = n^2/(n-1)$. Similarly, if we take $\Upsilon^* = ij|kl|mn$ and $\Upsilon = ik|jm|ln$, the pattern matrix is

$$\Upsilon^* \cap \Upsilon = \begin{matrix} 1 & 1 & 0 \\ 1 & 0 & 1 \\ 0 & 1 & 1 \end{matrix} \qquad (4.11)$$

corresponding to the algebraic expression $\sum \phi^{ij}\phi^{ik}\phi^{jk}$, as if the columns of the pattern matrix were labelled i, j and k. This pattern function appears in the final term of the expression for $\kappa_k(r, s|t, u|v, w)$ above and takes the value $n^3/(n-1)^2$: see Exercise 4.6.

Because of the orthogonality of the ϕ-arrays and the δ-arrays of coefficients, many pattern functions are identically zero and it is helpful to identify these at the outset. Evidently, from (4.10), if $\Upsilon^* \vee \Upsilon < 1$, the pattern function is zero. Moreover, additional vanishing pattern functions can be identified using (4.8), but this test must, in general, be applied iteratively. For example, if $\Upsilon^* = ij|kl|m$ and $\Upsilon = ijk|lm$, we find that

$$\phi^{ij}\phi^{kl}\phi^m \delta_{ijk}\delta_{lm} = \phi^{kl}\phi^m \delta_k \delta_{lm} = 0$$

on applying (4.8) twice. The pattern matrix is

$$\begin{array}{cc} 1 & 0 \\ 1 & 1 \\ 0 & 1 \end{array}$$

and the corresponding pattern function is zero because the columns of this matrix can be partitioned into two blocks that connect only through one row. The pattern matrix for $\Upsilon = ikm|jl$ cannot be partitioned in this way and the corresponding pattern function is $n^2/(n-1)$. Rows containing a single entry may be deleted since $\phi^i = 1$. Table 4.1 gives a list of some of the more useful pattern functions.

To conserve space, the patterns in Table 4.1 are coded numerically. For example, the pattern (4.11) is coded under the section marked *Three rows* as 12/13/23 and takes the value $n^3/(n-1)^2$. Patterns that can be derived from (4.11) by permuting rows or columns are not listed explicitly.

4.3 Related symmetric functions

4.3.1 *Generalized k-statistics*

Generalized *k*-statistics are the sample versions of the generalized cumulants, so that, for example, $k^{r,st}$ is a symmetric function and is an unbiased estimate of $\kappa^{r,st}$. A fairly simple extension of the argument used in Section 4.2.2 shows that

$$k^{r,st} = n^{-1} \sum \phi^{ij} Y_i^r Y_j^s Y_j^t \tag{4.12}$$

Table 4.1 *Some useful non-zero pattern functions*

Pattern	Pattern function
1/1/.../1	n
	Two rows
12/12	$n^2/(n-1)$
123/123	$n^3/(n-1)^{(2)}$
1234/1234	$n^3(n+1)/(n-1)^{(3)}$
12345/12345	$n^4(n+5)/(n-1)^{(4)}$
12356/123456	$n^3(n+1)(n^2+15n-4)/(n-1)^{(5)}$
	Three rows
12/12/12	$n^2(n-2)/(n-1)^2$
12/13/23, 123/12/13	$n^3/(n-1)^2$
123/123/12	$n^3(n-3)/\{(n-1)^2(n-2)\}$
123/123/123	$\dfrac{n^3(n^2-6n+10)}{(n-1)^2(n-2)^2}$
123/124/34, 1234/123/34	$\dfrac{n^4}{(n-1)^2(n-2)}$
123/124/134	$\dfrac{n^4(n-3)}{(n-1)^2(n-2)^2}$
123/24/1234	$\dfrac{n^4(n-4)}{(n-1)^2(n-2)^2}$
1234/1234/12	$\dfrac{n^3(n^2-4n-1)}{(n-1)^2(n-2)(n-3)}$
1234/1234/123	$\dfrac{n^3(n^3-18n^2+17n+2)}{(n-1)^2(n-2)^2(n-3)}$
1234/1234/1234	$\dfrac{n^3(n^4-12n^3+51n^2-74n-18)}{(n-1)^2(n-2)^2(n-3)^2}$

Table 4.1 *Non-zero pattern functions (continued)*

Pattern	Pattern function
Three rows (contd.)	
1234/125/345, 12345/215/435	$n^5/(n-1)^2(n-2)^2$
1234/1235/45, 12345/1235/34	$\dfrac{n^4(n+1)}{(n-1)^2(n-2)(n-3)}$
1234/1235/145	$\dfrac{n^4(n^2-4n-1)}{(n-1)^2(n-2)^2(n-3)}$
1234/1235/1245	$\dfrac{n^4(n^3-9n^2+19n+5)}{(n-1)^2(n-2)^2(n-3)^2}$
12345/1234/125	$\dfrac{n^4(n^2-5n-2)}{(n-1)^2(n-2)^2(n-3)}$
12345/1245/12	$\dfrac{n^4(n^2-4n-9)}{(n-1)(n-1)^{(4)}}$
1234/1256/3456	$\dfrac{n^4(n+1)(n^2-5n+2)}{(n-1)^2(n-1)^2(n-3)^2}$
12345/1236/456	$\dfrac{n^4(n^3-9n^2+19n+5)}{(n-1)^2(n-2)^2(n-3)^2}$
Four rows	
12/12/12/12	$n^2(n^2-3n+3)/(n-1)^3$
13/13/12/12	$n^3/(n-1)^2$
13/23/12/12, 123/23/12/12	$n^3(n-2)/(n-1)^3$
123/13/23/12	$n^3(n-3)/(n-1)^3$
123/123/12/12	$\dfrac{n^3(n^2-4n+5)}{(n-1)^3(n-2)}$
123/123/12/13	$\dfrac{n^3(n^2-5n+7)}{(n-1)^3(n-2)}$
123/123/123/12	$\dfrac{n^3(n-3)(n^2-4n+6)}{(n-1)^3(n-2)^2}$
123/123/123/123	$\dfrac{n^3(n^4-9n^3+33n^2-60n+48)}{(n-1)^3(n-2)^3}$

is the required symmetric function. Similarly,

$$k^{rs,tu} = n^{-1} \sum \phi^{ij} Y_i^r Y_i^s Y_j^t Y_j^u \qquad (4.13)$$

$$k^{r,s,tu} = n^{-1} \sum \phi^{ijk} Y_i^r Y_j^s Y_k^t Y_k^u \qquad (4.14)$$

and so on. Note that the coefficients ϕ^{ij} in (4.12) and (4.13) could
have been replaced by ϕ^{ijj} and ϕ^{iijj}, while that in (4.14) could
have been written ϕ^{ijkk}, matching the partition corresponding to
the required k-statistic.

To verify that these are indeed the appropriate estimators of
the generalized cumulants, we observe first that the generalized
k-statistics are symmetric functions of $Y_1, ..., Y_n$. Also, on taking
expectations, we have that

$$E\{k^{r,st}\} = n^{-1} \phi^{ij} \{ \kappa^{r,st} \delta_{ij} + \kappa^r \kappa^{st} \delta_i \delta_j \} = \kappa^{r,st}.$$

Similarly for $k^{rs,tu}$ and $k^{r,s,tu}$. More generally, if we define $Y_i^{rs} = Y_i^r Y_i^s$, it is immediately evident that (4.13) is an unbiased estimate
of $\mathrm{cov}(Y_i^{rs}, Y_i^{tu})$. Evidently, the sample moments are special cases
of generalized k-statistics, for we may write

$$k^{rst} = n^{-1} \sum_i \phi^i Y_i^r Y_i^s Y_i^t,$$

which is the same as the expression given in Section 4.2.2. In
fact, unbiased estimates exist for all generalized cumulants of order
(α, β), provided only that $\alpha \leq n$. In particular, unbiased estimates
exist for all moments of all orders but only for ordinary cumulants
whose order does not exceed n.

The importance of generalized k-statistics stems from the following properties:

(i) The generalized k-statistics are linearly independent (but not
functionally independent).

(ii) Every polynomial symmetric function can be expressed uniquely
as a *linear* combination of generalized k-statistics.

(iii) Any polynomial symmetric function whose expectation is independent of n can be expressed as a *linear* combination of
generalized k-statistics with coefficients independent of n.

These properties are not sufficient to identify the generalized k-statistics uniquely. In fact, as will be shown in the sections that follow, there are alternative systems of symmetric functions that are in some ways more convenient than generalized k-statistics. All such systems are invertible linear functions of generalized k-statistics with coefficients independent of the sample size, and the three properties listed above are preserved under such transformations.

In view of the results derived earlier, particularly in Section 3.8, the proofs of the above assertions are fairly elementary. Only broad outline proofs are provided here. To establish (i), we suppose that there exists a linear combination of generalized k-statistics that is identically zero for all Y and show that this assumption leads to a contradiction. The expectation of such a linear combination is the same linear combination of generalized cumulants, which, by assumption, must be zero for all distributions. But this is known to be impossible because the generalized cumulants are linearly independent. Hence (i) follows.

To prove (ii), we note first that if a linear combination exists, it must be unique because the generalized k-statistics are linearly independent. For the remainder of (ii), we need to show that there are enough generalized k-statistics to span the space of polynomial symmetric functions. Without loss of generality, we may restrict attention to homogeneous symmetric functions of degree one in each of the variables $Y^1, ..., Y^p$. Any such polynomial may be written in the form

$$a^{i_1 i_2 \cdots i_p} Y^1_{i_1} \cdots Y^p_{i_p}.$$

This is one of the few examples in this book of an array of coefficients that is not symmetric under index permutation. The following discussion is given for $p = 4$ but generalizes in an obvious way to arbitrary p. For $p = 4$, symmetry implies that the array a^{ijkl} can have at most $B_4 = 15$ distinct values, namely a^{1111}, $a^{1112}[4]$, $a^{1122}[3]$, $a^{1123}[6]$ and a^{1234}. To see that this is so, we note the quartic $\sum a^{ijkl} Y^r_i Y^s_j Y^t_k Y^u_l$ must be invariant under permutation of $Y_1, ..., Y_n$. Hence, for any $n \times n$ permutation matrix π^i_r, we must have

$$a^{ijkl} = \pi^i_r \pi^j_s \pi^k_t \pi^l_u a^{rstu} = a^{\pi_i \pi_j \pi_k \pi_l},$$

where $\pi_1, ..., \pi_n$ is a permutation of the first n integers. Hence, if

i,j,k,l are distinct integers, then

$$a^{iiij} = a^{1112} \neq a^{2111}$$
$$a^{iijk} = a^{1123} \neq a^{1233}$$

and so on. It follows that there are exactly 15 linearly independent symmetric functions of degree one in each of four distinct variables. Any convenient basis of 15 linearly independent symmetric functions, each of degree one in the four variables, is adequate to span the required space. For example, one possibility would be to set each of the distinct as to unity in turn, the remainder being kept at zero. However, the generalized cumulants provide an alternative and more convenient basis. Of course, if the four variables were not distinct, as, for example, in univariate problems, the number of linearly independent symmetric functions of total degree four and of specified degree in each of the component variables would be reduced. In the univariate case, there are five linearly independent symmetric functions of degree four, one for each of the partitions of the *number* four.

The argument just given holds for homogeneous symmetric functions of degree one in an arbitrary number of *distinct* random variables. Even when the number of distinct variables is less than the degree of the polynomial, it is often convenient in the algebra to sustain the fiction that there are as many distinct variables as the total degree of the polynomial. Effectively, we replicate an existing variable and the algebra treats the replicate as a distinct variable. This device of algebraic pseudo-replication often simplifies the algebra and avoids the need to consider numerous special cases.

The final assertion (iii) follows from completeness of the set of generalized k-statistics together with the fact that generalized k-statistics are unbiased estimates of the corresponding cumulant.

4.3.2 *Symmetric means and polykays*

The symmetric means have previously been introduced in Section 4.2.2 as the unique polynomial symmetric functions that are unbiased estimates of products of moments. Symmetric means, also called power products in the combinatorial literature dealing with the univariate case (Dressel, 1940), are most conveniently indexed in the multivariate case by a partition of a set of indices. Of course, this must be done in such a way that the notation does not give rise to confusion with generalized k-statistics, which are indexed in a similar manner. Our proposal here is to write $k^{(rs)(tuv)}$ for the estimate of the moment product $\kappa^{rs}\kappa^{tuv}$. The bracketing of superscripts is intended to suggest that some kind of product is involved. In a sense, the letter k is redundant or may be inferred from the context, and we might well write $(rs)(tuv)$ or $\langle(rs)(tuv)\rangle$ corresponding more closely to the conventions used in the univariate case where $k^{(11)(111)}$ would typically be written as 23 or $\langle 2\,3\rangle$ (MacMahon, 1915; Dressel, 1940; Tukey, 1950, 1956a). In this chapter, we use the notations $k^{(rs)(tuv)}$ and $(rs)(tuv)$ interchangeably: the latter notation has the advantage of greater legibility.

The expressions for the symmetric means are rather simple. For instance, it is easily verified that

$$(rs)(tuv) \equiv k^{(rs)(tuv)} = \sum_{i\neq j} Y_i^r Y_i^s Y_j^t Y_j^u Y_j^v / n^{(2)}$$

and

$$(rs)(tu)(v) \equiv k^{(rs)(tu)(v)} = \sum_{i\neq j\neq k} Y_i^r Y_i^s Y_j^t Y_j^u Y_k^v / n^{(3)},$$

where the sum extends over distinct subscripts and the divisor is just the number of terms in the sum. The extension to arbitrary partitions is immediate and need not be stated explicitly. Additional symmetric means are listed in Section 4.2.2.

The polykays are the unique polynomial symmetric functions that are unbiased estimates of cumulant products. It is natural therefore to write $k^{(r,s)(t,u,v)}$ or $(r,s)(t,u,v)$ to denote that unique symmetric function whose expectation is the cumulant product $\kappa^{r,s}\kappa^{t,u,v}$. Again, the bracketing suggests multiplication, and the commas indicate that cumulant products rather than moment products are involved. We first give a few examples of polykays and then show how the three systems of symmetric functions, generalized k-statistics, symmetric means and polykays are related.

It was shown in Section 4.2.2 that

$$k^{(r)(s)} = \sum \phi^{i|j} Y_i^r Y_j^s / n^{(2)}$$

is an unbiased estimate of the product $\kappa^r \kappa^s$, where $\phi^{i|j} = 1$ if $i \neq j$ and zero otherwise. By extension,

$$k^{(r)(s,t)} = \sum \phi^{i|jk} Y_i^r Y_j^s Y_k^t / n^{(2)},$$

with suitably chosen coefficients $\phi^{i|jk}$, is an unbiased estimate of $\kappa^r \kappa^{s,t}$. The required coefficients are

$$
\phi^{i|jk} = \begin{cases} 0 & \text{if } i = j \text{ or } i = k \\ 1 & \text{if } j = k \neq i \\ -1/(n-2) & \text{otherwise,} \end{cases}
$$

as can be seen by writing $k^{(r)(s,t)} = k^{(r)(st)} - k^{(r)(s)(t)}$ in the form

$$
\sum{}^{\#} Y_i^r Y_j^s Y_j^t /n^{(2)} - \sum{}^{\#} Y_i^r Y_j^s Y_k^t /n^{(3)}.
$$

In addition, we may write

$$
k^{(r,s)(t,u)} = \sum \phi^{ij|kl} Y_i^r Y_j^s Y_k^t Y_l^u /n^{(2)}
$$

for the unbiased estimate of the product $\kappa^{r,s} \kappa^{t,u}$, where

$$
\phi^{ij|kl} = \begin{cases} 0 & i \text{ or } j = k \text{ or } l \\ 1 & i = j \text{ and } k = l \neq i \\ -1/(n-2) & i = j, \, k \neq l \neq i \text{ or reverse} \\ 1/\{(n-2)(n-3)\} & \text{all distinct.} \end{cases}
$$

Also,

$$
k^{(r)(s)(t,u)} = \sum \phi^{i|j|kl} Y_i^r Y_j^s Y_k^t Y_l^u /n^{(3)}
$$

is an unbiased estimate of the product $\kappa^r \kappa^s \kappa^{t,u}$, where the coefficients are given by

$$
\phi^{i|j|kl} = \begin{cases} 0 & \text{same value occurs in different blocks} \\ 1 & k = l, \, i \neq j \neq k \\ -1/(n-3) & \text{all indices distinct.} \end{cases}
$$

Evidently, the coefficients for the polykays are more complicated than those for ordinary k-statistics, symmetric means or generalized k-statistics. These complications affect the algebra but do not necessarily have much bearing on computational difficulty. Neither the formulae given above nor the corresponding ones for generalized k-statistics are suitable as a basis for computation. Computational questions are discussed in Section 4.5.

Since the generalized k-statistics, symmetric means and poly-kays are three classes of symmetric functions indexed in the same manner, it is hardly surprising to learn that any one of the three classes can be expressed as a linear function of any other. In fact, all of the necessary formulae have been given in Section 3.6.2: we need only make the obvious associations of symmetric means with moment products, polykays with cumulant products and generalized k-statistics with generalized cumulants. The following are a few simple examples of symmetric means expressed in terms of polykays.

$$(rs)(t) = (r,s)(t) + (r)(s)(t) = \{(r,s) + (r)(s)\}(t)$$
$$(rs)(tu) = (r,s)(t,u) + (r,s)(t)(u) + (r)(s)(t,u) + (r)(s)(t)(u)$$
$$= \{(r,s) + (r)(s)\}\{(t,u) + (t)(u)\}$$
$$(rs)(tuv) = \{(r,s) + (r)(s)\}\{(t,u,v) + (t)(u,v)[3] + (t)(u)(v)\}.$$

Of course, $(rs) = (r,s) + (r)(s)$ and similarly

$$(tuv) = (t,u,v) + (t)(u,v)[3] + (t)(u)(v)$$

but this does not imply that $k^{(rs)(tuv)}$ is the same as $k^{(rs)}k^{(tuv)}$. The above multiplication formulae for the indices are purely symbolic and the multiplication must be performed first before the interpretation is made in terms of polykays. Thus $(rs)(tuv)$ is expressible as the sum of the 10 polykays whose indices are sub-partitions of $rs|tuv$.

The corresponding expressions for polykays in terms of symmetric means may be written symbolically as

$$(r,s)(t) = \{(rs) - (r)(s)\}(t)$$
$$(r,s)(t,u) = \{(rs) - (r)(s)\}\{(tu) - (t)(u)\} \qquad (4.15)$$
$$(r,s)(t,u,v) = \{(rs) - (r)(s)\}\{(tuv) - (t)(uv)[3] + 2(t)(u)(v)\}.$$

Again, it is intended that the indices should be multiplied algebraically before the interpretation is made in terms of symmetric means. Thus, $(r,s)(t,u,v)$ is a linear combination of the 10 symmetric means whose indices are sub-partitions of $rs|tuv$. The coefficients in this linear combination are values of the Möbius function for the partition lattice.

From the identity connecting generalized cumulants with products of ordinary cumulants, it follows that we may express generalized k-statistics in terms of polykays using (3.3) and conversely for polykays in terms of generalized cumulants using (3.18). By way of illustration, we find using (3.3) that

$$k^{rs,tu} = k^{(r,s,t,u)} + k^{(r)(s,t,u)}[4] + k^{(r,t)(s,u)}[2] + k^{(r)(t)(s,u)}[4],$$

where $k^{(r,s,t,u)} \equiv k^{r,s,t,u}$ and the sum extends over all partitions complementary to $rs|tu$. The inverse expression giving polykays in terms of generalized cumulants is a little more complicated but fortunately it is seldom needed. Application of (3.18) gives, after some arithmetic,

$$6k^{(r,s)(t,u)} = k^{r,stu}[4] - k^{rs,tu}[3] - 2k^{r,s,tu}[2] + k^{r,t,su}[4] - k^{r,s,t,u}.$$

This identity can be read off the sixth row of the 15×15 matrix given at the end of Section 3.6.2. The remaining expressions involving four indices are

$$6k^{(r)(s,t,u)} = -k^{r,stu}[4] + k^{rs,tu}[3] + 2k^{rs,t,u}[3] - k^{r,s,tu}[3] - 2k^{r,s,t,u}$$
$$6k^{(r)(s)(t,u)} = -k^{r,stu}[2] + 2k^{rst,u}[2] - 2k^{rs,tu} + k^{rt,su}[2]$$
$$\qquad\qquad + 2k^{r,s,tu} - k^{r,t,su}[4] - k^{rs,t,u} + k^{r,s,t,u}$$
$$6k^{(r)(s)(t)(u)} = 6k^{rstu} - 2k^{r,stu}[4] - k^{rs,tu}[3] + k^{r,s,tu}[6] - k^{r,s,t,u}.$$

These examples emphasize that the relationship between the polykays and the generalized k-statistics is linear, invertible and that the coefficients are independent of the sample size. Because of linearity, unbiasedness of one set automatically implies unbiasedness of the other.

4.4 Derived scalars

To each of the derived scalars discussed in Section 2.8 there corresponds a sample scalar in which the κs are replaced by ks. Denote by $k_{i,j}$ the matrix inverse of $k^{i,j}$ and write

$$p\bar{r}_{13}^2 = k^{r,s,t}k^{u,v,w}k_{r,s}k_{t,u}k_{v,w}$$
$$p\bar{r}_{23}^2 = k^{r,s,t}k^{u,v,w}k_{r,u}k_{s,v}k_{t,w}$$
$$p\bar{r}_4 = k^{r,s,t,u}k_{r,s}k_{t,u}$$

for the sample versions of $p\bar{\rho}_{13}^2$, $p\bar{\rho}_{23}^2$ and $p\bar{\rho}_4$ defined by (2.14)–(2.16). Although these three statistics are in a sense, the obvious estimators of the invariant parameters, they are not unbiased because, for example, $k_{r,s}$ is not unbiased for $\kappa_{r,s}$.

By their construction, the three statistics listed above are invariant under affine transformation of the components of X. Hence their expectations and joint cumulants must also be expressible in terms of invariants. To obtain such expressions, it is convenient to expand the matrix inverse $k_{r,s}$ in an asymptotic expansion about $\kappa_{r,s}$. If we write

$$k^{r,s} = \kappa^{r,s} + \epsilon^{r,s}$$

it follows that $\epsilon^{r,s} = O_p(n^{-1/2})$ in the sense that $n^{1/2}\epsilon^{r,s}$ has a nondegenerate limiting distribution for large n. The matrix inverse may therefore be expanded as

$$k_{r,s} = \kappa_{r,s} - \epsilon_{r,s} + \epsilon_{r,i}\epsilon_{s,j}\kappa^{i,j} - \epsilon_{r,i}\epsilon_{j,k}\epsilon_{l,s}\kappa^{i,j}\kappa^{k,l} + ...$$

where $\epsilon_{r,s} = \kappa_{r,i}\kappa_{s,j}\epsilon^{i,j}$ is not the matrix inverse of $\epsilon^{r,s}$. The advantage of working with this expansion is that it involves only $\epsilon^{r,s}$ whose cumulants, apart from the first, are the same as those of $k^{r,s}$. In the case of the scalar $p\bar{r}_4$, we may write

$$p\bar{r}_4 = k^{r,s,t,u}(\kappa_{r,s} - \epsilon_{r,s} + \epsilon_{r,i}\epsilon_{s,j}\kappa^{i,j} - ...)$$
$$\times(\kappa_{t,u} - \epsilon_{t,u} + \epsilon_{t,i}\epsilon_{u,j}\kappa^{i,j} - ...).$$

On taking expectation and including terms up to order $O(n^{-1})$ only, we find using the identity (3.3) that

$$E(p\bar{r}_4) = p\bar{\rho}_4 - 2\kappa_k(r,s,t,t|r,s) + \kappa_k(r,s,t,u)\kappa_k(r,s|t,u)$$
$$+ 2\kappa_k(r,s,t,t)\kappa_k(r,u|s,u), \tag{4.16}$$

where, for example, $\kappa_k(r,s,t,t|r,s)$ is a convenient shorthand notation for the scalar

$$\kappa_{r,v}\kappa_{s,w}\kappa_{t,u}\mathrm{cov}(k^{r,s,t,u}, k^{v,w})$$

and $\kappa_k(r,s,t,u) = \kappa^{r,s,t,u}$ by construction. Simplification of this particular scalar gives

$$\kappa_{r,v}\kappa_{s,w}\kappa_{t,u}\{\kappa^{r,s,t,u,v,w}/n + (\kappa^{r,s,t,v}\kappa^{u,w}[6] + \kappa^{r,t,u,v}\kappa^{s,w}[2])/(n-1)$$
$$+ (\kappa^{r,s,v}\kappa^{t,u,w}[4] + \kappa^{r,t,w}\kappa^{s,u,v}[2])/(n-1)\}$$
$$= p\bar{\rho}_6/n + (6p\bar{\rho}_4 + 2p^2\bar{\rho}_4 + 4p\bar{\rho}_{13}^2 + 2p\bar{\rho}_{23}^2)/(n-1).$$

After simplification of the remaining terms in (4.16), we are left with

$$E(\bar{r}_4) = \bar{\rho}_4(1 - 8/n - 2p/n) - 2\bar{\rho}_6/n - 8\bar{\rho}_{13}^2/n - 4\bar{\rho}_{23}^2/n$$
$$+ 2\bar{\rho}_{14}^2/n + \bar{\rho}_{24}^2/n + O(n^{-2}),$$

where $p\bar{\rho}_{14}^2$ and $p\bar{\rho}_{24}^2$ are defined in Section 2.8.

Similar expressions may be found for higher-order cumulants, though such expressions tend to be rather lengthy especially when carried out to second order. To first order we have, for example, that

$$\mathrm{var}(p\bar{r}_4) = \kappa_k(r,r,s,s|t,t,u,u) - 4\kappa_k(r,r,s,t)\kappa_k(u,u,v,v|s,t)$$
$$+ O(n^{-2}),$$

both terms being of order $O(n^{-1})$. This variance can be expressed directly in terms of invariants but the expression is rather lengthy and complicated and involves invariants of a higher degree than those so far discussed.

In the case of jointly normal random variables, the second term above vanishes and the first reduces to

$$\mathrm{var}(p\bar{r}_4) = (8p^2 + 16p)/n + O(n^{-2}). \qquad (4.17)$$

In fact, it can be shown (Exercise 4.10), that the limiting distribution of $n^{1/2}p\bar{r}_4$ is normal with zero mean and variance $8p^2 + 16p$. The normal limit is hardly surprising because all the k-statistics are asymptotically normal: the joint cumulants behave in the same way as the cumulants of a straightforward average of independent random variables.

In the case of the quadratic scalar, $\bar{r}_{13}^2$, we may make the following expansion

$$p\bar{r}_{13}^2 = k^{r,s,t}k^{u,v,w}(k_{r,s} - \epsilon_{r,s} + \epsilon_{r,i}\epsilon_{s,j}\kappa^{i,j} - ...)$$
$$\times (k_{t,u} - \epsilon_{t,u} + \epsilon_{t,i}\epsilon_{u,j}\kappa^{i,j} - ...)$$
$$\times (k_{v,w} - \epsilon_{v,w} + \epsilon_{v,i}\epsilon_{w,j}\kappa^{i,j} - ...)$$

On taking expectation, and including terms up to order $O(n^{-1})$, we find

$$E(p\bar{r}_{13}^2) = p\rho_{13}^2 + \kappa_k(r,r,s|s,t,t) - 2\kappa_k(r,s,t)\kappa_k(t,u,u|r,s)$$
$$- 2\kappa_k(u,v,v)\kappa_k(r,s,u|r,s) - 2\kappa_k(r,r,s)\kappa_k(u,t,t|s,u)$$
$$+ 2\kappa_k(r,r,t)\kappa_k(t,v,w)\kappa_k(v,u|u,w)$$
$$+ \kappa_k(r,r,t)\kappa_k(u,u,w)\kappa_k(t,s|s,w) + O(n^{-2}).$$

Again, the above formula may be expressed directly in terms of invariants. For example, the second term may be written as

$$p(\bar{\rho}_6 + p\bar{\rho}_4 + 8\bar{\rho}_4 + 5\bar{\rho}_{13}^2 + 4\bar{\rho}_{23}^2 + 2p + 4)/n + O(n^{-2}).$$

In the case of normal random variables, only the second term contributes, giving

$$E(p\bar{r}_{13}^2) = 2p(p+2)/n + O(n^{-2}).$$

In fact, it may be shown (Exercise 4.11), that for large n,

$$\frac{(n-1)(n-2)p\bar{r}_{13}^2}{2n(p+2)} \sim \chi_p^2$$

under the assumption of normality.

Similar calculations for $\bar{r}_{23}^2$ give

$$\frac{(n-1)(n-2)p\bar{r}_{23}^2}{6n} \sim \chi_{p(p+1)(p+2)/6}^2$$

for large n under the assumption of normality (Exercise 4.11).

Finally, it is worth pointing out that, although $\bar{r}_{13}^2$, $\bar{r}_{23}^2$ and $\bar{r}_4$ are the most commonly used scalars for detecting multivariate non-normality, they are not the only candidates for this purpose. Other scalars that have an equal claim to be called the sample versions of $p\bar{\rho}_{13}^2$ and $p\bar{\rho}_{23}^2$ include

$$k^{(r,s,t)(u,v,w)}k_{r,s}k_{t,u}k_{v,w} \quad \text{and} \quad k^{(r,s,t)(u,v,w)}k_{r,u}k_{s,v}k_{t,w}.$$

Another class of invariant scalars that has considerable geometrical appeal despite its incompleteness, may be defined by examining the directional standardized skewness and kurtosis and choosing the directions corresponding to maxima and minima. In two dimensions, if $\bar{\rho}_{13}^2 = 0$, there are three directions of equal maximum skewness separated by $2\pi/3$, and zero skewness in the orthognal direction. On the other hand, if $4\bar{\rho}_{23}^2 = 3\bar{\rho}_{13}^2$ there is one direction of maximum skewness and zero skewness in the orthogonal direction. More generally, a complete picture of the directional skewness involves a combination of the above: see Exercises 2.36 and 2.37. In the case of the directional kurtosis, it is necessary to distinguish between maxima and minima. Machado (1976, 1983) gives approximate percentage points of the distribution under normality of the sample versions of the maximized directional skewness and kurtosis and the minimized directional kurtosis.

4.5   Computation

4.5.1   *Practical issues*

Practical considerations suggest strongly that we are seldom likely to require k-statistics or polykays of degree more than about four. Without further major assumptions, this is enough to enable the statistician to make approximate interval estimates for the first two cumulants only and to make approximate point estimates for cumulants of orders three and four. As a general rule, the higher-order k-statistics tend to have large sampling variability and consequently, large quantities of data are required to obtain estimates that are sufficiently precise to be useful. By way of example, under the optimistically favourable assumption that the data are approximately normally distributed with unit variance, approximately 40000 observations are required to estimate the fourth cumulant accurately to one decimal place. In the case of observations distributed approximately as Poisson with unit mean, the corresponding sample size is just over half a million. More realistically, to estimate κ_4 accurately to the nearest whole number, we require 400 observations if circumstances are favourable, and more than 5000 if they are only a little less favourable. These rough calculations give some idea of the kind of precision achievable in practice.

Table 4.2   *Numbers of k-statistics and polykays of various orders*

Number of distinct k-statistics

Order	Number of variables			
	1	2	3	4
1	1	2	3	4
2	1	3	6	10
3	1	4	10	20
4	1	5	15	35
5	1	6	21	56
k	1	$k+1$	$\binom{k+2}{2}$	$\binom{k+3}{3}$

Number of distinct polykays

1	1	2	3	4
2	2	6	12	20
3	3	14	38	80
4	5	33	117	305
5	7	70	336	1072
k	$p_1(k)$	$p_2(k)$	$p_3(k)$	$p_4(k)$

In addition to the statistical considerations just mentioned, it should be pointed out that the number of k-statistics and, more emphatically, the number of polykays, grows rapidly with the number of variables and with the order of k-statistic or polykay considered. For example, the number of distinct k-statistics of total order k in up to p variables is $\binom{k+p-1}{k}$, while, if we include those of order less than k, the number becomes $\binom{k+p}{k}$. The latter formula includes the sample size itself as the k-statistic of order zero. On the other hand, the number of polykays of total order exactly k in up to q variables is $p_q(k)$, where, for example, $p_1(k)$ is the number of partitions of the number k and

$$p_2(k) = \sum_{\substack{a+b=k \\ a,b \geq 0}} p(a,b),$$

where $p(a,b)$ is the number of distinct partitions of a set containing a objects of type A and b objects of type B. For the purposes of this discussion, objects of the same type are assumed to be indistinguishable. In an obvious notation,

$$p_3(k) = \sum_{\substack{a+b+c=k \\ a,b,c \geq 0}} p(a,b,c)$$

involves a sum over various tri-partite partition numbers of three integers. Some values for these totals are given in Table 4.2.

Partly for the reasons just given, but mainly to preserve the sanity of author and reader alike, we discuss computation only for the case $k \leq 4$.

4.5.2 From power sums to symmetric means

As a first step in the calculation, we compute the following $\binom{p+4}{4}$ 'power sums' and 'power products': $k^0 = 1$,

$$k^r = n^{-1} \sum_i Y_i^r \qquad k^{rs} = n^{-1} \sum_i Y_i^r Y_i^s$$
$$k^{rst} = n^{-1} \sum_i Y_i^r Y_i^s Y_i^t \qquad k^{rstu} = n^{-1} \sum_i Y_i^r Y_i^s Y_i^t Y_i^u.$$

One simple way of organizing these calculations for a computer is to augment the data by adding the dummy variable $Y_i^0 = 1$ and by computing k^{rstu} for $0 \leq r \leq \ldots \leq u$. This can be accomplished in a single pass through the data. The three-index, two-index and one-index quantities can be extracted as k^{0rst}, k^{00rs} and k^{000r} as required.

The above quantities are special cases of symmetric means. All subsequent $k$statistics and polykays whose total degree does not exceed 4 may be derived from these symmetric functions. No further passes through the data matrix are required. The remaining symmetric means are $k^{(r)} = k^r$,

$$k^{(r)(s)} = \{nk^r k^s - k^{rs}\}/(n-1)$$
$$k^{(r)(s)(t)} = \{n^2 k^r k^s k^t - nk^r k^{st}[3] + 2k^{rst}\}/(n-1)^{(2)}$$
$$k^{(r)(s)(t)(u)} = \{n^3 k^r k^s k^t k^u - n^2 k^r k^s k^{tu}[6] + 2nk^r k^{stu}[4]$$
$$+ nk^{rs} k^{tu}[3] - 6k^{rstu}\}/(n-1)^{(3)}. \qquad (4.18)$$

Included, essentially as special cases of the above, are the following symmetric means:

$$k^{(r)(st)} = \{nk^r k^{st} - k^{rst}\}/(n-1)$$
$$k^{(rs)(tu)} = \{nk^{rs} k^{tu} - k^{rstu}\}/(n-1)$$
$$k^{(r)(s)(tu)} = \{n^2 k^r k^s k^{tu} - nk^r k^{stu} - nk^s k^{rtu}$$
$$- nk^{rs} k^{tu} + 2k^{rstu}\}/(n-1)^{(2)}. \qquad (4.19)$$

Of course, if the computations are organized as suggested in the previous paragraph, then the above formulae may all be regarded as special cases of expression (4.18) for $k^{(r)(s)(t)(u)}$. By way of example, direct substitution gives

$$k^{(0)(r)(s)(t)} = \left(n^3 k^r k^s k^t - 3n^2 k^r k^s k^t - n^2 k^r k^{st}[3]\right.$$
$$\left. + 2nk^{rst} + 2nk^r k^{st}[3] + nk^r k^{st}[3] - 6k^{rst}\right)/(n-1)^{(3)}$$
$$= k^{(r)(s)(t)}.$$

For $p = 2$, there are 55 such terms, all linearly independent but functionally dependent on the 15 basic power sums.

4.5.3 From symmetric means to polykays

In going from symmetric means to polykays, the expressions are all linear and the coefficients are integers independent of the sample size. These properties lead to simple recognizable formulae. Some examples are as follows:

$$k^{(r,s)} = k^{(rs)} - k^{(r)(s)}$$
$$k^{(r)(s,t)} = k^{(r)(st)} - k^{(r)(s)(t)}$$
$$k^{(r,s,t)} = k^{(rst)} - k^{(r)(st)}[3] + 2k^{(r)(s)(t)}$$
$$k^{(r,s)(t,u)} = k^{(rs)(tu)} - k^{(r)(s)(tu)} - k^{(rs)(t)(u)} + k^{(r)(s)(t)(u)}.$$

More generally, the Möbius coefficients that occur in the above formulae may be obtained using the symbolic index multiplication formulae (4.15). For example, in the final expression above, the indices and the coefficients are given by the expression

$$\{(rs) - (r)(s)\}\{(tu) - (t)(u)\}.$$

Similarly, in the expression for $k^{(r)(s,t,u)}$, the indices and the coefficients are given by

$$(r)\{(stu) - (s)(t\dot{u})[3] + 2(s)(t)(u)\}.$$

The foregoing two-stage operation, from power sums to symmetric means to polykays, produces the symmetric means as an undesired by-product of the computation. For most statistical purposes, it is the polykays that are most useful and, conceivably, there is some advantage to be gained in going from the power sums to the polykays directly. The machinery required to do this will now be described. First, we require the 15×15 upper triangular matrix $\mathbf{M} = m(\Upsilon_i, \Upsilon_j)$ whose elements are the values of the Möbius function for the lattice of partitions of four items. In addition, we require the vector $\mathbf{K}$ whose 15 elements are products of power sums.

In practice, since $\mathbf{M}$ is rather sparse, it should be possible to avoid constructing the matrix explicitly. To keep the exposition as simple as possible, however, we suppose here that $\mathbf{M}$ and $\mathbf{K}$ are constructed explicitly as shown in Table 4.3.

Evidently, (4.18) and (4.19) amount to a statement that the vector of symmetric means is given by the matrix product

$$\mathbf{D}_1\mathbf{M}^T\mathbf{D}_2\mathbf{K}$$

where

$$\mathbf{D}_1 = \text{diag}\{1/n^{(\nu_j)}\}, \quad \mathbf{D}_2 = \text{diag}\{n^{\nu_j}\}$$

and ν_j is the number of blocks in the jth partition. The polykays are obtained by Möbius inversion of the symmetric means, giving the vector

$$\mathbf{P} = \mathbf{M}\mathbf{D}_1\mathbf{M}^T\mathbf{D}_2\mathbf{K}. \qquad (4.20)$$

By this device, the computation of polykays is reduced to a simple linear operation on matrices and vectors. Despite this algebraic

Table 4.3 *The Möbius matrix and the vector of power sums used in* (4.20) *to compute the polykays*

M	K
1 −1 −1 −1 −1 −1 −1 −1 2 2 2 2 2 2 −6	k^{rstu}
1 −1 −1 −1 2	$k^{rst}k^u$
1 −1 −1 −1 2	$k^{rsu}k^t$
1 −1 −1 −1 2	$k^{rtu}k^s$
1 −1 −1 −1 2	$k^{stu}k^r$
1 −1 −1 1	$k^{rs}k^{tu}$
1 −1 −1 1	$k^{rt}k^{su}$
1 −1 −1 1	$k^{ru}k^{st}$
1 −1	$k^{rs}k^tk^u$
1 −1	$k^{rt}k^sk^u$
1 −1	$k^{ru}k^sk^t$
1 −1	$k^{st}k^rk^u$
1 −1	$k^{su}k^rk^t$
1 −1	$k^{tu}k^rk^s$
1	$k^rk^sk^tk^u$

simplicity, it is necessary to take care in the calculation to avoid rounding error, particularly where n is large or where some components have a mean value that is large.

The inverse relationship, giving K in terms of P, though not of much interest for computation, involves a matrix having a remarkably elegant form. The (i, j) element of the inverse matrix is n raised to the power $|\Upsilon_i \vee \Upsilon_j| - |\Upsilon_i|$, where $|\Upsilon|$ is the number of blocks in the partition. See Exercise 4.18. It follows that the inverse matrix has a unit entry where M is non-zero and negative powers of n elsewhere.

4.6 Application to sampling

4.6.1 *Simple random sampling*

In a population containing N units or individuals, there are $\binom{N}{n}$ distinct subsets of size n. The subsets are distinct with regard to their labels though not necessarily in their values. A sample of size n chosen in such a way that each of the distinct subsets occurs with equal probability, namely $\binom{N}{n}^{-1}$, is said to be a simple random sample of size n taken without replacement from the population. In particular, each unit in the population occurs in such a sample with probability n/N; each distinct pair occurs in the sample with probability $n(n-1)/\{N(N-1)\}$, and similarly for triplets and so on.

Suppose that on the ith unit in the sample, a p-dimensional variable $Y_i = Y_i^1, ..., Y_i^p$, $(i = 1, ..., n)$ is measured. Most commonly, $p = 1$, but it is more convenient here to keep the notation as general as possible. Let k^r be the rth component of the sample average and let K^r be the corresponding population average. Each unit in the population occurs in exactly $\binom{N-1}{n-1}$ of the $\binom{N}{n}$ distinct samples. Thus, if we denote by ave(k^r), the average value of the sample mean, averaged over all possible samples, we have that

$$
\begin{aligned}
\text{ave}(k^r) &= \binom{N}{n}^{-1} n^{-1} \sum_{\text{all subsets}} Y_1^r + ... + Y_n^r \\
&= \binom{N}{n}^{-1} n^{-1} \binom{N-1}{n-1} \{Y_1^r + ... + Y_N^r\} \\
&= N^{-1}\{Y_1^r + ... + Y_N^r\} = K^r.
\end{aligned}
$$

In other words, ave(k^r), averaged over all possible samples, is just the same function computed for the whole population of N units, here denoted by K^r.

The same argument immediately gives ave(k^{rs}) $= K^{rs}$, where

$$
K^{rs} = N^{-1}\{Y_1^r Y_1^s + ... + Y_N^r Y_N^s\}
$$

is a population average of products. It follows by direct analogy that ave(k^{rst}) $= K^{rst}$ and so on.

In order to show that symmetric means have the same property, namely

$$
\text{ave}\{k^{(r)(s)}\} = \sum{}^{\#} Y_i^r Y_j^s / \{N(N-1)\} = K^{(r)(s)}
$$

$$
\text{ave}\{k^{(r)(st)}\} = \sum{}^{\#} Y_i^r Y_j^s Y_j^t / \{N(N-1)\} = K^{(r)(st)},
$$

where summation runs from 1 to N over unequal indices, we need only replace each occurrence of 'unit' in the previous argument with 'pair of distinct units' and make appropriate cosmetic changes in the formulae. For example, each pair of distinct units in the population occurs in $\binom{N-2}{n-2}$ of the $\binom{N}{n}$ distinct samples. It follows that the average value of the symmetric mean, $k^{(r)(s)}$ is

$$
\begin{aligned}
\text{ave}\{k^{(r)(s)}\} &= \binom{N}{n}^{-1} \frac{1}{n(n-1)} \sum_{\text{all subsets}} Y_1^r Y_2^s + \ldots + Y_{n-1}^r Y_n^s \\
&= \binom{N}{n}^{-1} \frac{1}{n(n-1)} \binom{N-2}{n-2} \sum_{\substack{1 \le i,j \le N \\ i \ne j}} Y_i^r Y_j^s \\
&= K^{(r)(s)}.
\end{aligned}
$$

This argument easily extends to any symmetric mean and hence to any polykay. The conclusion can therefore be summarized as follows.

Under simple random sampling, the average value of any polykay or symmetric mean, averaged over all $\binom{N}{n}$ possible samples, is just the same function computed in the population of N values.

Tukey (1950), carefully avoiding the more conventional terminology of unbiasedness, refers to this property as 'inheritance on the average'. To see that this property does not hold for arbitrary symmetric functions, see Exercise 4.13.

The main advantage in the present context of working with polykays, rather than with the less numerous power sums and products, is that variances and higher-order cumulants of computed statistics are more aesthetically appealing when expressed linearly in terms of population polykays.

4.6.2 *Joint cumulants of k-statistics*

In this section, no attempt will be made at achieving total generality. Instead, we concentrate mainly on the joint cumulants likely to be of most use in applications, usually where the degree does not exceed four. First, we give a few simple derivations of the more useful formulae. Since $\text{ave}(k^r) = K^r$, it follows that the covariance of k^r and k^s is

$$
\text{ave}\{k^r k^s - K^r K^s\},
$$

averaged over all possible samples. From (4.18) it follows that

$$k^r k^s = k^{(r)(s)} + k^{(r,s)}/n$$
$$K^r K^s = K^{(r)(s)} + K^{(r,s)}/N.$$

Hence, since $\text{ave}(k^{(r)(s)}) = K^{(r)(s)}$, we have

$$\text{cov}(k^r, k^s) = K^{r,s}\left(\frac{1}{n} - \frac{1}{N}\right),$$

a well known result easily derived in other ways.

Similarly, the covariance of k^r and $k^{s,t}$ may be written as

$$\text{ave}\{k^r k^{s,t} - K^r K^{s,t}\}.$$

From the multiplication formula

$$k^r k^{s,t} = k^{(r)(s,t)} + k^{r,s,t}/n$$
$$K^r K^{s,t} = K^{(r)(s,t)} + K^{r,s,t}/N,$$

it follows that

$$\text{cov}(k^r, k^{s,t}) = K^{r,s,t}\left(\frac{1}{n} - \frac{1}{N}\right).$$

In the case of the covariance of two sample variances or co-variances, we use the multiplication formula

$$k^{r,s} k^{t,u} = k^{(r,s)(t,u)} + k^{r,s,t,u}/n + k^{(r,t)(s,u)}[2]/(n-1),$$

together with an identical expression for a product of Ks. This gives

$$\begin{aligned}
\text{cov}(k^{r,s}, k^{t,u}) &= K^{r,s,t,u}\left(\frac{1}{n} - \frac{1}{N}\right) \\
&\quad + K^{(r,t)(s,u)}[2]\left(\frac{1}{n-1} - \frac{1}{N-1}\right),
\end{aligned} \tag{4.21}$$

which should be compared with the corresponding infinite population expression

$$\kappa_k(r,s|t,u) = \kappa^{r,s,t,u}/n + \kappa^{r,t}\kappa^{s,u}[2]/(n-1).$$

Note that if (4.21) were expressed in terms of products of population k-statistics such as $K^{r,t}K^{s,u}[2]$, it would be necessary to introduce the additional product $K^{r,s}K^{t,u}$ thereby involving a partition not satisfying the connectivity condition in (3.3).

The key to the derivation of formulae such as those given above, is evidently to express multiple products of k-statistics and polykays as a *linear* combination of polykays. Formulae for multiple products of ordinary k-statistics are easy to write down, particularly with the help of the expressions in Exercises 4.5 and 4.7 for the joint cumulants of k-statistics. The following example helps to illustrate the method.

Consider, in an infinite population, the mean value of the product of three covariances. We find

$$E(k^{r,s}k^{t,u}k^{v,w}) =$$
$$\kappa^{r,s}\kappa^{t,u}\kappa^{v,w} + \kappa^{r,s}\kappa_k(t,u|v,w)[3] + \kappa_k(r,s|t,u|v,w)$$
$$=\kappa^{r,s}\kappa^{t,u}\kappa^{v,w} + \kappa^{r,s}\{\kappa^{t,u,v,w}/n + \kappa^{t,v}\kappa^{u,w}[2]/(n-1)\}[3]$$
$$+ \kappa^{r,s,t,u,v}/n^2 + \kappa^{r,t}\kappa^{s,u,v,w}[12]/\{n(n-1)\}$$
$$+ \kappa^{r,t,v}\kappa^{s,u,w}[4](n-2)/\{n(n-1)^2\}$$
$$+ \kappa^{r,t}\kappa^{s,v}\kappa^{u,w}[8]/(n-1)^2.$$

Evidently, the combination

$$k^{(r,s)(t,u)(v,w)} + k^{(r,s)(t,u,v,w)}[3]/n + k^{(r,s)(t,v)(u,w)}[6]/(n-1)$$
$$+ k^{r,s,t,u,v,w}/n^2 + k^{(r,t)(s,u,v,w)}[12]/\{n(n-1)\} \qquad (4.22)$$
$$+ k^{(r,t,v)(s,u,w)}[4](n-2)/\{n(n-1)^2\} + k^{(r,t)(s,v)(u,w)}[8]/(n-1)^2$$

is an unbiased estimate of $E(k^{r,s}k^{t,u}k^{v,w})$. It follows immediately by linear independence that (4.22) is identical to the product $k^{r,s}k^{t,u}k^{v,w}$, giving the required multiplication formula. Although it is certainly possible to write down general formulae for the product of two arbitrary polykays, such expressions tend to be rather complicated. For most purposes, a multiplication table is more useful: see Table 4.4, which gives complete products up to fourth degree and selected products up to sixth degree. In the univariate case, more extensive tables and general formulae are given by Wishart (1952), Dwyer & Tracy (1964) and Tracy (1968).

By definition, the third-order joint cumulant of $k^{r,s}$, $k^{t,u}$ and $k^{v,w}$ is equal to

$$\text{ave}(k^{r,s}k^{t,u}k^{v,w}) - K^{r,s}\text{ave}(k^{t,u}k^{v,w})[3] + 2K^{r,s}K^{t,u}K^{v,w}.$$

On substituting

$$\text{ave}(k^{t,u}k^{v,w}) = K^{(t,u)(v,w)} + K^{t,u,v,w}/n + K^{(t,v)(u,w)}[2]/(n-1)$$

and simplifying using (4.22), we find that the third cumulant may be written in the form

$$\begin{aligned}
\alpha_1 &\left(K^{(r,s)(t,u,v,w)} - K^{(r,s)}K^{(t,u,v,w)} \right)[3] \\
+ \beta_1 &\left(K^{(r,s)(t,v)(u,w)} - K^{(r,s)}K^{(t,v)(u,w)} \right)[6] \\
+ &(n^{-2} - N^{-2})K^{r,s,t,u,v,w} + \gamma_1 K^{(r,t)(s,u,v,w)}[12] \\
+ &\left(\frac{n-2}{n(n-1)^2} - \frac{N-2}{N(N-1)^2} \right) K^{(r,t,v)(s,u,w)}[4] \\
+ &\left((n-1)^{-2} - (N-1)^{-2} \right) K^{(r,t)(s,v)(u,w)}[8].
\end{aligned} \qquad (4.23)$$

The coefficients in this formula are given by

$$\begin{aligned}
\alpha_1 &= n^{-1} - N^{-1}, & \alpha_2 &= \alpha_1 - N^{-1}, \\
\beta_1 &= (n-1)^{-1} - (N-1)^{-1}, & \beta_2 &= \beta_1 - (N-1)^{-1}, \\
\gamma_1 &= \{n(n-1)\}^{-1} - \{N(N-1)\}^{-1}, & \gamma_2 &= \gamma_1 - \{N(N-1)\}^{-1}.
\end{aligned}$$

On replacing all products by population polykays, we find the following gratifyingly simple expression involving only connecting partitions, namely

$$\begin{aligned}
\text{cum}(k^{r,s}, k^{t,u}, k^{v,w}) = &K^{r,s,t,u,v,w}\alpha_1\alpha_2 + K^{(r,t)(s,u,v,w)}[12]\alpha_1\beta_2 \\
&+ K^{(r,t,v)(s,u,w)}[4]\{\alpha_1\beta_2 - \beta_1\gamma_2\} \\
&- K^{(r,s,t)(u,v,w)}[6]\alpha_1/(N-1) \\
&+ K^{(r,t)(s,v)(u,w)}[8]\beta_1\beta_2.
\end{aligned} \qquad (4.24)$$

The simplicity of this formula, together with the corresponding one (4.21) for the covariance of $k^{r,s}$ and $k^{t,u}$, should be sufficient to justify the emphasis on polykays.

Table 4.4 *Some multiplication formulae for polykays*

Product	Linear expression in polykays
$k^r k^s$	$k^{(r)(s)} + k^{(r,s)}/n$
$k^r k^{s,t}$	$k^{(r)(s,t)} + k^{r,s,t}/n$
$k^r k^{(s)(t)}$	$k^{(r)(s)(t)} + \{k^{(s)(r,t)} + k^{(t)(r,s)}\}/n$
$k^r k^s k^t$	$k^{(r)(s)(t)} + k^{(r)(s,t)}[3]/n + k^{r,s,t}/n^2$
$k^r k^s k^t k^u$	$k^{(r)(s)(t)(u)} + k^{(r)(s)(t,u)}[6]/n + k^{(r,s)(t,u)}[3]/n^2$
	$\quad + k^{(r)(s,t,u)}[4]/n^2 + k^{r,s,t,u}/n^3$
$k^r k^s k^{t,u}$	$k^{(r)(s)(t,u)} + k^{(r,s)(t,u)}/n + k^{(r)(s,t,u)}[2]/n$
	$\quad + k^{r,s,t,u}/n^2$
$k^r k^{s,t,u}$	$k^{(r)(s,t,u)} + k^{r,s,t,u}/n$
$k^r k^s k^{(t)(u)}$	$k^{(r)(s)(t)(u)} + k^{(r)(t)(s,u)}[4]/n + k^{(t)(u)(r,s)}/n$
	$\quad + k^{(r,t)(s,u)}[2]/n^2 + k^{(t)(r,s,u)}[2]/n^2$
$k^r k^{(s)(t)(u)}$	$k^{(r)(s)(t)(u)} + k^{(r,s)(t)(u)}[3]/n$
$k^r k^{(s)(t,u)}$	$k^{(r)(s)(t,u)} + k^{(r,s)(t,u)}/n + k^{(s)(r,t,u)}/n$
$k^{r,s} k^{(t)(u)}$	$k^{(r,s)(t)(u)} + k^{(t)(r,s,u)}[2]/n - k^{(r,t)(s,u)}[2]/\{n(n-1)\}$
$k^{(r)(s)} k^{(t)(u)}$	$k^{(r)(s)(t)(u)} + k^{(r,t)(s)(u)}[4]/n$
	$\quad + k^{(r,t)(s,u)}[2]/\{n(n-1)\}$
$k^{r,s} k^{t,u}$	$k^{(r,s)(t,u)} + k^{r,s,t,u}/n$
	$\quad + \{k^{(r,t)(s,u)} + k^{(r,u)(s,t)}\}/(n-1)$
$k^{r,s} k^{t,u} k^{v,w}$	$k^{(r,s)(t,u)(v,w)} + k^{(r,s)(t,u,v,w)}[3]/n$
	$\quad + k^{(r,s)(t,v)(u,w)}[6]/(n-1) + k^{r,s,t,u,v,w}/n^2$
	$\quad + k^{(r,t)(s,u,v,w)}[12]/\{n(n-1)\} + k^{(r,t)(s,v)(u,w)}[8]/(n-1)^2$
	$\quad + k^{(r,t,v)(s,u,w)}[4](n-2)/\{n(n-1)^2\}$
$k^{r,s} k^{t,u,v,w}$	$k^{(r,s)(t,u,v,w)} + k^{r,s,t,u,v,w}/n$
	$\quad + k^{(r,t)(s,u,v,w)}[8]/(n-1) + k^{(r,t,u)(s,v,w)}[6]/(n-1)$
$k^{r,s} k^{(t,u)(v,w)}$	$k^{(r,s)(t,u)(v,w)} + k^{(r,t)(s,u)(v,w)}[4]/(n-1)$
	$\quad + k^{(t,u)(r,s,v,w)}[2]/n - k^{(r,t,u)(s,v,w)}[2]/\{n(n-1)\}$

Wishart (1952), dealing with the univariate case, does not make this final step from (4.23) to (4.24). As a result, his formulae do not bring out fully the simplicity afforded by expressing the results in terms of polykays alone.

In this final step, it was necessary to find a linear expression in polykays for the product $K^{(r,s)}K^{(t,v)(u,w)}$, which appears in (4.23). The required multiplication formula is given in Table 4.4. Notice that as $N \to \infty$, the fourth term on the right vanishes and the remaining terms converge to the corresponding expression in the infinite population cumulant $\kappa_k(r, s|t, u|v, w)$.

Table 4.5 *Joint cumulants of k-statistics for finite populations*

k-statistics	Joint cumulant
k^r, k^s	$\alpha_1 K^{r,s}$
$k^r, k^{s,t}$	$\alpha_1 K^{r,s,t}$
k^r, k^s, k^t	$\alpha_1 \alpha_2 K^{r,s,t}$
$k^r, k^{s,t,u}$	$\alpha_1 K^{r,s,t,u}$
$k^{r,s}, k^{t,u}$	$\alpha_1 K^{r,s,t,u} + \beta_1 K^{(r,t)(s,u)}[2]$
$k^r, k^s, k^{t,u}$	$\alpha_1 \alpha_2 K^{r,s,t,u} - \alpha_1 K^{(r,t)(s,u)}[2]/(N-1)$
$k^{r,s}, k^{t,u,v}$	$\alpha_1 K^{r,s,t,u,v} + \beta_1 K^{(r,t,u)(s,v)}[6]$

$k^r, k^{s,t}, k^{u,v,w}$ $\alpha_1 \alpha_2 K^{r,s,t,u,v,w} + \alpha_1 \beta_2 K^{(r,s,u)(t,v,w)}[6]$

$$+ \alpha_1 \beta_2 K^{(r,s,u,v)(t,w)}[6] - \alpha_1 K^{(r,s)(t,u,v,w)}[2]/(N-1)$$

$$- \alpha_1 \{ K^{(r,u)(s,t,v,w)}[3] + K^{(r,u,v)(s,t,w)}[3] \}/(N-1)$$

$k^{r,s}, k^{t,u}, k^{v,w}$ Equation (4.24)

$k^{r,s,t}, k^{u,v,w}$ $\alpha_1 K^{r,s,t,u,v,w} + \beta_1 K^{(r,u)(s,t,v,w)}[9]$

$$+ \beta_1 K^{(r,s,u)(t,v,w)}[9]$$

$$+ \{ n/(n-1)^{(2)} - N/(N-1)^{(2)} \} K^{(r,u)(s,v)(t,w)}[6]$$

Table 4.5 gives a selection of joint cumulants of k-statistics for finite populations, including all combinations up to fourth order and selected combinations up to sixth order.

There is, naturally, a fundamental contradiction involved in using formulae such as those in Table 4.5. The purpose of sampling is presumably to learn something about the population of values, perhaps the average value or the variance or range of values, and the first two k-statistics are useful as point estimators. However, in order to set confidence limits, it becomes necessary to know the population k-statistics – something we had hoped simple random sampling would help us avoid computing. The usual procedure is to substitute the estimated k-statistic for the population parameter and to hope that any errors so induced are negligible. Such procedures, while not entirely satisfactory, are perfectly sensible and are easily justified in large samples. However, it would be useful to have a rule-of-thumb to know roughly what extra allowance for sampling variability might be necessary in small to medium samples.

Note that if the objective is to test some fully specified hypothesis, no such contradiction arises. In this case, all population k-statistics are specified by the hypothesis and the problem is to test whether the sample k-statistics are in reasonable agreement with the known theoretical values.

4.7 k-statistics based on least squares residuals

4.7.1   *Multivariate linear regression*

In the previous sections where we dealt with independent and identically distributed observations, it was appropriate to require summary statistics to be invariant under the permutation group. The rationale for this requirement is that the joint distribution is unaffected by permuting the observations and therefore any derived statistics should be similarly unaffected. This argument leads directly to consideration of the symmetric functions. Of course, if departures from the model are suspected, it is essential to look beyond the symmetric functions in order to derive a suitable test statistic. For example, suspected serial correlation might lead the statistician to examine sample autocorrelations: suspected dependence on an auxiliary variable might lead to an examination of specific linear combinations of the response variable. Neither of these statistics is symmetric in the sense understood in Section 4.2. In this section, we consider the case where the mean value of the response is known to depend linearly on given explanatory variables. We must then abandon the notion of symmetric functions as understood in Section 4.2 and look for statistics that are invariant in a different, but more appropriate, sense.

In previous sections we wrote Y_i^r for the rth component of the ith independent observation. Subscripts were used to identify the observations, more out of convenience and aesthetic considerations, than out of principle or necessity. Only permutation transformations were considered and the k-statistics are invariant under this group. Thus, there is no compelling argument for using superscripts rather than subscripts. In this section, however, we consider arbitrary linear transformations and it is essential to recognize that Y and its cumulants are contravariant both with respect to transformation of the individual observations and with respect to the components. In other words, both indices must appear as superscripts.

Suppose then, that the rth component of the ith response variable has expectation

$$E(Y^{r;i}) = \omega^{\alpha;i}\kappa_\alpha^r,$$

where $\omega^{\alpha;i}$ is an $n \times q$ array of known constants and κ_α^r is a $q \times p$ array of unknown parameters to be estimated. When we revert to matrix notation we write

$$E(\mathbf{Y}) = \mathbf{X}\beta,$$

this being the more common notation in the literature on linear models. Here, however, we choose to write κ_α^r for the array of regression coefficients in order to emphasize the connection with cumulants and k-statistics. In addition to the above assumption regarding the mean value, we assume that there are available known arrays $\omega^{i,j}$, $\omega^{i,j,k}$ and so on, such that

$$\text{cov}(Y^{r;i}, Y^{s;j}) = \kappa^{r,s}\omega^{i,j},$$
$$\text{cum}(Y^{r;i}, Y^{s;j}, Y^{t;k}) = \kappa^{r,s,t}\omega^{i,j,k}$$

and so on for the higher-order cumulants. The simplest non-trivial example having this structure arises when the ith observation is the sum of m_i independent and identically distributed, but unrecorded, random variables. In this case, the ω-arrays of order two and higher take the value m_i on the main diagonal and zero elsewhere.

The notation carries with it the implication that any nonsingular linear transformation

$$\bar{Y}^{r;i} = a^i_j Y^{r;j} \tag{4.25}$$

applied to Y has a corresponding effect on the arrays $\omega^{\alpha;i}$, $\omega^{i,j}$, $\omega^{i,j,k}$, ..., namely

$$\bar{\omega}^{\alpha;i} = a^i_j \omega^{\alpha;j}, \quad \bar{\omega}^{i,j} = a^i_k a^j_l \omega^{k,l}, \quad \bar{\omega}^{i,j,k} = a^i_l a^j_m a^k_n \omega^{l,m,n}$$

and so on. In other words, the ω-arrays are assumed to behave as contravariant tensors under the action of the general linear group (4.25).

In what follows, it is assumed that the estimation problem is entirely invariant under the group of transformations (4.25). In some ways, this may seem a very natural assumption because, for example,

$$E(\bar{Y}^{r;i}) = \bar{\omega}^{\alpha;i} \kappa^r_\alpha, \quad \text{cov}(\bar{Y}^{r;i}, \bar{Y}^{s;j}) = \bar{\omega}^{i,j} \kappa^{r,s}$$

and so on, so that the definition of the cumulants is certainly invariant under the group. However, it is important to emphasize the consequences of the assumption that *all* linear transformations of the observations are to be treated on an equal footing and that no particular scale has special status. We are forced by this assumption to abandon the concepts of interaction and replication, familiar in the analysis of variance, because they are not invariant under the group (4.25). See Exercise 4.22. This is not to be construed as criticism of the notions of interaction and replication, nor is it a criticism of our formulation in terms of the general linear group rather than some smaller group. Instead, it points to the limitations of our specification and emphasizes that, in any given application, the appropriateness of the criteria used here must be considered carefully. In simple cases, it may be possible to find a better formulation in terms of a group more carefully tailored to the problem.

Often it is appropriate to take $\omega^{i,j} = \delta^{ij}$, $\omega^{i,j,k} = \delta^{ijk}$ and so on, but, since no particular significance attaches to the diagonal arrays, we shall suppose instead that $\omega^{i,j}$, $\omega^{i,j,k}$,... are arbitrary arrays and that $\omega^{i,j}$ has full rank.

Our aim, then, is to construct unbiased estimates of κ^r_α, $\kappa^{r,s}$, $\kappa^{r,s,t}$,... that are invariant under transformations (4.25). Note that it was necessary to introduce the arrays $\omega^{\alpha;i}$, $\omega^{i,j}$,... as tensors in order that the definition of the cumulants be unaffected by linear transformation. Otherwise, if the cumulants were not invariant, it would not be sensible to require invariant estimates.

Let $\omega_{i,j}$ be the matrix inverse of $\omega^{i,j}$. In matrix notation, $\omega_{i,j}$ is written as $\mathbf{W}$. By assumption, $\omega^{i,j}$ has full rank, but even in the rank deficient case, the choice of inverse does not matter (see Exercise 2.24). Now define

$$\omega^\alpha_i = \omega_{i,j}\omega^{\alpha;j}, \qquad \omega_{i,j,k} = \omega_{i,l}\omega_{j,m}\omega_{k,n}\omega^{l,m,n}$$

and so on, lowering the indices in the usual way. In the calculations that follow, it is convenient to have a concise notation for the $q \times q$ array

$$\lambda^{\alpha,\beta} = \omega^{\alpha;i}\omega^{\beta;j}\omega_{i,j} = \omega^\alpha_i\omega^\beta_j\omega^{i,j},$$

and the $q \times q \times q$ array

$$\lambda^{\alpha,\beta,\gamma} = \omega^\alpha_i\omega^\beta_j\omega^\gamma_k\omega^{i,j,k}.$$

In matrix notation, $\lambda^{\alpha,\beta}$ is usually written as $\mathbf{X}^T\mathbf{W}\mathbf{X}$, and $\lambda_{\alpha,\beta}$ as $(\mathbf{X}^T\mathbf{W}\mathbf{X})^{-1}$, but there is no convenient matrix notation for $\lambda^{\alpha,\beta,\gamma}$.

A straightforward calculation shows that

$$E(\omega^\alpha_i Y^{r;i}) = \lambda^{\alpha,\beta}\kappa^r_\beta$$

and hence, by matrix inversion, that

$$k^r_\alpha = \lambda_{\alpha,\beta}\omega^\beta_i Y^{r;i} \qquad (4.26)$$

is an unbiased estimate of κ^r_α, invariant under linear transformations (4.25). In matrix notation, (4.26) becomes

$$\hat\beta = (\mathbf{X}^T\mathbf{W}\mathbf{X})^{-1}\mathbf{X}^T\mathbf{W}\mathbf{Y},$$

which is instantly recognized as the weighted least squares estimate of the regression coefficients. Note that the equally weighted estimate is not invariant under linear transformation.

The notation used in (4.26) is chosen to emphasize that the weighted least squares estimate is in fact a k-statistic of order one. Although it is in a sense the obvious invariant estimate of κ_α^r, the k-statistic, k_α^r is not unique except in rather special cases. See Exercise 4.19.

The second- and third-order cumulants of k_α^r are

$$\text{cov}(k_\alpha^r, k_\beta^s) = \lambda_{\alpha,\beta}\kappa^{r,s}$$
$$\text{cum}(k_\alpha^r, k_\beta^s, k_\gamma^t) = \lambda_{\alpha,\beta,\gamma}\kappa^{r,s,t}$$

where the indices of $\lambda^{\alpha,\beta,\gamma}$ have been lowered in the usual way by multiplying by $\lambda_{\alpha,\beta}$ three times. Similar expressions hold for the higher-order cumulants.

In the case of quadratic expressions, we write

$$E(\omega_{i,j}Y^{r;i}Y^{s;j}) = n\kappa^{r,s} + \lambda^{\alpha,\beta}\kappa_\alpha^r\kappa_\beta^s$$
$$E(\omega_i^\alpha\omega_j^\beta Y^{r;i}Y^{s;j}) = \lambda^{\alpha,\beta}\kappa^{r,s} + \lambda^{\alpha,\gamma}\lambda^{\beta,\delta}\kappa_\gamma^r\kappa_\delta^s.$$

These equations determine the k-statistic $k^{r,s}$ and the polykays $k_{(\alpha)(\beta)}^{(r)(s)}$, which are unbiased estimates of the products $\kappa_\alpha^r\kappa_\beta^s$. There are as many equations as there are k-statistics and polykays. We find that

$$(n-q)k^{r,s} = \{\omega_{i,j} - \lambda_{\alpha,\beta}\omega_i^\alpha\omega_j^\beta\}Y^{r;i}Y^{s;j},$$

which is the usual weighted residual sum of squares and products matrix, and

$$k_{(\alpha)(\beta)}^{(r)(s)} = k_\alpha^r k_\beta^s - \lambda_{\alpha,\beta}k^{r,s}.$$

These k-statistics and polykays are not unique except in very special cases: see Exercise 4.19. Among the estimates that are invariant under the general linear group, it appears that the ks do not have minimum variance unless the observations are normally distributed and, even then, additional conditions are required for exact optimality. See Exercise 4.23. Under a smaller group such as the permutation group, which admits a greater number of invariant estimates, they are neither unique nor do they have minimum variance among invariant estimates. Conditions under which $k^{r,s}$ has minimum variance in the univariate case are given by Plackett

(1960, p. 40) and Atiqullah (1962). See also Exercises 4.22–4.26 for a more general discussion relating to higher-order ks.

When dealing with higher-order k-statistics, it is more convenient to work with the unstandardized residuals

$$R^{r;i} = Y^{r;i} - \omega^{\alpha;i} k_\alpha^r = \{\delta_j^i - \omega^{\alpha;i} \lambda_{\alpha,\beta} \omega_j^\beta\} Y^{r;j}$$
$$= \rho_j^i Y^{r;j}.$$

In matrix notation, this becomes

$$\mathbf{R} = (\mathbf{I} - \mathbf{H})\mathbf{Y}, \quad \text{where} \quad \mathbf{H} = \mathbf{X}(\mathbf{X}^T \mathbf{W} \mathbf{X})^{-1} \mathbf{X}^T \mathbf{W},$$

the projection matrix producing fitted values, is not symmetrical unless $\mathbf{W}$ is a multiple of $\mathbf{I}$. Working with residuals is entirely analogous to working with central moments, rather than moments about the origin, in the single sample case. Note, however, that polykays having a unit part are not estimable from the residuals because the distribution of $R^{r;i}$ does not depend on κ_α^r.

The least squares residuals have cumulants

$$E(R^{r;i}) = 0,$$
$$\text{cov}(R^{r;i}, R^{s;j}) = \rho_k^i \rho_l^j \omega^{k,l} \kappa^{r,s} = \rho^{i,j} \kappa^{r,s}$$
$$\text{cum}(R^{r;i}, R^{s;j}, R^{t;k}) = \rho_l^i \rho_m^j \rho_n^k \omega^{l,m,n} \kappa^{r,s,t} = \rho^{i,j,k} \kappa^{r,s,t}$$

and so on. Effectively, the arrays $\omega^{i,j}$, $\omega^{i,j,k}, \ldots$ are transformed to $\rho^{i,j}$, $\rho^{i,j,k}, \ldots$ by projection on to the residual space and its associated direct products. Note that $\omega_{i,j}$ is a generalized inverse of $\rho^{i,j}$, and hence we may write

$$\rho_{i,j} = \omega_{i,k} \omega_{j,l} \rho^{k,l}, \qquad \rho_{i,j,k} = \omega_{i,l} \omega_{j,m} \omega_{k,n} \rho^{l,m,n}$$

and so on, in agreement with the definitions given previously. Evidently, $\rho_{i,j}$ is the Moore-Penrose inverse of $\rho^{i,j}$. In the discussion that follows, we may use $\rho_{i,j}$ and $\omega_{i,j}$ interchangeably to lower indices, both being generalized inverses of $\rho^{i,j}$. Thus, the invariant unbiased estimates of the second- and third-order cumulants are

$$\begin{aligned} n_2 k^{r,s} &= \rho_{i,j} R^{r;i} R^{s;j} = \omega_{i,j} R^{r;i} R^{s;j} \\ n_3 k^{r,s,t} &= \rho_{i,j,k} R^{r;i} R^{s;j} R^{t;k} = \omega_{i,j,k} R^{r;i} R^{s;j} R^{t;k} \end{aligned} \qquad (4.27)$$

where $n_2 = n - q$, $n_3 = \omega_{i,j,k}\omega^{l,m,n}\rho_l^i\rho_m^j\rho_n^k$, which reduces to $\sum_{ij}(\rho^{i,j})^3$ when $\omega^{i,j} = \delta^{ij}$ and $\omega^{i,j,k} = \delta^{i,j,k}$. In matrix notation, $n_2 k^{r,s}$ is just $\mathbf{R}^T \mathbf{W} \mathbf{R}$ or equivalently, $\mathbf{R}^T \mathbf{W} \mathbf{Y}$, but there is no simple matrix notation for $n_3 k^{r,s,t}$, the residual sum of cubes and products array.

In the case of cumulants of degree four, we write

$$E(\omega_{i,j,k,l}R^{r;i}R^{s;j}R^{t;k}R^{u;l}) = n_4\kappa^{r,s,t,u} + n_{22}\kappa^{r,s}\kappa^{t,u}[3]$$

$$E(\omega_{i,j}\omega_{k,l}R^{r;i}R^{s;j}R^{t;k}R^{u;l}) = n_{22}\kappa^{r,s,t,u} + n_2^2\kappa^{r,s}\kappa^{t,u}$$
$$+ n_2\kappa^{r,t}\kappa^{s,u}[2],$$

where the coefficients are defined by

$$n_4 = \rho^{i,j,k,l}\rho_{i,j,k,l} = \omega_{i,j,k,l}\rho^{i,j,k,l}$$
$$n_{22} = \rho^{i,j,k,l}\rho_{i,j}\rho_{k,l} = \omega_{i,j,k,l}\rho^{i,j}\rho^{k,l}.$$

In the particular case where $\omega^{i,j} = \delta^{ij}$, the coefficients reduce to $n_4 = \sum_{ij}(\rho^{i,j})^4$ and $n_{22} = \sum_i(\rho^{i,i})^2$.

Matrix inversion gives the following invariant unbiased estimates of $\kappa^{r,s,t,u}$ and $\kappa^{r,s}\kappa^{t,u}$.

$$\Delta_1 k^{r,s,t,u} = \{n_2(n_2 + 2)\omega_{i,j,k,l} - n_{22}\omega_{i,j}\omega_{k,l}[3]\}R^{r;i}R^{s;j}R^{t;k}R^{u;l}$$

$$\Delta_2 k^{(r,s)(t,u)} = \{-n_2(n_2 - 1)n_{22}\omega_{i,j,k,l} + (n_2(n_2 + 1)n_4 - 2n_{22}^2)\omega_{i,j}\omega_k,$$
$$+ (n_{22}^2 - n_2 n_4)\omega_{i,k}\omega_{j,l}[2]\}R^{r;i}R^{s;j}R^{t;k}R^{u;l}$$

where

$$\Delta_1 = n_2 n_4(n_2 - 1) + 3(n_2 n_4 - n_{22}^2) \quad \text{and} \quad \Delta_2 = n_2(n_2 - 1)\Delta_1.$$

The final expression above simplifies to some extent when $r = s = t = u$, as is always the case in univariate regression problems. In particular, the final two terms can be combined and $n_2(n_2 - 1)$ may be extracted as a factor.

Note that, in the case where the observations are independent with identical cumulants of order two and higher, these k-statistics and polykays are symmetric functions of the least squares residuals, even though the residuals are neither independent nor identically distributed. It is this feature that is the source of non-uniqueness of k-statistics and polykays in regression problems.

4.7.2   *Univariate linear regression*

Some condensation of notation is inevitable in the univariate case, but here we make only the minimum of adjustment from the notation of the previous section. However, we do assume that the observations are independent with cumulants

$$\text{cov}(Y^i, Y^j) = \kappa_2 \delta^{ij}$$
$$\text{cum}(Y^i, Y^j, Y^k) = \kappa_3 \delta^{ijk}$$

and so on. It follows that the projection matrix producing fitted values, $\mathbf{H} = \mathbf{X}(\mathbf{X}^T\mathbf{X})^{-1}\mathbf{X}^T$, unusually for a projection matrix, is symmetrical and $\kappa_2 \mathbf{H}$ is also the covariance matrix of the fitted values. The residual projection matrix is also symmetrical with elements $\rho_{i,j} = \delta_{ij} - h_{ij}$ and it is immaterial whether we use subscripts or superscripts.

If we write

$$S_r = \sum (R^i)^r$$

for the rth power sum of the residuals, it follows that

$$k_2 = S_2/n_2$$
$$k_3 = S_3/n_3$$
$$k_4 = \{n_2(n_2 + 2)S_4 - 3n_{22}S_2^2\}/\Delta_1$$
$$k_{22} = \{n_4 S_2^2 - n_{22}S_4\}/\Delta_1$$

where

$$n_2 = n - q, \quad n_r = \sum_{ij} (\rho_{i,j})^r \quad \text{and} \quad n_{22} = \sum_i (\rho_{i,i})^2,$$

and Δ_1 is given at the end of the previous section. These are the unbiased estimates of residual cumulants and products of residual cumulants up to fourth order. In other words, $E(k_4) = \kappa_4$ and $E(k_{22}) = \kappa_2^2$.

Little simplification of the above formulae seems to be possible in general. However, in the case of designs that are quadratically balanced, the residuals have identical marginal variances, though they are not usually exchangeable, and $h_{ii} = q/n$, $\rho_{i,i} = 1 - q/n$. It follows then that $n_{22} = n(1 - q/n)^2$. In addition, if n is large, we may wish to avoid the extensive calculations involved in computing n_3 and n_4. The following approximations are often satisfactory if n is large and if the design is not grossly unbalanced:

$$n_3 \simeq n(1 - q/n)^3, \quad n_4 \simeq n(1 - q/n)^4, \tag{4.28}$$

the errors of approximation being typically of order $O(n^{-1})$ and $O(n^{-2})$ respectively. Evidently, the above approximation is a lower bound for n_4, but not for n_3 The upper bounds, $n_3, n_4 \leq n - q$, are attained if $n - q$ of the observations are uninformative for the regression parameters. In practice, n_3 and n_4 are typically closer to the lower limits than to the upper limits. It follows then that

$$n_2 n_4 - n_{22}^2 \simeq -q n_4$$

and hence that

$$\Delta_1 \simeq n_4 (n_2(n_2 - 1) - 3q).$$

The above approximations are often adequate even if the design is not quadratically balanced and even if n is not particularly large. For example, in simple linear regression, with 20 observations and with a single quantitative covariate whose values are equally spaced, the error incurred in using the approximations is less than 2%. In less favourable circumstances, other approximations might be preferred. For example, for $r \geq 3$,

$$n_r \simeq \sum (\rho_{i,i})^r, \tag{4.29}$$

which reduces to the earlier approximation if the design is quadratically balanced. On the basis of limited numerical comparisons, it appears that, on balance, (4.29) is usually more accurate than (4.28), at least when q is small relative to n.

For further details, including more efficient estimates of κ_3 and κ_4, see McCullagh & Pregibon (1987) and also, Exercise 4.22.

4.8   Bibliographic notes

The estimation of means and variances has been a concern of statisticians from the earliest days, and consequently the literature is very extensive. No attempt will be made here to survey the early work concerned with sample moments and the joint moments of sample moments.

The idea of simplifying calculations by estimating cumulants rather than moments is usually attributed to Fisher (1929), although similar ideas were put forward earlier by Thiele (1897). Thiele's book was published in English in 1903 and the work was cited by Karl Pearson. It is curious that, in 1929, Fisher was apparently unaware of Thiele's earlier contribution.

Following the publication of Fisher's landmark paper, subsequent work by Wishart (Fisher & Wishart, 1931) and Kendall (1940a,b,c), seems directed at demonstrating the correctness of Fisher's combinatorial method for evaluating pattern functions. Fisher's algorithm must have appeared mystifying at the time: it has weathered the passage of half a century extraordinarily well.

Cook (1951) gives extensive tables of the joint cumulants of multivariate k-statistics. Kaplan (1952), using a form of tensor notation, gives the same formulae, but more succinctly. For further discussion, including explicit formulae in the univariate and bivariate case, see Kendall & Stuart (1977, Chapters 12,13).

The connection with finite population sampling was made by Tukey (1950, 1956a), who also bears responsibility for the term *polykay*: see Kendall & Stuart (1977, Section 12.22), where the less informative term l-statistic is used. Calculations similar to Tukey's were previously given by Dressel (1940), though without any suggestion that the results might be useful for sampling theory calculations. See also Irwin & Kendall (1943–45) For later developments, see Dwyer & Tracy (1964) and Tracy (1968).

The reader who finds the notation used in this chapter cumbersome or impenetrable should be warned that notational problems get much worse for symmetric functions derived from two-way or higher-order arrays. In a one-way layout, we talk of the *between groups variance* and the *within groups variance*. A similar decomposition exists for the higher-order statistics, so that we may talk of the *between groups skewness*, the *within groups skewness* and also the *cross skewness*. These are the three k-statistics of the third degree. In the case of a two-way design with one observation per cell, there are three variance terms and five third-order k-statistics, one of which is Tukey's (1949) non-additivity statistic. Such statistics are called *bi-kays*. There are of course, the corresponding *bi-poly-kays*, also derived from a two-way array, not to mention *poly-bi-kays* and *poly-poly-kays*, which are much worse. Any reader who is interested in such matters should read Hooke (1956a,b) or Tukey (1956b). For those intrepid souls whose appetites are whetted by this fare, we heartily recommend Speed (1986a,b).

Anscombe (1961) discusses the computation of and the applications of k-statistics in detecting departures from the usual linear model assumptions. There has been much subsequent work in this vein. See, for example, Bickel (1978), Hinkley (1985) or McCullagh & Pregibon (1987) for some recent developments.

4.9   Further results and exercises 4

4.1   Show that
$$k^{r,s} = n^{-1} \sum_{ij} \phi^{ij} Y_i^r Y_j^s$$

is the usual sample covariance matrix based on n independent and identically distributed observations. Show that $k^{r,s}$ is unbiased for $\kappa^{r,s}$.

4.2   By applying the criteria of unbiasedness and symmetry to $k^{r,s,t}$ and $k^{r,s,t,u}$, derive the expression (4.7) for the coefficients ϕ^{ijk} and ϕ^{ijkl}.

4.3   Find the mean value of the following expressions:

$$\sum_i Y_i^r Y_i^s Y_i^t, \qquad \sum_{i \neq j} Y_i^r Y_j^s Y_j^t, \quad \text{and} \quad \sum_{i \neq j \neq k} Y_i^r Y_j^s Y_k^t.$$

Hence find the symmetric functions that are unbiased estimates of κ^{rst}, $\kappa^r \kappa^{st}$ and $\kappa^r \kappa^s \kappa^t$. By combining these estimates in the appropriate way, find the symmetric functions that are unbiased estimates of $\kappa^{r,s,t}$, $\kappa^{r,st}$ and $\kappa^r \kappa^{s,t}$.

4.4   Show that the expressions for ordinary k-statistics in terms of power sums are

$$k^{r,s} = \{k^{rs} - k^r k^s\}n/(n-1)$$
$$k^{r,s,t} = \{k^{rst} - k^r k^{st}[3] + 2k^r k^s k^t\}n^2/(n-1)^{(2)}$$
$$k^{r,s,t,u} = \{(n+1)k^{rstu} - (n+1)k^r k^{stu}[4] - (n-1)k^{rs}k^{tu}[3]$$
$$+ 2nk^r k^s k^{tu}[6] - 6nk^r k^s k^t k^u\}n^2/(n-1)^{(3)},$$

Kaplan (1952).

4.5 Show that
$$\sum_{ij}(\phi^{ij})^2 = n^2/(n-1).$$

Hence derive the following joint cumulants of k-statistics:

$$\kappa_k(r|s) = \kappa^{r,s}/n, \qquad\qquad \kappa_k(r|s,t) = \kappa^{r,s,t}/n,$$
$$\kappa_k(r|s|t) = \kappa^{r,s,t}/n^2, \qquad \kappa_k(r|s,t,u) = \kappa^{r,s,t,u}/n,$$
$$\kappa_k(r,s|t,u) = \kappa^{r,s,t,u}/n + \kappa^{r,t}\kappa^{s,u}[2]/(n-1),$$
$$\kappa_k(r|s|t,u) = \kappa^{r,s,t,u}/n^2, \qquad \kappa_k(r|s,t,u,v) = \kappa^{r,s,t,u,v}/n,$$
$$\kappa_k(r,s|t,u,v) = \kappa^{r,s,t,u,v}/n + \kappa^{r,t,u}\kappa^{s,v}[6]/(n-1),$$
$$\kappa_k(r|s|t,u,v) = \kappa^{r,s,t,u,v}/n^2,$$
$$\kappa_k(r|s,t|u,v) = \kappa^{r,s,t,u,v}/n^2 + \kappa^{r,s,u}\kappa^{t,v}[4]/\{n(n-1)\}$$
$$\kappa_k(r|s|t|u,v) = \kappa^{r,s,t,u,v}/n^3$$

Show explicitly that the term $\kappa^{r,t,u}\kappa^{s,v}$ does not appear in the third cumulant, $\kappa_k(r|s|t,u,v)$.

4.6 By considering separately the five distinct index patterns, show that

$$\sum_{ijk}\phi^{ij}\phi^{ik}\phi^{jk} = n + \frac{3n(n-1)}{(n-1)^2} - \frac{n(n-1)(n-2)}{(n-1)^3}$$
$$= n^3/(n-1)^2.$$

4.7   Using the pattern functions given in Table 4.1, derive the following joint cumulants:

$$\kappa_k(r,s|t,u,v,w) = \kappa^{r,s,t,u,v,w}/n + \kappa^{r,t}\kappa^{s,u,v,w}[8]/(n-1)$$
$$+ \kappa^{r,t,u}\kappa^{s,v,w}[6]/(n-1),$$

$$\kappa_k(r,s|t,u|v,w) = \kappa^{r,s,t,u,v,w}/n^2 + \kappa^{r,t}\kappa^{s,u,v,w}[12]/\{n(n-1)\}$$
$$+ \kappa^{r,t,v}\kappa^{s,u,w}[4](n-2)/\{n(n-1)^2\}$$
$$+ \kappa^{r,t}\kappa^{s,v}\kappa^{u,w}[8]/(n-1)^2,$$

$$\kappa_k(r,s,t|u,v,w) = \kappa^{r,s,t,u,v,w}/n + \kappa^{r,u}\kappa^{s,t,v,w}[9]/(n-1)$$
$$+ \kappa^{r,s,u}\kappa^{t,v,w}[9]/(n-1)$$
$$+ \kappa^{r,u}\kappa^{s,v}\kappa^{t,w}[6]n/(n-1)^{(2)}$$

Kaplan (1952). Compare these formulae with those given in Table 4.5 for finite populations.

4.8   Show that joint cumulants of ordinary k-statistics in which the partition contains a unit block, e.g. $\kappa_k(r|s,t|u,v,w)$, can be found from the expression in which the unit block is deleted, by dividing by n and adding the extra index at all possible positions. Hence, using the expression given above for $\kappa_k(s,t|u,v,w)$, show that

$$\kappa_k(r|s,t|u,v,w) = \kappa^{r,s,t,u,v,w}/n^2 + \kappa^{r,s,u,v}\kappa^{t,w}[6]/\{n(n-1)\}$$
$$+ \kappa^{r,s,u}\kappa^{t,v,w}[6]/\{n(n-1)\}.$$

4.9   Show that as $n \to \infty$, every array of k-statistics of fixed order has a limiting joint normal distribution when suitably standardized. Hence show in particular, that, for large n,

$$n^{1/2}(k^r - \kappa^r) \sim N_p(0, \kappa^{r,s})$$
$$n^{1/2}(k^{r,s} - \kappa^{r,s}) \sim N_{p^2}(0, \tau^{rs,tu}),$$

where $\tau^{rs,tu} = \kappa^{r,s,t,u} + \kappa^{r,t}\kappa^{s,u}[2]$ has rank $p(p+1)/2$.

4.10   Under the assumption that the data have a joint normal distribution, show that $n^{1/2}k^{r,s,t,u}$ has a limiting covariance matrix given by

$$\text{ncov}(k^{i,j,k,l}, k^{r,s,t,u}) \to \kappa^{i,r}\kappa^{j,s}\kappa^{k,t}\kappa^{l,u}[4!].$$

Hence show that $n^{1/2}p\bar{r}_4$ has a limiting normal distribution with mean zero and variance $8p^2 + 16p$.

4.11   Under the assumptions of the previous exercise, show that

$$n \operatorname{cov}(k^{i,j,k}, k^{r,s,t}) \;\rightarrow\; \kappa^{i,r}\kappa^{j,s}\kappa^{k,t}[3!]$$

as $n \rightarrow \infty$. Hence show that, for large n,

$$\frac{np\bar{r}_{13}^2}{2(p+2)} \;\sim\; \chi_p^2$$

$$\frac{np\bar{r}_{23}^2}{6} \;\sim\; \chi_{p(p+1)(p+2)/6}^2.$$

4.12   By considering the symmetric group, i.e. the group comprising all $n \times n$ permutation matrices, acting on $Y_1, ..., Y_n$, show that every invariant polynomial function of degree k is expressible as a *linear* combination of polykays of degree k.

4.13   Show that

$$\sum_{i,j=1}^{n} Y_i^r Y_j^s = nk^{rs} + n(n-1)k^{(r)(s)}.$$

Hence deduce that, under simple random sampling, the average value over all samples of $n^{-2}\sum_{i,j=1}^{n} Y_i^r Y_j^s$ is

$$n^{-1}K^{rs} + (1-n^{-1})K^{(r)(s)} = n^{-1}K^{r,s} + K^{(r)(s)}$$

while the same function calculated in the population is

$$N^{-2}\sum_{i,j=1}^{N} Y_i^r Y_j^s = N^{-1}K^{r,s} + K^{(r)(s)}.$$

4.14   Derive the following multiplication formulae for k-statistics and polykays

$$k^r k^s = k^{(r)(s)} + k^{(r,s)}/n$$

$$k^r k^{s,t} = k^{(r)(s,t)} + k^{r,s,t}/n$$

$$k^r k^{(s)(t)} = k^{(r)(s)(t)} + \{k^{(s)(r,t)} + k^{(t)(r,s)}\}/n$$

$$k^r k^s k^t = k^{(r)(s)(t)} + k^{(r)(s,t)}[3]/n + k^{r,s,t}/n^2$$

4.15 Derive the following multiplication formulae for k-statistics and polykays

$$k^{r,s}k^{t,u} = k^{(r,s)(t,u)} + k^{r,s,t,u}/n$$
$$+ \{k^{(r,t)(s,u)} + k^{(r,u)(s,t)}\}/(n-1)$$
$$k^{r,s}k^{(t,u)(v,w)} = k^{(r,s)(t,u)(v,w)} + k^{(r,t)(s,u)(v,w)}[4]/(n-1)$$
$$+ k^{(t,u)(r,s,v,w)}[2]/n - k^{(r,t,u)(s,v,w)}[2]/\{n(n-1)\}$$

4.16 Using the multiplication formulae given in the previous two exercises, derive the finite population joint cumulants listed in Table 4.3.

4.17 By considering the sample covariance matrix of the observed variables and their pairwise products, show that

$$k^{rs,tu} - k^{rs,i}k^{tu,j}k_{i,j},$$

regarded as a $p^2 \times p^2$ symmetric matrix, is non-negative definite. Hence deduce that

$$k^{r,s,t,u} - k^{r,s,i}k^{t,u,j}k_{i,j} + \{k^{(r,t)(s,u)} + k^{(r,u)(s,t)}\}n/(n-2)$$

is also non-negative definite when regarded as a matrix whose rows are indexed by (r,s) and columns by (t,u). Hence derive lower bounds for $\bar{r}_4 - \bar{r}_{13}^2$ and $\bar{r}_4 - \bar{r}_{23}^2$.

4.18 By expanding the product of several power sums along the lines suggested in Section 4.6.2 for products of k-statistics, show that

$$\mathbf{K} = \mathbf{HP}$$

in the notation of (4.20), where $\mathbf{P}$ is a vector of polykays, $\mathbf{K}$ is a vector of power sum products and $\mathbf{H}$ has components

$$h_{ij} = n^l \quad (l = |\Upsilon_i \vee \Upsilon_j| - |\Upsilon_i|),$$

where $|\Upsilon|$ denotes the number of blocks of the partition. Show also that $l \leq 0$ and $l = 0$ if and only if Υ_j is a sub-partition of Υ_i. Hence prove that as $n \to \infty$, the components of $\mathbf{P}$ are the Möbius transform of the components of $\mathbf{K}$.

4.19 Show, in the notation of Section 4.7.1, that if the scalar $n_{22} = \rho_{i,j,k,l}\rho^{i,j}\rho^{k,l}$ is non-zero, then

$$n_{22}^{-1}\rho_{i,j,k,l}\rho^{k,l}R^{r;i}R^{s;j}$$

is an invariant unbiased estimate of $\kappa^{r,s}$. Under what circumstances is the above estimate the same as $k^{r,s}$?

4.20 Show that the covariance of $k^{r,s}$ and $k^{t,u}$, as defined in (4.27), is

$$\text{cov}(k^{r,s}, k^{t,u}) = n_{22}\kappa^{r,s,t,u}/n_2^2 + \kappa^{r,t}\kappa^{s,u}[2]/n_2,$$

which reduces to

$$\kappa^{r,s,t,u}/n + \kappa^{r,t}\kappa^{s,u}[2]/(n-q)$$

for quadratically balanced designs. Hence show that, in the univariate notation of Section 4.7.2,

$$n_{22}k_4/n_2^2 + 2k_{22}/n_2$$

is an unbiased estimate of $\text{var}(k_2)$.

4.21 Show that, for the ordinary linear regression problem with an intercept and one dependent variable, x, that

$$n_{22} = (n-1)(n-3)/n + \sum(x_i - \bar{x})^4 / \left(\sum(x_i - \bar{x})^2\right)^2$$

and that, for equally spaced x-values, this reduces to

$$n_{22} = (n-2)^2/n + 4/(5n) + O(n^{-2})$$

in reasonable agreement with the approximations of Section 4.7.2. Show more generally, that the discrepancy between n_{22} and the approximation $(n-2)^2/n$, is a function of the standardized fourth cuumulant of the x-values. Find this function explicitly.

4.22 In the notation of Section 4.7.2, show that, when third- and higher-order cumulants are neglected, the cubes of the least squares residuals have covariance matrix $\text{cov}(R^iR^jR^k, R^lR^mR^n)$ given by

$$\kappa_2^3\{\rho^{i,j}\rho^{k,l}\rho^{m,n}[9] + \rho^{i,l}\rho^{j,m}\rho^{k,n}[6]\},$$

here taken to be of order $n^3 \times n^3$. Show that, if $\nu = n - p$ is the rank of $\rho^{i,j}$, then

$$w_{ijk,lmn} = \rho_{i,l}\rho_{j,m}\rho_{k,n}[6]/36 - \rho_{i,j}\rho_{k,l}\rho_{m,n}[9]/\{18(\nu + 4)\}$$

is the Moore-Penrose inverse matrix. Hence prove that

$$l_3 = \frac{k_3 - 3k_2\bar{R}\,n(n - p)/\{n_3(n - p + 4)\}}{1 - 3\sum_{ij}\rho_{i,j}\rho_{i,i}\rho_{j,j}/\{n_3(n - p + 4)\}}$$

where $n\bar{R} = \sum \rho_{i,i}R^i$, is unbiased for κ_3 and, under normality, has minimum variance among homogeneous cubic forms in the residuals, (McCullagh & Pregibon, 1987).

4.23 Deduce from the previous exercise that if the vector having components $\rho_{i,i}$ lies in the column space of the model matrix $\mathbf{X}$, then $l_3 \equiv k_3$. More generally, prove that if the constant vector lies in the column space of $\mathbf{X}$, then

$$n^{1/2}(l_3 - k_3) = O_p(n^{-1})$$

for large n under suitably mild limiting conditions on $\mathbf{X}$. Hence, deduce that k_3 is nearly optimal under normality.

4.24 Suppose $n = 4$, $p = 1$, $y = (1.2, 0.5, 1.3, 2.7)$, $x = (0, 1, 2, 3)$. Using the results in the previous two exercises, show that $k_2 = 0.5700$, $k_3 = 0.8389$, with variance $6.59\sigma^6$ under normality, and $l_3 = 0.8390$ with variance $4.55\sigma^6$ under normality. Compare with Anscombe (1961, p.15) and Pukelsheim (1980, p.110). Show also that $k_{22} = 0.6916$ and $k_4 = 0.7547$.

4.25 Repeat the calculations of Exercise 4.22, but now for l_4 and l_{22}, the optimal unbiased estimates of the fourth cumulant and the square of the second cumulant. Show that l_4 and l_{22} are linear combinations of

$$S_2^2 \quad \text{and} \quad S_4 - 6S_2\sum \rho_i^i R_i^2/(n_2 + 6).$$

Deduce that, in the quadratically balanced case, k_4 and k_{22} are the optimal unbiased estimates.
[Hint: It might be helpful to consult McCullagh & Pregibon (1987) to find the inverse of the $n^4 \times n^4$ covariance matrix of the quartic functions of the residuals.] Note that, in the case of the fourth cumulant, this calculation is different from Pukelsheim (1980, Section 4), who initially assumes κ_2 to be known and finally replaces κ_2 by k_2.

4.26 Let $\mathbf{Y}$ have components Y^i satisfying
$$E(Y^i) = \mu^i = \omega^{\alpha;i}\beta_\alpha$$
or, in matrix notation, $E(\mathbf{Y}) = \mathbf{X}\beta$, where $\mathbf{X}$ is $n \times q$ of rank q. Let $\omega^{i,j}$, $\omega^{i,j,k}, \dots$ be given tensors such that $Y^i - \mu^i$ has cumulants $\kappa_2\omega^{i,j}$, $\kappa_3\omega^{i,j,k}$ and so on. The first-order interaction matrix, $\mathbf{X}^*$, is obtained by appending to $\mathbf{X}$, $q(q+1)/2$ additional columns having elements in the ith row given by $\omega^{i;r}\omega^{i;s}$ for $1 \le r \le s \le q$. Let $\mathbf{H} = \mathbf{X}(\mathbf{X}^T\mathbf{W}\mathbf{X})^{-1}\mathbf{X}^T\mathbf{W}$ and let $\mathbf{H}^*$, defined analogously, have rank $q^* < n$. Show that
$$k_2 = \mathbf{Y}^T\mathbf{W}(\mathbf{I} - \mathbf{H})\mathbf{Y}/(n - q)$$
$$k_2^* = \mathbf{Y}^T\mathbf{W}(\mathbf{I} - \mathbf{H}^*)\mathbf{Y}/(n - q^*)$$
are both

  (i) unbiased for κ_2, (ii) invariant under the symmetric group (applied simultaneously to the rows of $\mathbf{Y}$ and $\mathbf{X}$),

 (iii) invariant under the general linear group applied to the columns of $\mathbf{X}$ (i.e. such that the column space of $\mathbf{X}$ is preserved).

Show also that k_2 is invariant under the general linear group (4.25) applied to the rows of $\mathbf{X}$ and $\mathbf{Y}$, but that k_2^* is not so invariant.

4.27 Justify the claims made in Section 4.7.1 that interaction and replication are not invariant under the general linear group (4.25). Show that these concepts are preserved under the permutation group.

4.28 In the notation of Section 4.7.2, let
$$F_i = h_{ij}Y^j, \qquad R_i = \rho_{ij}Y^j = (\delta_{ij} - h_{ij})Y^j$$
be the fitted value and the residual respectively. Define the derived statistics
$$T_1 = \sum R_j F_j^2 \quad \text{and} \quad T_2 = \sum R_j^2 F_j.$$
Show that
$$E(T_1) = \kappa_3(n_2 - 2n_{22} + n_3)$$
$$E(T_2) = \kappa_3(n_2 - n_3).$$
Show also, under the usual normal theory assumptions, that conditionally on the fitted values,
$$\mathrm{var}(T_1) = \kappa_2 \sum_{ij} \rho_{i,j} F_i^2 F_j^2$$
$$\mathrm{var}(T_2) = 2\kappa_2^2 \sum \rho_{i,j}^2 F_i F_j.$$

4.29 A population of size $N = N_0 + N_1$ comprises N_1 unit values and N_0 zero values. Show that the population Ks are

$$K_1 = N_1/N,$$
$$K_2 = N_0 N_1/N^{(2)},$$
$$K_3 = N_0 N_1 (N_0 - N_1)/N^{(3)},$$
$$K_4 = N_0 N_1 (N(N + 1) - 6N_0 N_1)/N^{(4)}$$

Hence derive the first four cumulants of the central hypergeometric distribution (Barton, David & Fix, 1960).

4.30 In a population consisting of the first N natural numbers, show that the population k-statistics are

$$K_1 = (N + 1)/2,$$
$$K_2 = N(N + 1)/12,$$
$$K_3 = 0,$$
$$K_4 = -N^2 (N + 1)^2/120.$$

(Barton, David & Fix, 1960).

CHAPTER 5

Edgeworth series

5.1 Introduction

While the lower-order cumulants of X are useful for describing, in both qualitative and quantitative terms, the shape of the joint distribution, it often happens that we require either approximations for the density function itself or, more commonly, approximations for tail probabilities or conditional tail probabilities. For example, if X^1 is a goodness-of-fit statistic and $X^2, ..., X^p$ are estimates of unknown parameters, it would often be appropriate to assess the quality of the fit by computing the tail probability

$$\text{pr}(X^1 \geq x^1 | X^2 = x^2, ..., X^p = x^p),$$

small values being taken as evidence of a poor fit. The cumulants themselves do not provide such estimates directly and it is necessary to proceed via an intermediate step where the density, or equivalently, the cumulative distribution function, is approximated by a series expansion. Series expansions of the Edgeworth type all involve an initial first-order approximation multiplied by a sum of correction terms whose coefficients are simple combinations of the cumulants of X. In the case of the Edgeworth expansion itself, the first-order approximation is based on the normal density having the same mean vector and covariance matrix as X. The correction terms then involve cumulants and cumulant products as coefficients of the Hermite polynomials. More generally, however, in order to minimize the number of correction terms, it may be advantageous to use a first-order approximation other than the normal. The correction terms then involve derivatives of the approximating density and are not necessarily polynomials.

In Section 5.2 we discuss the nature of such approximations for arbitrary initial approximating densities. These series take on different forms depending on whether we work with the density function, the log density function, the cumulative distribution function or some transformation of the cumulative distribution function. For theoretical calculations the log density is often the most convenient: for the computation of significance levels, the distribution function may be preferred because it is more directly useful.

The remainder of the chapter is devoted to the Edgeworth expansion itself and to derived expansions for conditional distributions.

5.2   A formal series expansion

5.2.1   *Approximation for the density*

Let $f_X(x; \kappa)$ be the joint density of the random variables $X^1, ..., X^p$, where the notation emphasizes the dependence of the density on the cumulants of X. Suppose that the initial approximating density, $f_0(x) \equiv f_0(x; \lambda)$ has cumulants $\lambda^i, \lambda^{i,j}, \lambda^{i,j,k}, ...$, which differ from the cumulants of X by

$$\eta^i = \kappa^i - \lambda^i, \quad \eta^{i,j} = \kappa^{i,j} - \lambda^{i,j}, \quad \eta^{i,j,k} = \kappa^{i,j,k} - \lambda^{i,j,k}$$

and so on. Often, it is good policy to choose $f_0(x; \lambda)$ so that $\eta^i = 0$ and $\eta^{i,j} = 0$, but it would be a tactical error to make this assumption at an early stage in the algebra, in the hope that substantial simplification would follow. While it is true that this choice makes many terms vanish, it does so at the cost of ruining the symmetry and structure of the algebraic formulae.

The cumulant generating functions for $f_X(x; \kappa)$ and $f_0(x; \lambda)$ are assumed to have their usual expansions

$$K_X(\xi) = \xi_i \kappa^i + \xi_i \xi_j \kappa^{i,j}/2! + \xi_i \xi_j \xi_k \kappa^{i,j,k}/3! + ...$$

and

$$K_0(\xi) = \xi_i \lambda^i + \xi_i \xi_j \lambda^{i,j}/2! + \xi_i \xi_j \xi_k \lambda^{i,j,k}/3! + ...$$

Subtraction gives

$$K_X(\xi) = K_0(\xi) + \xi_i \eta^i + \xi_i \xi_j \eta^{i,j}/2! + \xi_i \xi_j \xi_k \eta^{i,j,k}/3! + ...$$

and exponentiation gives

$$M_X(\xi) = M_0(\xi)\{1 + \xi_i\eta^i + \xi_i\xi_j\eta^{ij}/2! + \xi_i\xi_j\xi_k\eta^{ijk}/3! + ...\} \quad (5.1)$$

where

$$\eta^{ij} = \eta^{i,j} + \eta^i\eta^j,$$
$$\eta^{ijk} = \eta^{i,j,k} + \eta^i\eta^{j,k}[3] + \eta^i\eta^j\eta^k$$
$$\eta^{ijkl} = \eta^{i,j,k,l} + \eta^i\eta^{j,k,l}[4] + \eta^{i,j}\eta^{k,l}[3] + \eta^i\eta^j\eta^{k,l}[6] + \eta^i\eta^j\eta^k\eta^l$$

and so on, are the formal 'moments' obtained by treating η^i, $\eta^{i,j}$, $\eta^{i,j,k}$, ... as formal 'cumulants'. It is important here to emphasize that $\eta^{ij} \neq \kappa^{ij} - \lambda^{ij}$, the difference between second moments of X and those of $f_0(x)$. Furthermore, the ηs are not, in general, the cumulants of any real random variable. For instance, $\eta^{i,j}$ need not be positive definite.

To obtain a series approximation for $f_X(x; \kappa)$, we invert the approximate integral transform, (5.1), term by term. By construction, the leading term, $M_0(\xi)$, transforms to $f_0(x)$. The second term, $\xi_i M_0(\xi)$, transforms to $f_i(x) = -\partial f_0(x)/\partial x^i$ as can be seen from the following argument. Integration by parts with respect to x^r gives

$$\int_{-\infty}^{\infty} \exp(\xi_i x^i)\frac{\partial f_0(x)}{\partial x^r}\,dx^r = \exp(\xi_i x^i)f_0(x)\,\Big|_{x^r=-\infty}^{x^r=\infty}$$
$$- \xi_r \int_{-\infty}^{\infty} \exp(\xi_i x^i)f_0(x)\,dx^r.$$

For imaginary ξ, $\exp(\xi_i x^i)$ is bounded, implying that the first term on the right is zero. Further integration with respect to the remaining $p - 1$ variables gives

$$\int_{R^p} \exp(\xi_i x^i)f_r(x)dx = \xi_r M_0(\xi).$$

The critical assumption here is that $f_0(x)$ should have continuous partial derivatives everywhere in R^p. For example, if $f_0(x)$ had discontinuities, these would give rise to additional terms in the above integration.

By the same argument, if $f_0(x)$ has continuous second-order partial derivatives, it may be shown that $\xi_r\xi_s M_X(\xi)$ is the integral transform of $f_{rs}(x) = (-1)^2\partial^2 f_0(x)/\partial x^r \partial x^s$ and so on. In other words, (5.1) is the integral transform of

$$f_X(x) = f_0(x) + \eta^i f_i(x) + \eta^{ij} f_{ij}(x)/2! \\ + \eta^{ijk} f_{ijk}(x)/3! + \dots \qquad (5.2)$$

where the alternating signs on the derivatives have been incorporated into the notation.

Approximation (5.2) looks initially very much like a Taylor expansion in the sense that it involves derivatives of a function divided by the appropriate factorial. In fact, this close resemblance can be exploited, at least formally, by writing

$$X = Y + Z$$

where Y has density $f_0(y)$ and Z is a pseudo-variable independent of Y with 'cumulants' η^i, $\eta^{i,j}$, $\eta^{i,j,k}$,... and 'moments' η^i, η^{ij}, η^{ijk},... . Conditionally on $Z = z$, X has density $f_0(x - z)$ and hence, the marginal density of X is, formally at least,

$$f_X(x) = E_Z\{f_0(x - Z)\}.$$

Taylor expansion about $Z = 0$ and averaging immediately yields (5.2). This formal construction is due to Davis (1976).

To express (5.2) as a multiplicative correction to $f_0(x)$, we write

$$f_X(x; \kappa) = f_0(x)\{1 + \eta^i h_i(x) + \eta^{ij} h_{ij}(x)/2! \\ + \eta^{ijk} h_{ijk}(x)/3! + \dots\} \qquad (5.3)$$

where

$$h_i(x) = h_i(x; \lambda) = f_i(x)/f_0(x), \quad h_{ij}(x) = h_{ij}(x; \lambda) = f_{ij}(x)/f_0(x)$$

and so on. In many instances, $h_i(x)$, $h_{ij}(x)$, ... are simple functions with pleasant mathematical properties. For example, if $f_0(x) = \phi(x)$, the standard normal density, then $h_i(x)$, $h_{ij}(x)$,... are the standard Hermite tensors, or polynomials in the univariate case. See Sections 5.3 and 5.4.

5.2.2   *Approximation for the log density*

For a number of reasons, both theoretical and applied, it is often better to consider series approximations for the log density $\log f_X(x;\kappa)$ rather than for the density itself. One obvious advantage, particularly where polynomial approximation is involved, is that $f_X(x;\kappa) \geq 0$, whereas any polynomial approximation is liable to become negative for certain values of x. In addition, even if the infinite series (5.3) could be guaranteed positive, there is no similar guarantee for the truncated series that would actually be used in practice. For these reasons, better approximations may be obtained by approximating $\log f_X(x;\kappa)$ and exponentiating.

Expansion of the logarithm of (5.3) gives

$$
\begin{aligned}
\log f_X(x;\kappa) \simeq{} & \log f_0(x) + \eta^i h_i(x) \\
& + \{\eta^{i,j} h_{ij}(x) + \eta^i \eta^j h_{i,j}(x)\}/2! \\
& + \{\eta^{i,j,k} h_{ijk}(x) + \eta^i \eta^{j,k} h_{i,jk}(x)[3] + \eta^i \eta^j \eta^k h_{i,j,k}(x)\}/3! \\
& + \{\eta^{i,j,k,l} h_{ijkl} + \eta^i \eta^{j,k,l} h_{i,jkl}[4] + \eta^{i,j} \eta^{k,l} h_{ij,kl}[3] \\
& \quad + \eta^i \eta^j \eta^{k,l} h_{i,j,kl}[6] + \eta^i \eta^j \eta^k \eta^l h_{i,j,k,l}\}/4! \qquad (5.4)
\end{aligned}
$$

where

$$
\begin{aligned}
h_{i,j} &= h_{ij} - h_i h_j \\
h_{i,j,k} &= h_{ijk} - h_i h_{jk}[3] + 2h_i h_j h_k \\
h_{i,jk} &= h_{ijk} - h_i h_{jk} \\
h_{ij,kl} &= h_{ijkl} - h_{ij} h_{kl}
\end{aligned}
$$

and so on. Compare equations (2.7) and (3.2). Note that, apart from sign, the hs with fully partitioned indices are derivatives of the log density

$$
\begin{aligned}
h_i(x) &= -\partial \log f_0(x)/\partial x^i \\
h_{i,j}(x) &= \partial^2 \log f_0(x)/\partial x^i \partial x^j \\
h_{i,j,k}(x) &= -\partial^3 \log f_0(x)/\partial x^i \partial x^j \partial x^k
\end{aligned}
$$

and so on.

The h-functions with indices partially partitioned are related to each other in exactly the same way as generalized cumulants. Thus, for example, from (3.2) we find

$$
h_{i,jk} = h_{i,j,k} + h_j h_{i,k} + h_k h_{i,j}
$$

and

$$h_{i,jkl} = h_{i,j,k,l} + h_j h_{i,k,l}[3] + h_{i,j} h_k h_l[3],$$

using the conventions of Chapter 3.

Often it is possible to choose $f_0(x)$ so that its first-order and second-order moments agree with those of X. In this special case, (5.4) becomes

$$\log f_X(x) - \log f_0(x) = \eta^{i,j,k} h_{ijk}(x)/3! + \eta^{i,j,k,l} h_{ijkl}(x)/4!$$
$$+ \eta^{i,j,k,l,m} h_{ijklm}(x)/5! + \eta^{i,j,k,l,m,n} h_{ijklmn}(x)/6! \qquad (5.5)$$
$$+ \eta^{i,j,k} \eta^{l,m,n} h_{ijk,lmn}(x)[10]/6! + ...$$

This simplification greatly reduces the number of terms required but at the same time it manages to conceal the essential simplicity of the terms in (5.4). The main statistical reason for keeping to the more general expansion (5.4) is that, while it is generally possible to chose $f_0(x)$ to match the first two moments of X under some null hypothesis H_0, it is generally inconvenient to do so under general alternatives to H_0, as would be required in calculations related to the power of a test.

5.2.3   Approximation for the cumulative distribution function

One of the attractive features of expansions of the Edgeworth type is that the cumulative distribution function

$$F_X(x; \kappa) = \mathrm{pr}(X^1 \leq x^1, ..., X^p \leq x^p; \kappa)$$

can easily be approximated by integrating (5.2) term by term. If $F_0(x)$ is the cumulative distribution function corresponding to $f_0(x)$, this gives

$$F_X(x; \kappa) \simeq F_0(x) + \eta^i F_i(x) + \eta^{ij} F_{ij}(x)/2!$$
$$+ \eta^{ijk} F_{ijk}(x)/3! + ... \qquad (5.6)$$

where

$$F_i(x) = (-1)\partial F_0(x)/\partial x^i; \quad F_{ij}(x) = (-1)^2 \partial^2 F_0(x)/\partial x^i \partial x^j$$

and so on with signs alternating.

Of course, we might prefer to work with the multivariate survival probability

$$S_X(x) = \operatorname{pr}(X^1 \geq x^1, X^2 \geq x^2, ..., X^p \geq x^p; \kappa)$$

and the corresponding probability, $S_0(x)$, derived from $f_0(x)$. Integration of (5.2) term by term gives

$$S_X(x; \kappa) \simeq S_0(x) + \eta^i S_i(x) + \eta^{ij} S_{ij}(x)/2! + \eta^{ijk} S_{ijk}(x)/3! + ...$$

where

$$S_i(x) = -\partial S_0(x)/\partial x^i; \qquad S_{ij}(x) = (-1)^2 \partial^2 S_0(x)/\partial x^i \partial x^j$$

and so on with signs alternating as before.

In the univariate case, $p = 1$, but not otherwise, $F_X(x; \kappa) + S_X(x; \kappa) = 1$ if the distribution is continuous, and the correction terms in the two expansions sum to zero giving $F_0(x) + S_0(x) = 1$.

Other series expansions can be found for related probabilities. We consider here only one simple example, namely

$$\bar{S}_X(x; \kappa) = \operatorname{pr}(X^1 \leq x^1 \text{ or } X^2 \leq x^2 \text{ or } ... \text{ or } X^p \leq x^p; \kappa)$$

as opposed to $F_X(x; \kappa)$, whose definition involves replacing 'or' with 'and' in the expression above. By definition, in the continuous case, $\bar{S}_X(x; \kappa) = 1 - S_X(x; \kappa)$. It follows that

$$\bar{S}_X(x; \kappa) \simeq \bar{S}_0(x) + \eta^i \bar{S}_i(x) + \eta^{ij} \bar{S}_{ij}(x)/2! + \eta^{ijk} \bar{S}_{ijk}(x)/3! + ...$$

where

$$\bar{S}_i(x) = -\partial \bar{S}_0(x)/\partial x^i; \quad \bar{S}_{ij}(x) = (-1)^2 \partial^2 \bar{S}_0(x)/\partial x^i \partial x^j$$

and so on, again with alternating signs on the derivatives.

5.3   Expansions based on the normal density

5.3.1   *Multivariate case*

Suppose now, as a special case of (5.2) or (5.3) that the initial approximating density is chosen to be $f_0(x) = \phi(x; \lambda)$, the normal density with mean vector λ^i and covariance matrix $\lambda^{i,j}$. The density may be written

$$\phi(x; \lambda) = (2\pi)^{-p/2} |\lambda^{i,j}|^{-1/2} \exp\{-\tfrac{1}{2}(x^i - \lambda^i)(x^j - \lambda^j)\lambda_{i,j}\}$$

where $\lambda_{i,j}$ is the matrix inverse of $\lambda^{i,j}$ and $|\lambda^{i,j}|$ is the determinant of the covariance matrix. The functions $h_i(x) = h_i(x; \lambda)$, $h_{ij}(x) = h_{ij}(x; \lambda), ...$, obtained by differentiating $\phi(x; \lambda)$, are known as the Hermite tensors. The first six are given below

$$
\begin{aligned}
h_i &= \lambda_{i,j}(x^j - \lambda^j) \\
h_{ij} &= h_i h_j - \lambda_{i,j} \\
h_{ijk} &= h_i h_j h_k - h_i \lambda_{j,k}[3] \\
h_{ijkl} &= h_i h_j h_k h_l - h_i h_j \lambda_{k,l}[6] + \lambda_{i,j}\lambda_{k,l}[3] \\
h_{ijklm} &= h_i h_j h_k h_l h_m - h_i h_j h_k \lambda_{l,m}[10] + h_i \lambda_{j,k}\lambda_{l,m}[15] \\
h_{ijklmn} &= h_i ... h_n - h_i h_j h_k h_l \lambda_{m,n}[15] + h_i h_j \lambda_{k,l}\lambda_{m,n}[45] \\
&\quad - \lambda_{i,j}\lambda_{k,l}\lambda_{m,n}[15]
\end{aligned}
\tag{5.7}
$$

The general pattern is not difficult to describe: it involves summation over all partitions of the indices, unit blocks being associated with h_i, double blocks with $-\lambda_{i,j}$ and partitions having blocks of three or more elements being ignored.

In the univariate case, these reduce to the Hermite polynomials, which form an orthogonal basis with respect to $\phi(x; \lambda)$ as weight function, for the space of functions continuous over $(-\infty, \infty)$. The Hermite tensors are the polynomials that form a similar orthogonal basis for functions continuous over R^p. Their properties are discussed in detail in Section 5.4.

The common practice is to chose $\lambda^i = \kappa^i$, $\lambda^{i,j} = \kappa^{i,j}$, so that

$$\eta^i = 0, \quad \eta^{i,j} = 0, \quad \eta^{i,j,k} = \kappa^{i,j,k}, \quad \eta^{i,j,k,l} = \kappa^{i,j,k,l}$$

$$\eta^{ijk} = \kappa^{i,j,k}, \quad \eta^{ijkl} = \kappa^{i,j,k,l}, \quad \eta^{ijklm} = \kappa^{i,j,k,l,m},$$

$$\eta^{ijklmn} = \kappa^{i,j,k,l,m,n} + \kappa^{i,j,k}\kappa^{l,m,n}[10]$$

and so on, summing over all partitions ignoring those that have blocks of size 1 or 2. Thus (5.3) gives

$$
\begin{aligned}
f_X(x; \kappa) &\simeq \phi(x; \kappa)\{1 + \kappa^{i,j,k} h_{ijk}(x)/3! + \kappa^{i,j,k,l} h_{ijkl}(x)/4! \\
&\quad + \kappa^{i,j,k,l,m} h_{ijklm}(x)/5! \\
&\quad + (\kappa^{i,j,k,l,m,n} + \kappa^{i,j,k}\kappa^{l,m,n}[10]) h_{ijklmn}(x)/6! + ...\}.
\end{aligned}
\tag{5.8}
$$

This form of the approximation is sometimes known as the Gram-Charlier series. To put it in the more familiar and useful Edgeworth form, we note that if X is a standardized sum of n independent random variables, then $\kappa^{i,j,k} = O(n^{-1/2})$, $\kappa^{i,j,k,l} = O(n^{-1})$ and so on, decreasing in power of $n^{-1/2}$. Thus, the successive correction terms in the Gram-Charlier series are of orders $O(n^{-1/2})$, $O(n^{-1})$, $O(n^{-3/2})$ and $O(n^{-1})$ and these are not monotonely decreasing in n. The re-grouped series, formed by collecting together terms that are of equal order in n,

$$
\begin{aligned}
\phi(x; \kappa) \Big[&1 + \kappa^{i,j,k} h_{ijk}(x)/3! \\
&+ \{\kappa^{i,j,k,l} h_{ijkl}(x)/4! + \kappa^{i,j,k}\kappa^{l,m,n} h_{ijklmn}(x)[10]/6!\} \\
&+ \{\kappa^{i,j,k,l,m} h_{ijklm}(x)/5! + \kappa^{i,j,k}\kappa^{l,m,n,r} h_{i...r}(x)[35]/7! \\
&\quad + \kappa^{i,j,k}\kappa^{l,m,n}\kappa^{r,s,t} h_{i...t}(x)[280]/9!\} + ... \Big]
\end{aligned}
\tag{5.9}
$$

is called the Edgeworth series and is often preferred for statistical calculations. The infinite versions of the two series are formally identical and the main difference is that truncation of (5.8) after a fixed number of terms gives a different answer than truncation of (5.9) after a similar number of terms.

At the origin of x, the approximating density takes the value

$$
\phi(0; \kappa) \left[1 + 3\rho_4/4! - \{9\rho_{13}^2 + 6\rho_{23}^2\}/72 + O(n^{-2}) \right]
\tag{5.10}
$$

where ρ_4, ρ_{13}^2 and ρ_{23}^2 are the invariant standardized cumulants of X. Successive terms in this series decrease in whole powers of n as opposed to the half-powers in (5.9).

5.3.2 Univariate case

The univariate case is particularly important because the notion of a tail area, as used in significance testing, depends on the test statistic being one-dimensional. In the univariate case it is convenient to resort to the conventional power notation by writing

$$h_1(x; \kappa) = \kappa_2^{-1}(x - \kappa_1)$$
$$h_2(x; \kappa) = h_1^2 - \kappa_2^{-1}$$
$$h_3(x; \kappa) = h_1^3 - 3\kappa_2^{-1}h_1$$

and so on for the Hermite polynomials. These are derived in a straightforward way from the Hermite tensors (5.7). The standard Hermite polynomials, obtained by putting $\kappa_1 = 0$, $\kappa_2 = 1$ are

$$h_1(z) = z$$
$$h_2(z) = z^2 - 1$$
$$h_3(z) = z^3 - 3z$$
$$h_4(z) = z^4 - 6z^2 + 3 \qquad (5.11)$$
$$h_5(z) = z^5 - 10z^3 + 15z$$
$$h_6(z) = z^6 - 15z^4 + 45z^2 - 15.$$

The first correction term in (5.9) is

$$\kappa_3 h_3(x; \kappa)/6 = \kappa_3\{\kappa_2^{-3}(x - \kappa_1)^3 - 3\kappa_2^{-2}(x - \kappa_1)\}/6$$
$$= \rho_3 h_3(z)/6 = \rho_3(z^3 - 3z)/6$$

where $z = \kappa_2^{-1/2}(x - \kappa_1)$ and $Z = \kappa_2^{-1/2}(X - \kappa_1)$ is the standardized version of X. Similarly, the $O(n^{-1})$ correction term becomes

$$\rho_4 h_4(z)/4! + 10\rho_3^2 h_6(z)/6!$$

Even in theoretical calculations it is rarely necessary to use more than two correction terms from the Edgeworth series. For that reason, we content ourselves with corrections up to order $O(n^{-1})$, leaving an error that is of order $O(n^{-3/2})$.

For significance testing, the one-sided tail probability may be approximated by

$$\mathrm{pr}(Z \geq z) \simeq 1 - \Phi(z)$$
$$+ \phi(z)\{\rho_3 h_2(z)/3! + \rho_4 h_3(z)/4! + \rho_3^2 h_5(z)/72\},$$
$$(5.12)$$

which involves one correction term of order $O(n^{-1/2})$ and two of order $O(n^{-1})$. The two-sided tail probability is, for $z > 0$,

$$
\begin{aligned}
\mathrm{pr}(|Z| \geq z) = {} & 2\{1 - \Phi(z)\} \\
& + 2\phi(z)\{\rho_4 h_3(z)/4! + \rho_3^2 h_5(z)/72\}
\end{aligned}
\tag{5.13}
$$

and this involves only correction terms of order $O(n^{-1})$. In essence, what has happened here is that the $O(n^{-1/2})$ corrections are equal in magnitude but of different sign in the two tails and they cancel when the two tails are combined.

5.3.3 Regularity conditions

So far, we have treated the series expansions (5.8) to (5.13) in a purely formal way. No attempt has been made to state precisely the way in which these series expansions are supposed to approximate either the density or the cumulative distribution function. In this section, we illustrate in an informal way, some of the difficulties encountered in making this notion precise.

First, it is not difficult to see that the infinite series expansion (5.3) or (5.8) for the density is, in general, divergent. For example, if the cumulant differences, η^i, $\eta^{i,j}$, $\eta^{i,j,k}$,... are all equal to 1, then $\eta^{ij} = 2$, $\eta^{ijk} = 5$, $\eta^{ijkl} = 15$,..., otherwise known as the Bell numbers. These increase exponentially fast so that (5.3) fails to converge. For this reason we need a justification of an entirely different kind for using the truncated expansion, truncated after some fixed number of terms.

To this end, we introduce an auxiliary quantity, n, assumed known, and we consider the formal mathematical limit $n \to \infty$. The idea now is to approximate $K_X(\xi)$ by a truncated series such that after $r + 2$ terms the error is $O(n^{-r/2})$, say. Formal inversion gives a truncated series expansion for the density or distribution function which, under suitable smoothness conditions, has an error also of order $O(n^{-r/2})$. In all the applications we have in mind, X is a standardized sum and $f_0(x)$ is the normal density having the same mean and variance as X. Then $\eta^i = 0$, $\eta^{i,j} = 0$, $\eta^{i,j,k} = O(n^{-1/2})$, $\eta^{i,j,k,l} = O(n^{-1})$ and so on. Thus the truncated expansion for $K_X(\xi)$, including terms up to degree 4 in ξ, has error $O(n^{-3/2})$. To obtain similar accuracy in the approximations for $M_X(\xi)$ or $f_X(x)$ it is necessary to go as far as terms of degree 6.

This rather informal description ignores one fairly obvious point. The truncated series approximation (5.5) for $F_X(x; \kappa)$ is continuous. If X has a discrete distribution, $F_X(x; \kappa)$ will have discontinuities with associated probabilities typically of order $O(n^{-1/2})$. Any continuous approximation for $F_X(x)$ will therefore have an error of order $O(n^{-1/2})$, however many correction terms are employed.

In the univariate case, if X is the standardized sum of n independent and identically distributed random variables, Feller (1971, Chapter XVI) gives conditions ensuring the validity of expansions (5.9) for the density and (5.12) for the cumulative distribution function. The main condition, that

$$\limsup_{\varsigma \to \infty} |M_X(i\varsigma)| < 1,$$

excludes lattice distributions but ensures that the error in (5.12) is $o(n^{-1})$ uniformly in z. With weaker conditions, but again excluding lattice distributions, the error is known to be $o(n^{-1/2})$ uniformly in z (Esseen, 1945; Bahadur & Ranga-Rao, 1960). In the lattice case, (5.12) has error $o(n^{-1/2})$ provided that it is used with the usual correction for continuity. An asymptotic expansion given by Esseen (1945) extends the Edgeworth expansion to the lattice case: in particular, it produces correction terms for (5.12) so that the renaining error is $O(n^{-3/2})$ uniformly in z.

Similar conditions dealing with the multivariate case are given by Barndorff-Nielsen & Cox (1979).

The case where X is a standardized sum of non-independent or non-identically distributed random variables is considerably more complicated. It is necessary, for example, to find a suitably strengthened version of the Lindeberg condition involving, perhaps, the higher-order cumulants. In addition, if X is a sum of discrete and continuous random variables, it would be necessary to know whether the discrete or the continuous components dominated in the overall sum. No attempt will be made to state such conditions here.

For details of regularity conditions, see Bhattacharya & Rao (1976) or Skovgaard, (1981a,b, 1986b).

5.4   Some properties of Hermite tensors

5.4.1 *Tensor properties*

We now investigate the extent to which the arrays of polynomials
$h_i(x; \lambda)$, $h_{ij}(x; \lambda)$, $h_{ijk}(x; \lambda)$, ..., defined apart from choice of sign,
as the coefficient of $\phi(x; \lambda)$ in the partial derivatives of $\phi(x; \lambda)$, de-
serve to be called tensors. To do so, we must examine the effect on
the polynomials of transforming from x to new variables $\bar{x}$. In doing
so, we must also take into account the induced transformation on
the λs, which are the formal cumulants of x.

Consider the class of non-singular affine transformations on x

$$\bar{x}^i = a^i + a^i_r x^r \tag{5.14}$$

together with the induced transformation on the λs

$$\bar{\lambda}^i = a^i + a^i_r \lambda^r \quad \text{and} \quad \bar{\lambda}^{i,j} = a^i_r a^j_s \lambda^{r,s}.$$

It is easily checked that

$$(x^i - \lambda^i)(x^i - \lambda^j)\lambda_{i,j} = (\bar{x}^i - \bar{\lambda}^i)(\bar{x}^j - \bar{\lambda}^j)\bar{\lambda}_{i,j} \tag{5.15}$$

so that this quadratic form is invariant under nonsingular affine
transformation. If a^i_r does not have full rank, this argument breaks
down. Apart from the determinantal factor $|\lambda^{i,j}|^{-1/2}$, which has
no effect on the definition of the Hermite tensors, $\phi(x; \lambda)$ is invariant under the transformation (5.14) because it is a function of
the invariant quadratic form (5.15). It follows therefore that the
derivatives with respect to $\bar{x}$ are

$$\frac{\partial \phi}{\partial x^r} \frac{\partial x^r}{\partial \bar{x}^i}, \quad \frac{\partial^2 \phi}{\partial x^r \partial x^s} \frac{\partial x^r}{\partial \bar{x}^i} \frac{\partial x^s}{\partial \bar{x}^j} + \frac{\partial \phi}{\partial x^r} \frac{\partial^2 x^r}{\partial \bar{x}^i \partial \bar{x}^j}$$

and so on. Since $\partial^2 x^r / \partial \bar{x}^i \partial \bar{x}^j \equiv 0$, these derivatives may be written
in the form

$$b^r_i \phi_r, \quad b^r_i b^s_j \phi_{rs}, \quad b^r_i b^s_j b^t_k \phi_{rst}$$

and so on where b^r_i is the matrix inverse of a^i_r. Thus the derivatives
of ϕ are Cartesian tensors and they transform in the covariant
manner. It follows immediately that $h_i(x; \lambda)$, $h_{ij}(x; \lambda)$ and so on
are also Cartesian tensors, justifying the terminology of Section 5.3.

The above result applies equally to the more general case discussed in Section 5.2 where the form of the initial approximating
density $f_0(x; \lambda)$ is left unspecified. See Exercise 5.5.

5.4.2 *Orthogonality*

The principal advantage of index notation is its total transparency: the principal disadvantage, nowhere more evident than in Chapter 1 where tensors were defined, is that general expressions for tensors of arbitrary order are difficult to write down without introducing unsightly subscripted indices. For this reason, we have avoided writing down general expressions for Hermite tensors, even though the general pattern is clear and is easily described in a few words (Section 5.3.1). To prove orthogonality it is necessary either to devise a suitable general notation or to use a less direct method of proof.

Our method of proof uses the generating function

$$\exp\{\xi_i(x^i - \lambda^i) - \tfrac{1}{2}\xi_i\xi_j\lambda^{i,j}\} = 1 + \xi_i h^i + \xi_i\xi_j h^{ij}/2! \\ + \xi_i\xi_j\xi_k h^{ijk}/3! + \dots \tag{5.16}$$

where

$$h^i = \lambda^{i,r} h_r(x; \lambda), \qquad h^{ij} = \lambda^{i,r}\lambda^{j,s} h_{rs}(x; \lambda),$$
$$h^{ijk} = \lambda^{i,r}\lambda^{j,s}\lambda^{k,t} h_{rst}(x; \lambda)$$

and so on, are the contravariant expressions of the Hermite tensors. To show that (5.16) is indeed a generating function for the *h*s, we observe that

$$\phi(x^i - \lambda^{i,j}\xi_j; \lambda) = \phi(x; \lambda)\exp\{\xi_i(x^i - \lambda^i) - \xi_i\xi_j\lambda^{i,j}/2\}.$$

Taylor expansion about $\xi = 0$ of $\phi(x^i - \lambda^{i,j}\xi_j; \lambda)$ gives

$$\phi(x; \lambda)\{1 + \xi_i h^i + \xi_i\xi_j h^{ij}/2! + \xi_i\xi_j\xi_k h^{ijk}/3! + \dots\},$$

which we have simplified using the definition of Hermite tensors. This completes the proof that $\exp\{\xi_i(x^i - \lambda^i) - \xi_i\xi_j\lambda^{i,j}/2\}$ is the generating function for the Hermite tensors.

To prove orthogonality, consider the product

$$\phi(x^i - \lambda^{i,j}\varsigma_j; \lambda)\exp\{\xi_i(x^i - \lambda^i) - \xi_i\xi_j\lambda^{i,j}/2\}$$
$$= \phi(x; \lambda)\{1 + \varsigma_i h^i + \varsigma_i\varsigma_j h^{ij}/2! + \dots\}\{1 + \xi_i h^i + \xi_i\xi_j h^{ij}/2! + \dots\}.$$

Simplification of the exponent gives

$$\phi(x^i - \lambda^{i,j}\xi_j - \lambda^{i,j}\varsigma_j; \lambda)\exp(\xi_i\varsigma_j\lambda^{i,j})$$

and integration with respect to x over R^p gives $\exp(\xi_i\varsigma_j\lambda^{i,j})$. Orthogonality follows because $\exp(\xi_i\varsigma_j\lambda^{i,j})$ involves only terms of equal degree in ξ and ς. Moreover, from the expansion

$$\exp(\xi_i\varsigma_j\lambda^{i,j}) = 1 + \xi_i\varsigma_j\lambda^{i,j} + \xi_i\xi_j\varsigma_k\varsigma_l\lambda^{i,k}\lambda^{j,l}/2! + ...,$$

it follows that the inner products over R^p are

$$\int h^i h^j \phi \, dx = \lambda^{i,j}$$

$$\int h^{ij} h^{kl} \phi \, dx = \lambda^{i,k}\lambda^{j,l} + \lambda^{i,l}\lambda^{j,k} \qquad (5.17)$$

$$\int h^{ijk} h^{lmn} \phi \, dx = \lambda^{i,l}\lambda^{j,m}\lambda^{k,m}[3!]$$

and so on for tensors of higher order.

In the univariate case where the hs are standard Hermite polynomials, the orthogonality relations are usually written in the form $\int h_r(x)h_s(x)\phi(x)\,dx = r!\delta_{rs}$.

The extension of the above to scalar products of three or more Hermite polynomials or tensors is given in Exercises 5.9–5.14.

5.4.3   *Generalized Hermite tensors*

From any sequence of arrays h^i, h^{ij}, h^{ijk}, ..., each indexed by an unordered set of indices, there may be derived a new sequence h^i, $h^{i,j}$, $h^{i,j,k}$, ..., each indexed by a fully partitioned set of indices. If $M_h(\xi)$ is the generating function for the first set, then $K_h(\xi) = \log M_h(\xi)$ is the generating function for the new set and the relationship between the two sequences is identical to the relationship between moments and cumulants. In the case of Hermite tensors, we find on taking logs in (5.16) that

$$K_h(\xi) = \xi_i(x^i - \lambda^i) - \xi_i\xi_j\lambda^{i,j}/2 \qquad (5.18)$$

so that $h^i = x^i - \lambda^i$, $h^{i,j} = -\lambda^{i,j}$ and all other arrays in the new sequence are identically zero. In fact, apart from choice of sign, the arrays in this new sequence are the partial derivatives of $\log \phi(x; \lambda)$ with respect to the components of x.

Just as ordinary moments and ordinary cumulants are conveniently considered as special cases of generalized cumulants, so too the sequences h^i, h^{ij}, h^{ijk}, ... and h^i, $h^{i,j}$, $h^{i,j,k}$ are special cases of generalized Hermite tensors. Some examples with indices fully partitioned are given below.

$$
\begin{aligned}
h^{i,jk} &= h^{ijk} - h^i h^{jk} \\
&= h^{i,j,k} + h^j h^{i,k} + h^k h^{i,j} \\
&= -(x^j - \lambda^j)\lambda^{i,k} - (x^k - \lambda^k)\lambda^{i,j} \\
h^{i,jkl} &= h^{ijkl} - h^i h^{jkl} \\
&= h^{i,j,k,l} + h^j h^{i,k,l}[3] + h^{i,j} h^{k,l}[3] + h^{i,j} h^k h^l[3] \\
&= -h^k h^l \lambda^{i,j}[3] + \lambda^{i,j} \lambda^{k,l}[3] \\
h^{ij,kl} &= h^{ijkl} - h^{ij} h^{kl} \\
&= h^{i,j,k,l} + h^i h^{j,k,l}[4] + h^{i,k} h^{j,l}[2] + h^{i,k} h^j h^l[4] \\
&= -h^j h^l \lambda^{i,k}[4] + \lambda^{i,k} \lambda^{j,l}[2] \\
h^{i,j,kl} &= h^{ijkl} - h^i h^{jkl}[2] - h^{ij} h^{kl} + 2h^i h^j h^{kl} \\
&= h^{i,j,k,l} + h^k h^{i,j,l}[2] + h^{i,k} h^{j,l}[2] \\
&= \lambda^{i,k} \lambda^{j,l}[2]
\end{aligned}
$$

In these examples, the first line corresponds to the expressions (3.2) for generalized cumulants in terms of moments. The second line corresponds to the fundamental identity (3.3) for generalized cumulants in terms of ordinary cumulants and the final line is obtained on substituting the coefficients in (5.18).

A generalized Hermite tensor involving β indices partitioned into α blocks is of degree $\beta - 2\alpha + 2$ in x provided that $\beta - 2\alpha + 2 \geq 0$. Otherwise the array is identically zero.

The importance of these generalized Hermite tensors lies in the formal series expansion (5.4) for the log density, which is used in computing approximate conditional cumulants.

5.4.4 *Factorization of Hermite tensors*

Suppose that X is partitioned into two vector components $X^{(1)}$ and $X^{(2)}$ of dimensions q and $p - q$ respectively. We suppose also that the normal density $\phi(x; \lambda)$, from which the Hermite tensors are derived, is expressible as the product of two normal densities, one for $X^{(1)}$ and one for $X^{(2)}$. Using the indices $i, j, k, \ldots$ to refer to components of $X^{(1)}$ and $r, s, t, \ldots$ to refer to components of $X^{(2)}$, we may write $\lambda^{i,r} = 0$ and $\lambda_{i,r} = 0$. Since the two components in the approximating density are uncorrelated, there is no ambiguity in writing $\lambda_{i,j}$ and $\lambda_{r,s}$ for the matrix inverses of $\lambda^{i,j}$ and $\lambda^{r,s}$ respectively. More generally, if the two components were correlated, this notation would be ambiguous because the leading $q \times q$ sub-matrix of

$$\begin{bmatrix} \lambda^{i,j} & \lambda^{i,s} \\ \lambda^{r,j} & \lambda^{r,s} \end{bmatrix}^{-1},$$

namely $\{\lambda^{i,j} - \lambda^{i,r}\{\lambda^{r,s}\}^{-1}\lambda^{s,j}\}^{-1}$, is not the same as the matrix inverse of $\lambda^{i,j}$.

Since $h^{i,r} = -\lambda^{i,r} = 0$ and all higher-order generalized Hermite tensors, with indices fully partitioned, are zero, it follows immediately from (3.3) that

$$h^{i,rs} = 0; \quad h^{ij,r} = 0; \quad h^{i,rst} = 0; \quad h^{ij,rs} = 0$$
$$h^{ijk,r} = 0; \quad h^{i,j,rs} = 0$$

and so on for any sub-partition of $\{(i, j, k, \ldots), (r, s, t, \ldots)\}$. This is entirely analogous to the statement that mixed cumulants involving two independent random variables and no others are zero. The corresponding statement for moments involves multiplication or factorization so that

$$h^{irs} = h^i h^{rs}; \qquad h^{ijr} = h^{ij} h^r; \qquad h^{ijrs} = h^{ij} h^{rs};$$
$$h^{i,jrs} = h^{i,j} h^{rs}; \qquad h^{i,r,js} = h^{i,j} h^{r,s};$$
$$h^{ir,js} = h^{i,j} h^{r,s} + h^i h^j h^{r,s} + h^{i,j} h^r h^s = h^{ij} h^{r,s} + h^{i,j} h^r h^s$$

and so on.

5.5   Linear regression and conditional cumulants

5.5.1   *Covariant representation of cumulants*

For calculations involving conditional distributions or conditional cumulants, it is often more convenient to work not with the cumulants of X^i directly but rather with the cumulants of $X_i = \kappa_{i,j} X^j$, which we denote by κ_i, $\kappa_{i,j}$, $\kappa_{i,j,k}$, ... We refer to these as the covariant representation of X and the covariant representation of its cumulants. This transformation may, at first sight, seem inconsequential but it should be pointed out that, while the notation X^i is unambiguous, the same cannot be said of X_i because the value of X_i depends on the entire set of variables $X^1, ..., X^p$. Deletion of X^p, say, leaves the remaining components $X^1, ..., X^{p-1}$ and their joint cumulants unaffected but the same is not true of $X_1, ..., X_{p-1}$.

5.5.2   *Orthogonalized variables*

Suppose now that X is partitioned into two vector components $X^{(1)}$ and $X^{(2)}$ of dimensions p and $q - p$ respectively and that we require the conditional cumulants of $X^{(1)}$ after linear regression on $X^{(2)}$. This is a simpler task than finding the conditional cumulants of $X^{(1)}$ given $X^{(2)} = x^{(2)}$, because it is assumed implicitly that only the conditional mean of $X^{(1)}$ and none of the higher-order conditional cumulants depends on the value of $x^{(2)}$. Further, each component of the conditional mean is assumed to depend linearly on $x^{(2)}$.

We first make a non-singular linear transformation from the original $(X^{(1)}, X^{(2)})$ to new variables $Y = (Y^{(1)}, Y^{(2)})$ in such a way that $Y^{(2)} = X^{(2)}$ and $Y^{(1)}$ is uncorrelated with $Y^{(2)}$. Extending the convention established in Section 5.4.4, we let the indices $i, j, k, ...$ refer to components of $X^{(1)}$ or $Y^{(1)}$, indices $r, s, t, ...$ refer to components of $X^{(2)}$ or $Y^{(2)}$ and indices $\alpha, \beta, \gamma, ...$ refer to components of the joint variable X or Y. The cumulants of X are κ^α partitioned into κ^i and κ^r, $\kappa^{\alpha,\beta}$ partitioned into $\kappa^{i,j}$ $\kappa^{i,r}$ and $\kappa^{r,s}$ and so on. The same convention applies to the covariant representation, κ_i, κ_r, $\kappa_{i,j}$, $\kappa_{i,r}$ and $\kappa_{r,s}$.

At this stage, it is necessary to distinguish between $\kappa_{r,s}$, the $(p - q) \times (p - q)$ sub-matrix of $[\kappa^{\alpha,\beta}]^{-1}$ and the $(p - q) \times (p - q)$ matrix inverse of $\kappa^{r,s}$, the covariance matrix of $Y^{(2)}$. The usual way of doing this is to distinguish the cumulants of Y from those of X by means of an overbar. Thus $\bar{\kappa}^{r,s} = \kappa^{r,s}$ is the covariance matrix of $Y^{(2)}$. In addition, $\bar{\kappa}^{i,j}$ is the covariance matrix of $Y^{(1)}$ and $\bar{\kappa}^{i,r} = 0$ by construction. The inverse matrix has elements $\bar{\kappa}_{r,s}$, $\bar{\kappa}_{i,j}$, $\bar{\kappa}_{i,r} = 0$, where $\bar{\kappa}^{r,s}\,\bar{\kappa}_{s,t} = \delta^r_t$.

The linear transformation to orthogonalized variables $Y^{(1)}$ and $Y^{(2)}$ may be written

$$Y^i = X^i - \beta^i_r X^r; \qquad Y^r = X^r \qquad (5.19)$$

where $\beta^i_r = \kappa^{i,s}\bar{\kappa}_{r,s}$ is the regression coefficient of X^i on X^r. This is the contravariant representation of the linear transformation but, for our present purposes, it will be shown that the covariant representation is the more convenient to work with. The covariance matrix of $Y^{(1)}$ is

$$\bar{\kappa}^{i,j} = \kappa^{i,j} - \kappa^{i,r}\kappa^{j,s}\bar{\kappa}_{r,s} = \{\kappa_{i,j}\}^{-1}. \qquad (5.20)$$

Further, using formulae for the inverse of a partitioned matrix, we find

$$\kappa_{i,j} = \bar{\kappa}_{i,j}$$
$$\kappa_{i,r} = -\kappa_{i,j}\beta^j_r$$
$$\kappa_{r,s} = \bar{\kappa}_{r,s} + \beta^i_r\beta^j_s\kappa_{i,j}.$$

Hence the expressions for Y_i and Y_r are

$$Y_i = \bar{\kappa}_{i,j}Y^j = \kappa_{i,j}Y^j = \kappa_{i,j}(X^j - \beta^j_r X^r)$$
$$= \kappa_{i\alpha}X^\alpha = X_i$$

and

$$Y_r = \bar{\kappa}_{r,s}Y^s = \bar{\kappa}_{r,s}X^s = \bar{\kappa}_{r,s}\kappa^{s,\alpha}X_\alpha$$
$$= X_r + \beta^i_r X_i.$$

Thus, the covariant representation of the linear transformation (5.19) is

$$Y_i = X_i; \qquad Y_r = X_r + \beta^i_r X_i. \qquad (5.21)$$

It follows that the joint cumulants of $(Y^{(1)}, Y^{(2)})$ expressed in covariant form in terms of the cumulants of $(X^{(1)}, X^{(2)})$, are

$$\bar{\kappa}_i = \kappa_i, \qquad\qquad \bar{\kappa}_{i,j} = \kappa_{i,j}, \qquad\qquad \bar{\kappa}_{i,j,k} = \kappa_{i,j,k}, \dots$$
$$\bar{\kappa}_{i,r} = 0, \qquad\qquad \bar{\kappa}_{r,s} = \kappa_{r,s} + \beta^i_r\kappa_{i,s}[2] + \beta^i_r\beta^j_s$$
$$\kappa_{i,j} = \kappa_{r,s} - \beta^i_r\beta^j_s\kappa_{i,j}$$

$$\bar{\kappa}_{i,j,r} = \kappa_{i,j,r} + \beta_r^k \kappa_{i,j,k}$$
$$\bar{\kappa}_{i,r,s} = \kappa_{i,r,s} + \beta_r^j \kappa_{i,j,s}[2] + \beta_r^j \beta_s^k \kappa_{i,j,k} \qquad (5.22)$$
$$\bar{\kappa}_{r,s,t} = \kappa_{r,s,t} + \beta_r^i \kappa_{i,s,t}[3] + \beta_r^i \beta_s^j \kappa_{i,j,t}[3] + \beta_r^i \beta_s^j \beta_r^k \kappa_{i,j,k}.$$

The main point that requires emphasis here is that, in covariant form, the cumulants of $Y^{(1)}$ are the same as those of $X^{(1)}$ and they are unaffected by the Gram-Schmidt orthogonalization (5.19).

The contravariant expressions for the cumulants of $(Y^{(1)}, Y^{(2)})$ are

$$\bar{\kappa}^i = \kappa^i - \beta_r^i \kappa^r$$
$$\bar{\kappa}^{i,j} = \kappa^{i,j} - \beta_r^i \kappa^{r,j}[2] + \beta_r^i \beta_s^j \kappa^{r,s} = \kappa^{i,j} - \beta_r^i \beta_s^j \kappa^{r,s}$$
$$\bar{\kappa}^{i,j,k} = \kappa^{i,j,k} - \beta_r^i \kappa^{r,j,h}[3] + \beta_r^i \beta_s^j \kappa^{r,s,k}[3] - \beta_r^i \beta_s^j \beta_r^k \kappa^{r,s,t}$$
$$\bar{\kappa}^{i,r} = 0, \qquad \bar{\kappa}^{r,s} = \kappa^{r,s} \qquad (5.23)$$
$$\bar{\kappa}^{i,j,r} = \kappa^{i,j,r} - \beta_s^j \kappa^{i,r,s}[2] + \beta_s^i \beta_r^j \kappa^{r,s,t}$$
$$\bar{\kappa}^{i,r,s} = \kappa^{i,r,s} - \beta_r^i \kappa^{r,s,t}$$
$$\bar{\kappa}^{r,s,t} = \kappa^{r,s,t}$$

It is important to emphasize at this stage that $Y^{(1)}$ and $Y^{(2)}$ are not independent unless the strong assumptions stated in the first paragraph of this section apply. Among the cumulants given above, exactly two of the mixed third-order expressions are non-zero, namely $\kappa^{i,r,s}$ and $\kappa^{i,j,r}$. A non-zero value of $\kappa^{i,r,s}$ means that, although $E(Y^i|X^{(2)})$ has no linear dependence on $X^{(2)}$, there is some quadratic dependence on products $X^r X^s$. There is a close connection here with Tukey's (1949) test for non-additivity. Similarly, a non-zero value of $\kappa^{i,j,r}$ implies heterogeneity of covariance, namely that $\mathrm{cov}(Y^i, Y^j|X^{(2)})$ depends linearly or approximately linearly on $X^{(2)}$. See (5.28).

5.5.3 Linear regression

Suppose now, in the notation of the previous section, that the orthogonalized variables $Y^{(1)}$ and $Y^{(2)}$ are independent. It follows that the second- and higher-order conditional cumulants of $X^{(1)}$ given $X^{(2)}$ are the same as the unconditional cumulants of $Y^{(1)}$. These are given in covariant form in the first set of equations (5.22) and in contravariant form in (5.23). To complete the picture, we require only the conditional mean of $X^{(1)}$. We find from (5.19) that

$$E\{X^i|X^{(2)}\} = \kappa^i + \beta_r^i(x^r - \kappa^r) = \kappa^i + \kappa^{i,r}h_r(x^{(2)};\bar{\kappa})$$

where $h_r(x^{(2)};\bar{\kappa}) = \bar{\kappa}_{r,s}(x^s - \kappa^s)$ is the first Hermite polynomial in the components of $X^{(2)}$.

The above expression may be written in covariant form as

$$E(X^i|X^{(2)}) = \bar{\kappa}^{i,j}\{\kappa_j - \kappa_{j,r}x^r\}$$

so that $\kappa_j - \kappa_{j,r}x^r$ is the covariant representation of the conditional mean of $X^{(1)}$.

These results may be summarized by a simple recipe for computing the conditional cumulants of $X^{(1)}$ after linear regression on $X^{(1)}$

(i) First compute the cumulants of $\kappa_{\alpha,\beta}X^\beta$, giving κ_i κ_r; $\kappa_{i,j}$, $\kappa_{i,r}$, $\kappa_{r,s}$; $\kappa_{i,j,k}$, ...

(ii) Compute $\bar{\kappa}^{i,j}$ the $p \times p$ matrix inverse of $\kappa_{i,j}$.

(iii) Replace κ_i by $\kappa_i - \kappa_{i,r}x^r$.

(iv) Raise all indices by multiplying by $\bar{\kappa}^{i,j}$ as often as necessary, giving

$$\bar{\kappa}^{i,j}\{\kappa_j - \kappa_{j,r}x^r\}$$
$$\bar{\kappa}^{i,k}\bar{\kappa}^{j,l}\kappa_{k,l} = \bar{\kappa}^{i,j}$$
$$\bar{\kappa}^{i,j,k} = \bar{\kappa}^{i,l}\bar{\kappa}^{j,m}\bar{\kappa}^{k,n}\kappa_{l,m,n}$$

and so on for the transformed cumulants.

5.6   Conditional cumulants

5.6.1   *Formal expansions*

In the previous section we considered the simple case where only the conditional mean of $X^{(1)}$ given $X^{(2)} = x^{(2)}$ and none of the higher-order cumulants depends on the value of $x^{(2)}$. In this special case it was possible to compute exact expressions for the conditional cumulants. More generally, all of the conditional cumulants may depend, to some extent, on the value of $x^{(2)}$. Our aim in this section is to use the series expansion (5.4), together with the results of the previous three sections, to find formal series expansions for at least the first four conditional cumulants and, in principle at least, for all of the conditional cumulants.

To simplify matters, we assume that the initial approximating density for $(X^{(1)}, X^{(2)})$ is the product of two normal densities, $\phi_1(x^{(1)}; \lambda^{(1)})\phi_2(x^{(2)}; \lambda^{(2)})$. The advantage derived from this choice is that the Hermite tensors factor into products as described in Section 5.4.4. In practice, it is often sensible to choose the argument $\lambda^{(1)}$ of ϕ_1 to be the mean vector and covariance matrix of $X^{(1)}$ but, for reasons given in Section 5.2.1, we shall not do so for the moment.

Expansion (5.4) for the logarithm of the joint density of $X^{(1)}$ and $X^{(2)}$ may be written

$$
\log \phi_1(x^{(1)}; \lambda^{(1)}) + \log \phi_2(x^{(2)}; \lambda^{(2)}) + \eta^i h_i + \eta^r h_r
$$
$$
+ \left\{ \eta^{i,j} h_{ij} + \eta^{i,r} h_i h_r[2] + \eta^{r,s} h_{rs} + \eta^i \eta^j h_{i,j} + \eta^r \eta^s h_{r,s} \right\}/2!
$$
$$
+ \left\{ \left(\eta^{i,j,k} h_{ijk} + \eta^{i,j,r} h_{ij} h_r[3] + \eta^{i,r,s} h_i h_{rs}[3] + \eta^{r,s,t} h_{rst} \right) \right.
$$
$$
\left. + \left(\eta^i \eta^j {}^k h_{i,jk} + \eta^i \eta^{j,r} h_{i,j} h_r[2] + \eta^r \eta^{i,s} h_i h_{r,s}[2] + \eta^r \eta^{s,t} h_{r,st} \right)[3] \right\}/3!
$$
$$
+ \left\{ \left(\eta^{i,j,k,l} h_{ijkl} + \eta^{i,j,k,r} h_{ijk} h_r[4] + \eta^{i,j,r,s} h_{ij} h_{rs}[6] + \eta^{i,r,s,t} h_i h_{rst}[4] \right. \right.
$$
$$
\left. + \eta^{r,s,t,u} h_{rstu} \right)
$$
$$
+ \left(\eta^i \eta^{j,k,l} h_{i,jkl} + \eta^i \eta^{j,k,r} h_{i,jk} h_r[3] + \eta^i \eta^{j,r,s} h_{i,j} h_{rs}[3] \right.
$$
$$
\left. + \eta^r \eta^{i,j,s} h_{ij} h_{r,s}[3] + \eta^r \eta^{i,s,t} h_i h_{r,st}[3] + \eta^r \eta^{s,t,u} h_{r,stu} \right)[4]
$$
$$
+ \left(\eta^{i,j} \eta^{k,l} h_{ij,kl} + \eta^{i,j} \eta^{k,r} h_{ij,k} h_r[4] + ... + \eta^{r,s} \eta^{t,u} h_{rs,tu} \right)[3]
$$
$$
+ \left(\eta^i \eta^j \eta^{k,l} h_{i,j,kl} + \eta^i \eta^r \eta^{j,s} h_{i,j} h_{r,s}[4] + \eta^r \eta^s \eta^{t,u} h_{r,s,tu} \right)[6] \right\}/4!
$$
$$
+ ... \, . \tag{5.24}
$$

This series may look a little complicated but in fact it has a very simple structure and is easy to extrapolate to higher-order terms. Such terms will be required later. Essentially every possible combination of terms appears, except that we have made use of the identities $h_{i,j,k} = 0$, $h_{i,j,k,l} = 0$ and so on.

On subtracting the logarithm of the marginal distribution of $X^{(2)}$, namely

$$
\log \phi_2(x^{(2)}; \lambda^{(2)}) + \eta^r h_r + \{ \eta^{r,s} h_{rs} + \eta^r \eta^s h_{r,s} \}/2! + ...
$$

we find after collecting terms that the conditional log density of $X^{(1)}$ given $X^{(2)}$ has a formal Edgeworth expansion in which the first four cumulants are

$$E(X^i|X^{(2)}) = \kappa^i$$

$$+ \eta^{i,r} h_r + \eta^{i,r,s} h_{rs}/2! + \eta^{i,r,s,t} h_{rst}/3! + \eta^{i,r,s,t,u} h_{rstu}/4! + \dots$$

$$+ \eta^r \eta^{i,s} h_{r,s} + \eta^r \eta^{i,s,t} h_{r,st}/2! + \eta^r \eta^{i,s,t,u} h_{r,stu}/3! + \dots$$

$$+ \eta^{i,r} \eta^{s,t} h_{r,st}/2! + \eta^{i,r} \eta^{s,t,u} h_{r,stu}/3! + \eta^{r,s} \eta^{i,t,u} h_{rs,tu}/(2!\,2!) + \dots$$

$$+ \eta^r \eta^{s,t} \eta^{i,u} h_{r,st,u}/2! + \dots$$

$$\mathrm{cov}(X^i, X^j|X^{(2)}) = \kappa^{i,j} + \eta^{i,j,r} h_r + \eta^{i,j,r,s} h_{rs}/2! + \eta^{i,j,r,s,t} h_{rst}/3! + \dots$$

$$+ \eta^r \eta^{i,j,s} h_{r,s} + \eta^r \eta^{i,j,s,t} h_{r,st}/2! + \dots$$

$$+ \eta^{i,r} \eta^{j,s} h_{r,s} + \eta^{i,r} \eta^{j,st}[2] h_{r,st}/2! + \eta^{r,s} \eta^{i,j,t} h_{rs,t}/2! + \dots$$

$$\mathrm{cum}(X^i, X^j, X^k|X^{(2)}) = \kappa^{i,j,k} + \eta^{i,j,k,r} h_r + \eta^{i,j,k,r,s} h_{rs}/2! + \dots$$

$$+ \eta^r \eta^{i,j,k,s} h_{r,s} + \eta^r \eta^{i,j,k,s,t} h_{r,st}/2! + \dots$$

$$+ \eta^{i,r} \eta^{j,k,s}[3] h_{r,s} + \eta^{i,r} \eta^{j,k,s,t}[3] h_{r,st}/2! + \eta^{r,s} \eta^{i,j,k,t} h_{rs,t}/2! + \dots$$

$$+ \eta^{r,s,i} \eta^{j,k,t}[3] h_{rs,t}/2! + \dots$$

$$\mathrm{cum}(X^i, X^j, X^k, X^l|X^{(2)}) = \kappa^{i,j,k,l} + \eta^{i,j,k,l,r} h_r + \eta^{i,j,k,l,r,s} h_{rs}/2! + \dots$$

$$+ \eta^r \eta^{i,j,k,l,s} h_{r,s} + \eta^{i,r} \eta^{j,k,l,s}[4] h_{r,s} + \eta^{i,j,r} \eta^{k,l,s}[3] h_{r,s} + \dots$$

$$+ \dots \,. \tag{5.25}$$

Of course, $\eta^{r,s,t} = \kappa^{r,s,t}$, $\eta^{r,s,i} = \kappa^{r,s,i}$, $\eta^{r,s} = \kappa^{r,s} - \lambda^{r,s}$ and so on, but the ηs have been retained in the above expansions in order to emphasize the essential simplicity of the formal series. We could, in fact, choose $\lambda^r = \kappa^r$, $\lambda^{r,s} = \kappa^{r,s}$ giving $\eta^r = 0$, $\eta^{r,s} = 0$. This choice eliminates many terms and greatly simplifies computations but it has the effect of destroying the essential simplicity of the pattern of terms in (5.25). Details are given in the following section.

5.6.2 *Asymptotic expansions*

If X is a standardized sum of n independent random variables, we may write the cumulants of X as 0, $\kappa^{\alpha,\beta}$, $n^{-1/2}\kappa^{\alpha,\beta,\gamma}$, $n^{-1}\kappa^{\alpha,\beta,\gamma,\delta}$ and so on. Suppose for simplicity that the components $X^{(1)}$ and $X^{(2)}$ are uncorrelated so that $\kappa^{i,r} = 0$. Then, from (5.25), the conditional cumulants of $X^{(1)}$ given $X^{(2)} = x^{(2)}$ have the following expansions up to terms of order $O(n^{-1})$.

$$E(X^i | X^{(2)}) = \kappa^i + n^{-1/2}\kappa^{i,r,s}h_{rs}/2!$$
$$+ n^{-1}\{\kappa^{i,r,s}\kappa^{t,u,v}h_{rs,tuv}/(3!\,2!) + \kappa^{i,r,s,t}h_{rst}/3!\}$$

$$\mathrm{cov}(X^i, X^j | X^{(2)}) = \kappa^{i,j} + n^{-1/2}\kappa^{i,j,r}h_r + n^{-1}\{\kappa^{i,j,r,s}h_{rs}/2!$$
$$+ \kappa^{i,j,r}\kappa^{s,t,u}h_{r,stu}/3! + \kappa^{i,r,s}\kappa^{j,t,u}h_{rs,tu}/(2!\,2!)\}$$

$$\mathrm{cum}(X^i, X^j, X^k | X^{(2)}) = n^{-1/2}\kappa^{i,j,k}$$
$$+ n^{-1}\{\kappa^{i,j,k,r}h_r + \kappa^{i,j,r}\kappa^{k,s,t}[3]h_{r,st}/2!\}$$

$$\mathrm{cum}(X^i, X^j, X^k, X^l | X^{(2)}) = n^{-1}\{\kappa^{i,j,k,l} + \kappa^{i,j,r}\kappa^{k,l,s}[3]h_{r,s}\}.$$

In the above expansions, the Hermite tensors are to be calculated using the exact mean vector κ^r and covariance matrix $\kappa^{r,s}$ of $X^{(2)}$. Note that, to the order given, the conditional mean is a cubic function of $X^{(2)}$, the conditional covariances are quadratic, the conditional skewnesses are linear and the conditional kurtosis is constant, though not the same as the unconditional kurtosis. All higher-order cumulants are $O(n^{-3/2})$ or smaller.

5.7 Normalizing transformation

In the multivariate case, there is an infinite number of smooth transformations, $g(.)$, that make the distribution of $Y = g(X)$ normal to a high order of approximation. Here, in order to ensure a unique solution, at least up to choice of signs for the components, we ask that the transformation be *triangular*. In other words, Y^1 is required to be a function of X^1 alone, Y^2 is required to be a function of the pair X^1, X^2, and so on. Algebraically, this condition may be expressed by writing

$$Y^1 = g_1(X^1)$$
$$Y^2 = g_2(X^1, X^2)$$
$$Y^r = g_r(X^1, ..., X^r) \qquad (r = 1, ..., p).$$

To keep the algebra as simple as possible without making the construction trivial, it is assumed that X is a standardized random variable with cumulants

$$0, \qquad \delta^{ij}, \qquad n^{-1/2}\kappa^{i,j,k}, \qquad n^{-1}\kappa^{i,j,k,l}$$

and so on, decreasing in powers of $n^{1/2}$. Thus, X is standard normal to first order: the transformed variable, Y is required to be standard normal with error $O(n^{-3/2})$.

The derivation of the normalizing transformation is unusually tedious and rather unenlightening. For that reason, we content ourselves with a statement of the result, which looks as follows.

$$Y^i = X^i - n^{-1/2}\left\{3\kappa^{i,r,s}h_{rs} + 3\kappa^{i,i,r}h_i h_r + \kappa^{i,i,i}h_{ii}\right\}/3!$$

$$- n^{-1}\left\{4\kappa^{i,r,s,t}h_{rst} + 6\kappa^{i,i,r,s}h_{rs}h_i + 4\kappa^{i,i,i,r}h_{ii}h_r + \kappa^{i,i,i,i}h_{iii}\right\}/4!$$

$$+ n^{-1}\left\{(36\kappa^{\alpha,i,r}\kappa^{\alpha,s,t} + 18\kappa^{i,i,r}\kappa^{i,s,t})h_{rst}\right.$$

$$+ (18\kappa^{\alpha,i,i}\kappa^{\alpha,r,s} + 12\kappa^{i,i,i}\kappa^{i,r,s} + 36\kappa^{\alpha,i,r}\kappa^{\alpha,i,s} + 27\kappa^{i,i,r}\kappa^{i,i,s})h_{rs}h$$

$$+ (36\kappa^{\alpha,i,i}\kappa^{\alpha,i,r} + 30\kappa^{i,i,i}\kappa^{i,i,r})h_{ii}h_r$$

$$+ (9\kappa^{\alpha,i,i}\kappa^{\alpha,i,i} + 8\kappa^{i,i,i}\kappa^{i,i,i})h_{iii}\left.\right\}/72$$

$$+ n^{-1}\left\{(36\kappa^{\alpha,\beta,i}\kappa^{\alpha,\beta,r} + 36\kappa^{\alpha,i,i}\kappa^{\alpha,i,r} + 12\kappa^{i,i,i}\kappa^{i,i,r})h_r\right.$$

$$+ (18\kappa^{\alpha,\beta,i}\kappa^{\alpha,\beta,i} + 27\kappa^{\alpha,i,i}\kappa^{\alpha,i,i} + 10\kappa^{i,i,i}\kappa^{i,i,i})h_i\left.\right\}/72 \qquad (5.2()$$

In the above expression, all sums run from 1 to $i-1$. Greek letters repeated as superscripts are summed but Roman letters are not. In other words, $\kappa^{\alpha,\beta,i}\kappa^{\alpha,\beta,i}$ is a shorthand notation for

$$\sum_{\alpha=1}^{i-1}\sum_{\beta=1}^{i-1}\kappa^{\alpha,\beta,i}\kappa^{\alpha,\beta,i}.$$

In addition, the index i is regarded as a fixed number so that

$$\kappa^{i,i,i,r}h_{ii}h_r = \sum_{r=1}^{i-1}\kappa^{i,i,i,r}h_{ii}h_r.$$

Fortunately, the above polynomial transformation simplifies considerably in the univariate case because most of the terms are null. Reverting now to power notation, we find

$$Y = X - \rho_3(X^2 - 1)/6 - \rho_4(X^3 - 3X)/24 + \rho_3^2(4X^3 - 7X)/36$$

to be the polynomial transformation to normality. In this expression, we have inserted explicit expressions for the Hermite polynomials. In particular, $4X^3 - 7X$ occurs as the combination $4h_3 + 5h_1$.

5.8 Bibliographic notes

The terms 'Edgeworth series' and 'Edgeworth expansion' stem from the paper by Edgeworth (1905). Similar series had previously been investigated by Chebyshev, Charlier, and Thiele (1897): Edgeworth's innovation was to group the series inversely by powers of the sample size rather than by the degree of the Hermite polynomial.

For a historical perspective on Edgeworth's contribution to Statistics, see the discussion paper by Stigler (1978).

Jeffreys (1966, Section 2.68) derives the univariate Edgeworth expansion using techniques similar to those used here.

Wallace (1958) gives a useful discussion in the univariate case, of Edgeworth series for the density and Cornish-Fisher series for the percentage points. See also Cornish & Fisher (1937).

Proofs of the validity of Edgeworth series can be found in the books by Cramér (1937) and Feller (1971). Esseen (1945) and Bhattacharya & Ranga-Rao (1976) give extensions to the lattice case. See also Chambers (1967) or Bhattacharya & Ghosh (1978).

Skovgaard (1981a) discusses the conditions under which a transformed random variable has a density that can be approximated by an Edgeworth series.

Michel (1979) discusses regularity conditions required for the validity of Edgeworth expansions to conditional distributions.

The notation used here is essentially the same as that used by Amari & Kumon (1983): see also Amari (1985). Skovgaard (1986) prefers to use coordinate-concealing notation for conceptual reasons and to deal with the case where the eigenvalues of the covariance matrix may not tend to infinity at equal rates.

5.9   Further results and exercises 5

5.1 Show, under conditions to be stated, that if
$$f_X(x;\kappa) = f_0(x) + \eta^i f_i(x) + \eta^{ij} f_{ij}(x)/2! + \eta^{ijk} f_{ijk}(x)/3! + \dots$$
then the moment generating function of $f_X(x;\kappa)$ is
$$M_0(\xi)\{1 + \xi_i\eta^i + \xi_i\xi_j\eta^{ij}/2! + \xi_i\xi_j\xi_k\eta^{ijk}/3! + \dots\}$$
where $M_0(\xi)$ is the moment generating function of $f_0(x)$.

5.2 Using expansion (5.2) for the density, derive expansion (5.4) for the log density.

5.3 Give a heuristic explanation for the formal similarity of expansions (5.2) and those in Section 5.2.3.

5.4 Show that any generalized Hermite tensor involving β indices partitioned into α blocks, is of degree $\beta - 2\alpha - 2$ in x or is identically zero if $\beta - 2\alpha - 2 < 0$.

5.5 If $f_X(x)$ is the density function of $X^1, \dots, X^p$, show that the density of
$$Y^r = a^r + a_i^r X^i$$
is
$$f_Y(y) = J f_X\{b_r^i(y^r - a^r)\},$$
where $b_r^i a_j^r = \delta_j^i$ and J is the determinant of b_r^i. Hence show that the partial derivatives of $\log f_X(x)$ are Cartesian tensors.

5.6 Show that the mode of a density that can be approximated by an Edgeworth series occurs at
$$\hat{x}^i = -\kappa^{i,j,k}\kappa_{j,k}/2 + O(n^{-3/2}).$$

5.7 Show that the median of a univariate density that can be approximated by an Edgeworth series occurs approximately at the point
$$\hat{x} = \frac{-\kappa_3}{6\kappa_2}.$$
Hence show that, to the same order of approximation, in the univariate case,
$$\frac{(\text{mean} - \text{median})}{(\text{mean} - \text{mode})} = \frac{1}{3}$$
(Haldane, 1942). See also Haldane (1948) for a discussion of medians of multivariate distributions.

5.8 Let X be a normal random variable with mean vector λ^r and covariance matrix $\lambda^{r,s}$. Define

$$h^r = h^r(x; \lambda), \quad h^{rs}(x; \lambda), \ldots$$

to be the Hermite tensors based on the same normal distribution, i.e.,

$$h^r = x^r - \lambda^r$$
$$h^{rs} = h^r h^s - \lambda^{r,s}$$

and so on as in (5.7). Show that the random variables

$$h^r(X), \quad h^{rs}(X), \quad h^{rst}(X), \ldots$$

have zero mean and are uncorrelated.

5.9 Using the notation established in the previous exercise, show that

$$\text{cum}\left(h^{rs}(X), h^{tu}(X), h^{vw}(X)\right) = \lambda^{r,r}\lambda^{s,v}\lambda^{t,w}[8]$$
$$\text{cum}\left(h^r(X), h^s(X), h^{tu}(X)\right) = \lambda^{r,t}\lambda^{s,u}[2]$$
$$\text{cum}\left(h^r(X), h^{st}(X), h^{uvw}(X)\right) = \lambda^{r,u}\lambda^{s,v}\lambda^{t,w}[6].$$

Give an expression for the cumulant corresponding to an arbitrary partition of the indices.

5.10 Suppose now that X has cumulants κ^r, $\kappa^{r,s}$, $\kappa^{r,s,t}$,... and that the Hermite tensors are based on the normal density with mean λ^r and covariance matrix $\lambda^{r,s}$. Show that

$$E\{h^r(X)\} = \eta^r$$
$$E\{h^{rs}(X)\} = \eta^{rs}$$
$$E\{h^{rst}(X)\} = \eta^{rst}$$

and so on, where the ηs are defined in Section 5.2.1.

5.11 Using the notation established in the previous exercise, show that

$$\text{cov}\left(h^r(X), h^s(X)\right) = \kappa^{r,s}$$
$$\text{cov}\left(h^r(X), h^{st}(X)\right) = \kappa^{r,s,t} + \eta^s\kappa^{r,t}[2]$$
$$\text{cov}\left(h^{rs}(X), h^{tu}(X)\right) = \kappa^{r,s,t,u} + \eta^r\kappa^{s,t,u}[4] + \kappa^{r,t}\kappa^{s,u}[2]$$
$$+ \eta^r\eta^t\kappa^{s,u}[4]$$
$$\text{cov}\left(h^r(X), h^{stu}(X)\right) = \kappa^{r,s,t,u} + \eta^s\kappa^{r,t,u}[3] + \kappa^{r,s}\eta^{t,u}[3]$$
$$+ \eta^s\eta^t\kappa^{r,u}[3].$$

5.12 Generalize the result of the previous exercise by showing that the joint cumulant corresponding to an arbitrary set of Hermite tensors involves a sum over connecting partitions. Describe the rule that determines the contribution of each connecting partition.

5.13 Show that

$$\int h_1(x)h_2(x)h_3(x)\phi(x)\,dx = 6$$

$$\int h_1(x)h_2(x)h_3(x)h_4(x)\phi(x)\,dx = 264$$

where $h_r(x)$ is the standard univariate Hermite polynomial of degree r and $\phi(x)$ is the standard normal density. [Hint: use the tables of connecting partitions.]

5.14 More generally, using the notation of the previous exercise, show that, for $i > j > k$,

$$\int h_i(x)h_j(x)h_k(x)\phi(x)\,dx = \frac{i!\,j!\,k!}{\{\frac{1}{2}(j+k-i)\}!\,\{\frac{1}{2}(i+k-j)\}!\,\{\frac{1}{2}(i+j-k)\}!}$$

when $j+k-i$ is even and non-negative, and zero otherwise, (Jarrett, 1973, p. 26).

5.15 Using (5.26) or otherwise, show that in the univariate case, where X is a standardized sum with mean zero, unit variance and so on, then

$$Y^* = X - \rho_3 X^2/6 - \rho_4 X^3/24 + \rho_3^2 X^3/9$$

has mean $-\rho_3/6$ and standard deviation

$$1 - \rho_4/8 + 7\rho_3^2/36$$

when terms of order $O(n^{-3/2})$ are ignored. Show also that

$$\frac{Y^* + \rho_3/6}{1 - \rho_4/8 + 7\rho_3^2/36} \sim N(0,1) + O(n^{-3/2})$$

5.16 Taking the definition of Y^* as given in the previous exercise, show that

$$W/2 = (Y^*)^2 = X^2/2 - \rho_3 X^3/3! - \{\rho_4 - 3\rho_3^2\}X^4/4!$$

has mean given by

$$E(W) = 1 + (5\rho_3^2 - 3\rho_4)/12 = 1 + b/n.$$

Deduce that

$$\frac{W}{1 + b/n} \sim \chi_1^2 + O(n^{-3/2}).$$

5.17 Using the equation following (5.26), show by reversal of series, that X may be expressed as the following polynomial in the normally distributed random variable Y

$$X = Y + \rho_3(Y^2 - 1)/6 + \rho_4(Y^3 - 3Y)/24 - \rho_3^2(2Y^3 - 5Y)/36.$$

Hence, express the approximate percentage points of X in terms of standard normal percentage points (Cornish & Fisher, 1937).

5.18 Let $X = Y + Z$ where Y has density function $f_0(y)$ and Z is independent of Y with moments η^i, η^{ij}, η^{ijk},.... . Show formally, that the density of X is given by

$$f_X(x) = E_Z\{f_0(x - Z)\}.$$

Hence derive the series (5.2) by Taylor expansion of $f_0(x)$. By taking $\eta^i = 0$, $\eta^{i,j} = 0$, and $f_0(x) = \phi(x; \kappa)$, derive the usual Edgeworth expansion for the density of X, taking care to group terms in the appropriate manner. (Davis, 1976).

CHAPTER 6

Saddlepoint approximation

6.1 Introduction

One difficulty that arises as a result of approximating the density function or log density function is that the density is not invariant under affine transformation of X. Any approximation, therefore, ought to have similar non-invariant properties and this requirement raises difficulties when we work exclusively with tensors. For example, if $Y^r = a^r + a_i^r X^i$ is an affine transformation of X, then the density function of Y at y is

$$f_Y(y) = |A|^{-1} f_X(x)$$

where $x^i = b_r^i(y^r - a^r)$, $b_r^i a_j^r = \delta_j^i$ and $|A|$ is the determinant of a_i^r, assumed to be non-zero. In the terminology of Thomas (1965), the density is said to be an invariant of *weight* 1: ordinary invariants have weight zero.

One way to exploit the advantages of working with invariants, at least under affine transformation, is to work with probabilities of sets rather than with probability densities. The difficulty then is to specify the sets in an invariant manner, for example as functions of the invariant polynomials

$$(x^i - \kappa^i)(x^j - \kappa^j)\kappa_{i,j}, \quad (x^i - \kappa^i)(x^j - \kappa^j)(x^k - \kappa^k)\kappa_{i,j,k},$$
$$(x^i - \kappa^i)\kappa^{j,k}\kappa_{i,j,k}, \quad (x^i - \kappa^i)(x^j - \kappa^j)\kappa^{k,l}\kappa_{i,j,k,l}$$

and so on. Another way, more convenient in the present circumstances, is to specify the probability density with respect to a so-called carrier measure on the sample space. This can always be done in such a way that the approximating density is invariant and the carrier measure transforms in such a way as to absorb the Jacobian of the transformation.

172

To be more explicit, suppose that the density function of X is written as the product

$$f_X(x) = |\kappa^{i,j}|^{-1/2} g(x).$$

Now let Y be an affine transformation of X as before. The covariance matrix, being a contravariant tensor, transforms to

$$\bar{\kappa}^{r,s} = a_i^r a_j^s \kappa^{i,j}.$$

Then the density of Y at y is simply

$$f_Y(y) = |\bar{\kappa}^{r,s}|^{-1/2} g(x).$$

Thus, with the inverse square root of the determinant of the covariance matrix playing the role of carrier measure, the density $g(x)$ is invariant under affine transformation of coordinates.

From this viewpoint, the usual normal-theory approximation uses the carrier measure $(2\pi)^{-p/2} |\kappa^{i,j}|^{-1/2}$ together with the invariant quadratic approximation

$$(x^i - \kappa^i)(x^j - \kappa^j)\kappa_{i,j}/2$$

for the negative log density. The Edgeworth approximation retains the carrier measure but augments the approximation for the negative log density by the addition of further invariant polynomial terms, namely

$$-\kappa^{i,j,k} h_{ijk}/3! - \kappa^{i,j,k,l} h_{ijkl}/4! - \kappa^{i,j,k}\kappa^{l,m,n} h_{ijk,lmn}[10]/6! - \ldots$$

In the Edgeworth system of approximation, the carrier measure is taken as constant throughout the sample space. Thus the whole burden of approximation lies on the invariant series approximation. Furthermore, this property of constancy of the carrier measure is preserved only under transformations for which the Jacobian is constant, i.e. under affine transformation alone. It is therefore appropriate to investigate the possibility of using alternative systems of approximation using non-polynomial invariants together with carrier measures that are not constant throughout the sample space. In all cases considered, the carrier measure is the square root of the determinant of a covariant tensor or, equivalently so far as transformation properties are concerned, the inverse square root of the determinant of a contravariant tensor.

6.2   Legendre transformation of $K(\xi)$

6.2.1   *Definition*

In what follows, it is convenient to consider the set of ξ-values, denoted by Ξ, for which $K(\xi) < \infty$ as being, in a sense, complementary or dual to the sample space of possible averages of identically distributed Xs. Thus, $\mathcal{X}$ is the interior of the convex hull of the sample space appropriate to a single X. In many cases, the two sample spaces are identical, but, particularly in the case of discrete random variables there is an important technical distinction. Both $\mathcal{X}$ and Ξ are subsets of p-dimensional space. It is essential, however, to think of the spaces as distinct and qualitatively different: if we are contemplating the effect of linear transformation on X, then vectors in $\mathcal{X}$ are contravariant whereas vectors in Ξ are covariant and inner products are invariant. To keep the distinction clear, it is sometimes helpful to think of Ξ as a parameter space even though we have not yet introduced any parametric models in this context.

Corresponding to the cumulant generating function $K(\xi)$ defined on Ξ, there is a dual function $K^*(x)$ defined on $\mathcal{X}$ such that the derivatives $K^r(\xi)$ and $K_r^*(x)$ are functional inverses. In other words, the solution in ξ to the p equations

$$K^r(\xi) = x^r \qquad (6.1)$$

is

$$\xi_i = K_i^*(x), \qquad (6.2)$$

where $K_i^*(x)$ is the derivative of $K^*(x)$. The existence of a solution to (6.1) has been demonstrated by Daniels (1954); see also Barndorff-Nielsen (1978, Chapter 5), where $K^*(x)$ is called the *conjugate function* of $K(\xi)$. Uniqueness follows from the observation that $K(\xi)$ is a strictly convex function (Exercise 6.2).

The function $K^*(x)$ is known as the Legendre or Legendre-Fenchel transformation of $K(\xi)$: it occurs in the theory of large deviations (Ellis, 1985, p.220), where $-K^*(x)$ is also called the *entropy* or *point entropy* of the distribution $f_X(x)$.

In the literature on convex analysis, the term convex conjugate is also used (Fenchel, 1949; Rockafellar, 1970, Sections 12, 26).

An alternative, and in some ways preferable definition of $K^*(x)$ is

$$K^*(x) = \sup_{\xi}\{\xi_i x^i - K(\xi)\}. \qquad (6.3)$$

To see that these two definitions are mutually consistent, we note
that (6.1) or (6.2) determines the stationary points of the function
in (6.3). Now write $h(x)$ for the maximum value, namely

$$h(x) = x^i K_i^*(x) - K(K_i^*(x)).$$

Differentiation with respect to x^i gives $h_i(x) = K_i^*(x)$, showing
that $K^*(x) = h(x) + \text{const}$. The constant is identified by (6.3) but
not by the previous definition. The turning point gives a maximum
because $K(\xi)$ is convex on Ξ. Note also that $K^*(x)$ is convex on
X and achieves its minimum value of zero at $x^i = \kappa^i$.

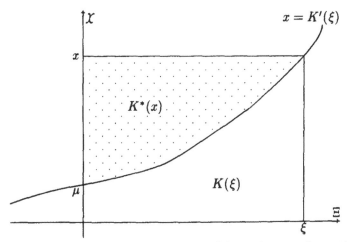

Figure 6.1: *Graphical construction of Legendre transformation
in the univariate case.*

Figure 6.1 illustrates graphically the construction of the Legendre transformation in the univariate case. The solid line gives the
graph of $K'(\xi)$ against ξ and has positive gradient since $K'' > 0$.
In the illustration, the intercept, which is equal to $E(X)$, is taken
to be positive. The area under the curve from zero to ξ is just the
cumulant generating function $K(\xi) = \int_0^\xi K'(t)dt$. For any given
value of x, ξx is the area of the rectangle whose opposite corners
are at the origin and (ξ, x). Evidently, from the geometry of the diagram, $\xi x - K(\xi)$ is maximized at the value ξ satisfying $K'(\xi) = x$.

The shaded area above the curve is equal to $\xi K'(\xi) - K(\xi)$. Regarded as a function of x, this is the Legendre transformation of $K(\xi)$. Equivalently, the graph reflected about the 45° line gives the inverse function of $K'(\xi)$. Integration from μ to x gives the area shaded above the line, showing that the two definitions given earlier are equivalent.

A similar geometrical description applies in the p-dimensional case with the two axes in Figure 6.1 replaced by the appropriate p-dimensional dual spaces. The graph is replaced by a mapping from Ξ to $\mathcal{X}$, but unfortunately, this is awkward to visualize even for $p = 2$ because 4 dimensions are involved.

Equations (6.1) or (6.2) identify corresponding points in Ξ and $\mathcal{X}$. In fact, every point in Ξ has a unique image in $\mathcal{X}$ given by (6.1) and conversely, every point in $\mathcal{X}$ has a unique image in Ξ given by (6.2). For example, the point $x^i = \kappa^i$ in $\mathcal{X}$ is identified with the point $\xi = 0$ in Ξ. The values of the two functions at these points are $K(0) = 0$ and $K^*(\kappa^i) = 0$ respectively.

Schematically, the relation between the first few derivatives of $K(\xi)$ and the dual function, $K^*(x)$, is as follows.

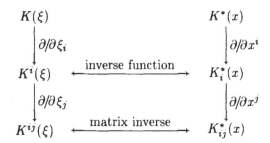

Apart from the choice of constant at the integration step, it is evident that the procedure can be reversed and that the Legendre transformation of $K^*(x)$ is just $K(\xi)$, i.e. $K^{**} = K$. Equivalently, we may show algebraically that

$$K(\xi) = \sup_{x}\{x^i\xi_i - K^*(x)\},$$

which is obvious from the diagram in Figure 6.1.

6.2.2 Applications

The following is a brief informal description mostly without proofs of the role of the Legendre transformation in approximating distributions. For a more formal and rigorous account of the theory in the context of large deviations, see Ellis (1985). For the connection with exponential family models, see Barndorff-Nielsen (1978).

Large deviations: The following inequality, which is central to much of the theory of large deviations in the univariate case, helps to explain the role played by the Legendre transformation. Let X be a real-valued random variable with mean μ and let $x > \mu$ be a given number. We show that

$$\text{pr}(X \geq x) \leq \exp\{-K^*(x)\}. \tag{6.4}$$

To derive this inequality, we first write the required probability in the form

$$\text{pr}(X - x \geq 0) = \text{pr}\{\exp(\xi(X - x)) \geq 1\},$$

which is valid for all $\xi > 0$. Thus, since $\exp(\xi x) \geq H(x)$, where $H(.)$ is the Heaviside function, it follows that

$$\text{pr}(X \geq x) \leq \inf_{\xi>0} \exp\{-\xi x + K(\xi)\}$$
$$= \exp\{-K^*(x)\}$$

and the inequality is proved.

More generally, for vector-valued X, it may be shown that, for any set A in $\mathcal{X}$,

$$\text{pr}(X \in A) \leq \exp\{I(A)\}$$

where

$$I(A) = \sup_{x \in A}\{-K^*(x)\}$$

is called the entropy of the set A. Evidently, this is a generalization of the univariate inequality. The multivariate inequality follows from the univariate inequality together with the observation that $A \subset B$ implies $I(A) \leq I(B)$.

To see the relevance of the above inequalities to the theory of large deviations, let $X_1, ..., X_n$ be independent and identically distributed with cumulant generating function $K(\xi)$ and let $\bar{X}_n$ be the sample average. Large deviation theory is concerned mainly with approximations for the probability of the event $\bar{X}_n \geq x$ where $x > \mu$ is a fixed value independent of n. It is not difficult to see from the law of large numbers that the event in question has negligible probability for large n: in fact, for fixed x, the probability decreases exponentially fast as $n \to \infty$. The central result due to Cramér (1938) and Chernoff (1952), which determines the exponential rate of decrease, is that

$$n^{-1} \log \text{pr}(\bar{X}_n \geq x) \to -K^*(x) \qquad (6.5)$$

as $n \to \infty$. Note that $nK^*(x)$ is the Legendre transformation of $\bar{X}_n$, implying, in effect, that the inequality (6.4) becomes increasingly sharp as $n \to \infty$. As a numerical device for approximating tail probabilities however, the above limit is rarely of sufficient accuracy for statistical purposes. Better approximations are described in Section 6.2.6.

The Cramér-Chernoff key limit theorem may be extended to vector-valued random variables in the following way. Let $\bar{X}_n \in R^p$ and $A \subset R^p$ be open and convex. The condition that A be open can often be dropped. Then the required limit may be written

$$n^{-1} \log \text{pr}(\bar{X}_n \in A) \to I(A) \qquad (6.6)$$

as $n \to \infty$. Note that if $E(X)$ lies in the interior of A then $I(A)$ is zero, which is consistent with the law of large numbers, $\text{pr}(\bar{X}_n \in A) \to 1$. Similarly, if $E(X)$ lies on the boundary of A, $I(A)$ is zero and $\text{pr}(\bar{X}_n \in A) \to \text{const} \leq \frac{1}{2}$. In other words, if $\text{pr}(\bar{X}_n \in A)$ tends to a finite limit, the limiting value cannot be deduced from the entropy limit (6.6).

Proofs of the limits (6.5) and (6.6) may be found in the books by Bahadur (1971) and Ellis (1985). Bahadur & Zabell (1979) give a number of generalizations.

Likelihood ratio statistics: Consider now the exponential family of distributions parameterized by θ in the form

$$f_X(x; \theta) = \exp\{\theta_i x^i - K(\theta)\} f_X(x). \qquad (6.7)$$

First note that since $K(.)$ is the cumulant generating function of the distribution $f_X(x)$, it follows that $f_X(x; \theta)$ is a probability distribution on X for all θ in Ξ. The cumulant generating function of $f_X(x; \theta)$ is $K(\xi + \theta) - K(\theta)$.

Suppose we wish to test the hypothesis that $\theta = 0$ based on an observed value x on X. The log likelihood ratio is

$$\log f_X(x; \theta) - \log f_X(x) = \theta_i x^i - K(\theta).$$

The maximized log likelihood ratio statistic, maximized over θ, gives

$$\sup_{\theta}\{\theta_i x^i - K(\theta)\} = K^*(x)$$

where the maximum occurs at $\hat{\theta}_i = K_i^*(x)$. In this setting, the Legendre transformation is none other than the maximized log likelihood ratio statistic. It follows then, by the usual asymptotic property of likelihood ratio statistics, that

$$2nK^*(\bar{X}_n) \sim \chi_p^2 + o(1)$$

and typically the error of approximation is $O(n^{-1})$. For a derivation of this result including an explicit expression for the $O(n^{-1})$ term, see Section 6.2.4.

Saddlepoint approximation: One reason for considering the Legendre transformation of $K(\xi)$ is that it is a function on X, invariant under affine transformation, that is useful for approximating the density of X, and hence, the density of any one-to-one function of X. In fact, the saddlepoint approximation, derived in Section 6.3, for the density of X may be written in the form

$$(2\pi)^{-p/2}|K_{rs}^*(x)|^{1/2}\exp\{-K^*(x)\}. \tag{6.8}$$

This approximation arises from applying the inversion formula to $K(\xi)$, namely

$$f_X(x) = (2\pi i)^{-p}\int_{c-i\infty}^{c+i\infty}\exp\{K(\xi) - \xi_i x^i\}\,d\xi.$$

The saddlepoint of the integrand is given by (6.1) and occurs at a point for which ξ_i is real. Approximation of the exponent in the neighbourhood of the saddlepoint by a quadratic function and integrating gives (6.8). The approximation can be justified as an asymptotic approximation if X is an average or standardized sum of n independent random variables where n is assumed to be large. From the point of view discussed in Section 6.1, approximation (6.8) uses the carrier measure $(2\pi)^{-p/2}|K^*_{rs}(x)|^{1/2}$ on $\mathcal{X}$, together with the invariant approximation $K^*(x)$ for the negative log density. Note that the carrier measure in this case is not constant on $\mathcal{X}$. This property permits us to consider non-linear transformations of X incorporating the non-constant Jacobian into the determinant and leaving the exponent in (6.8) unaffected. Of particular importance is the distribution of the random variable $Y_i = K^*_i(X)$. Direct substitution using (6.8) gives as the required approximation

$$(2\pi)^{-p/2}|K^{rs}(y)|^{1/2}\exp\{-K^*(x)\}.$$

In this case, $Y_i = \hat{\theta}_i$, the maximum likelihood estimate of the parameter in the exponential family derived from $f_X(x)$.

Greater accuracy in the asymptotic sense can be achieved by retaining the carrier measure and replacing the exponent in (6.8) by

$$-K^*(x) - (3\rho^*_4(x) - 4\rho^{*2}_{23}(x))/4! \qquad (6.9)$$

where

$$\rho^*_4 = K^*_{ijkl}K^{*ij}K^{*kl}$$
$$\rho^{*2}_{23} = K^*_{ijk}K^*_{rst}K^{*ir}K^{*js}K^{*kt}$$

and K^{*ij}, the matrix inverse of K^*_{ij}, is identical to $K^{ij}(\xi)$ with ξ satisfying (6.2). The additional correction terms in (6.9) arise from expanding the exponent, $K(\xi) - \xi_i x^i$, about the saddlepoint as far as terms of degree four in ξ. These correction terms may alternatively be expressed in terms of the derivatives of $K(\xi)$ evaluated at the saddlepoint. Thus (6.9) becomes

$$-K^*(x) + (3\rho_4(\xi) - 3\rho^2_{13}(\xi) - 2\rho^2_{23}(\xi))/4!$$

where, for example,

$$\rho^2_{13}(\xi) = K_{ijk}K_{rst}K^{ij}K^{kr}K^{st}$$

and the functions are evaluated at the saddlepoint. Evidently, $\rho^2_{13}(\xi)$ evaluated at $\xi = 0$ is identical to ρ^2_{13}, and similarly for the remaining invariants.

6.2.3 Some examples

In general, it is not easy to compute the Legendre transformation of an arbitrary cumulant generating function in terms of known functions. The following examples show that the calculation is feasible in a number of cases.

Multivariate normal density: The cumulant generating function for the normal distribution is

$$K(\xi) = \kappa^i \xi_i + \kappa^{i,j} \xi_i \xi_j / 2!.$$

The derivative, which is linear in ξ, may be inverted giving the Legendre transformation, which is

$$K^*(x) = (x^i - \kappa^i)(x^j - \kappa^j)\kappa_{i,j}/2!.$$

Evidently, since $K^*(x)$ is quadratic in x, K^*_{rs} is a constant equal to $\kappa_{r,s}$ and the saddlepoint approximation (6.8) is exact.

Gamma distribution: Suppose that X has the univariate gamma distribution with mean μ and variance μ^2/ν, where ν is the index or precision parameter. The cumulant generating function is

$$K(\xi) = -\nu \log(1 - \mu\xi/\nu).$$

Taylor expansion about $\xi = 0$ yields the higher-order cumulants $\kappa_r = (r-1)! \, \mu^r/\nu^{r-1}$. On solving the equation $K'(\xi) = x$, we find

$$\xi(x) = \frac{\nu}{\mu} - \frac{\nu}{x}.$$

Integration gives

$$K^*(x) = \nu \left\{ \frac{x - \mu}{\mu} - \log \left(\frac{x}{\mu} \right) \right\},$$

which is exactly one half of the deviance contribution for gamma models (McCullagh & Nelder, 1983, p.153).

Since the second derivative of $K^*(x)$ is ν/x^2, it follows that the saddlepoint approximation for the density is

$$\frac{\left(\dfrac{\nu x}{\mu}\right)^\nu \exp\left(\dfrac{-\nu x}{\mu}\right) \dfrac{1}{x} \, dx}{(2\pi)^{1/2} \nu^{\nu - 1/2} \exp(-\nu)}.$$

This approximation differs from the exact density only to the extent that Stirling's approximation is used in the denominator in place of $\Gamma(\nu)$. In this example, this kind of defect can be corrected by choosing a multiplicative constant in such a way that the total integral is exactly one. This re-normalized saddlepoint approximation is exact for all gamma distributions.

Poisson distribution: The cumulant generating function is $\mu(\exp(\xi) - 1)$ showing that all cumulants are equal to μ. Following the procedure described above, we find

$$K^*(x) = x \log(x/\mu) - (x - \mu).$$

In the literature on log linear models, $2K^*(x)$ is more familiar as the deviance contribution or contribution to the likelihood ratio test statistic. The second derivative of K^* is just x^{-1}. Thus, the saddlepoint approximation gives

$$\frac{\exp(-\mu)\mu^x}{(2\pi)^{1/2}x^{x+1/2}\exp(-x)}.$$

Again, this differs from the exact distribution in that Stirling's approximation has been used in place of $x!$. Unlike the previous example, however, re-normalization cannot be used to correct for this defect because x is not a constant and the *relative* weights given to the various values of x by the approximation are not exact.

6.2.4 *Transformation properties*

The Legendre transformation possesses a number of invariance properties that help to explain its role in approximating densities. First, let $Y^r = a^r + a_i^r X^i$ be an affine transformation of X. Any point in $\mathcal{X}$ can be identified either by its x-coordinates or by its y-coordinates, and the two are, in a sense, equivalent. The inverse transformation may be written $X^i = b_r^i(Y^r - a^r)$. It is easy to show that the Legendre transformation of $K_Y(\xi)$ is identical to the Legendre transformation of $K_X(\xi)$. To demonstrate this fact, we note first that

$$K_Y(\xi) = K_X(\xi_i a_j^i) + \xi_i a^i$$

Differentiation with respect to ξ_i and equating the derivative to y^i gives

$$K_X^i(\xi_i a_j^i) = b_j^i(y^j - a^j) = x^i.$$

Hence
$$\xi_i = b_i^j K_j^*(x),$$

which is the derivative with respect to y^i of $K^*(x)$.

For an alternative proof using definition (6.3) directly, see Exercise 6.9.

More generally, if a_i^r does not have full rank, it is easily shown that
$$K_X^*(x) \geq K_Y^*(y)$$

for all $x \in \mathcal{X}$ and y in the image set. To derive this inequality, we work directly from definition (6.3), giving

$$
\begin{aligned}
K_Y(y) &= \sup_{\varsigma}\{\varsigma_r y^r - K_Y(\varsigma)\} \\
&= \sup_{\varsigma}\{(\varsigma_r a_i^r)x^i - K_X(\varsigma_r a_i^r)\} \\
&\leq \sup_{\xi}\{\xi_i x^i - K_X(\xi)\} = K^*(x).
\end{aligned}
$$

This inequality is intuitively obvious from the interpretation of $K^*(x)$ as a log likelihood ratio statistic or *total deviance* in the sense of McCullagh & Nelder (1983).

A second important property of the Legendre transformation concerns its behaviour under exponential tilting of the density $f_X(x)$. The exponentially tilted density is

$$f_X(x;\theta) = \exp\{\theta_i x^i - K(\theta)\}f_X(x)$$

where θ is the tilt parameter, otherwise known as the canonical parameter of the exponential family. The effect of exponential tilting on the negative log density is to transform from $-\log f_X(x)$ to

$$-\log f_X(x) - \theta_i x^i + K(\theta).$$

To see the effect on the Legendre transform, we write

$$K^*(x;\theta) = \sup_{\xi}\{\xi_i x^i - K(\xi + \theta) + K(\theta)\} \qquad (6.10)$$

where $K(\xi + \theta) - K(\theta)$ is the cumulant generating function of the tilted density (6.7). An elementary calculation gives

$$K^*(x;\theta) = K^*(x) - \theta_i x^i + K(\theta) \qquad (6.11)$$

Table 6.1   *Transformation properties of the Legendre transform*

Transformation	Cumulant generating function	Legendre transform
Identity	$K(\xi)$	$K^*(y)$
Convolution (sum)	$nK(\xi)$	$nK^*(y/n)$
Average	$nK(\xi/n)$	$nK^*(y)$
Location shift	$K(\xi) + \xi_i a^i$	$K^*(y-a)$
Exponential tilt	$K(\xi + \theta) - K(\theta)$	$K^*(y) - \theta_i y^i + K(\theta)$
Affine	$K(a_j^i \xi_i) + \xi_i a^i$	$K^*(b_j^i(y^j - a^j))$

so that, under this operation, the Legendre transformation behaves exactly like the negative log density.

The above transformation properties have important consequences when the Legendre transform is used to approximate the negative log density function. In particular, whatever the error incurred in using the saddlepoint approximation to $f_X(x)$, the same error occurs uniformly for all θ in the saddlepoint approximation to $f_X(x; \theta)$, the exponentially tilted density.

Table 6.1 provides a list of some of the more important transformation properties of the Legendre transformation.

6.2.5   *Expansion of the Legendre transformation*

In this section, we derive the Taylor expansion of $K^*(x)$ about $x^i = \kappa^i$. The expansion is useful for finding the approximate distribution of the random variables $K^*(X)$ and $K_i^*(X)$. As shown in Section 6.2.2, the first of these is a likelihood ratio statistic: the second is the maximum likelihood estimate of the canonical parameter in the exponential family generated by $f_X(x)$.

The first step is to expand the derivative of the cumulant generating function in a Taylor series about the origin giving

$$K^r(\xi) = \kappa^r + \kappa^{r,i}\xi_i + \kappa^{r,i,j}\xi_i \xi_j/2! + \kappa^{r,i,j,k}\xi_i \xi_j \xi_k/3! + \dots .$$

On solving the equation $K^r(\xi) = x^r$ by reversal of series and substituting $z^i = x^i - \kappa^i$, we find

$$\xi_i = \kappa_{i,r} z^r - \kappa_{i,r,s} z^r z^s/2!$$
$$- \{\kappa_{i,r,s,t} - \kappa_{i,r,u}\kappa_{s,t,v}\kappa^{u,v}[3]\}z^r z^s z^t/3! + \dots .$$

After integrating term by term and after applying the boundary condition $K^*(\kappa^r) = 0$, we find

$$K^*(x) = \kappa_{i,j} z^i z^j/2! - \kappa_{i,j,k} z^i z^j z^k/3!$$
$$- \{\kappa_{i,j,k,l} - \kappa_{i,j,r}\kappa_{k,l,s}\kappa^{r,s}[3]\} z^i z^j z^k z^l/4!$$
$$- \{\kappa_{i,j,k,l,m} - \kappa_{i,j,k,r}\kappa_{l,m,s}\kappa^{r,s}[10]$$
$$+ \kappa_{i,j,r}\kappa_{k,l,s}\kappa_{m,t,u}\kappa^{r,t}\kappa^{s,u}[15]\} z^i...z^m/5!$$
$$- ... . \tag{6.12}$$

Note that the invariance of $K^*(x)$ follows immediately from the above expansion.

The close formal similarity between (6.12) and expansion (5.5) for the negative log density is remarkable: in fact, the two expansions are identical except that certain polynomial terms in (6.12) are replaced by generalized Hermite tensors in (5.5). For example, in (5.5) the coefficient of $\kappa_{i,j,k}$ is $-h^{ijk}/3!$ while the coefficient of $\kappa_{i,j,r}\kappa_{k,l,s}$ is $-h^{ijr,kls}[10]/6!$, which is quartic in x.

If $\bar{X}_n$ is the mean of n identically distributed Xs, then $Z^r = \bar{X}_n^r - \kappa^r$ is $O_p(n^{-1/2})$. A straightforward calculation using (6.12) reveals that the likelihood ratio statistic has expectation

$$E(2nK^*(\bar{X})) = p\{1 + (3\bar{\rho}_{13}^2 + 2\bar{\rho}_{23}^2 - 3\bar{\rho}_4)/(12n)\} + O(n^{-2})$$
$$= p\{1 + b/n\} + O(n^{-2}), \tag{6.13}$$

where the ρs are the invariant cumulants of X_1. It is evident from expansion (6.12) that $2nK^*(\bar{X}_n)$ has a limiting χ_p^2 distribution because the terms beyond the first are negligible. With a little extra effort, it can be shown that all cumulants of $\{1+b/n\}^{-1}2nK^*(\bar{X}_n)$ are the same as those of χ_p^2 when terms of order $O(n^{-2})$ are neglected. This adjustment to the likelihood ratio statistic is known as the Bartlett factor.

The key idea in the proof is to write the likelihood ratio statistic as a quadratic form in derived variables Y. Thus

$$2nK^*(\bar{X}_n) = Y^r Y^s \kappa_{r,s}$$

where Y is a polynomial in Z. Rather intricate, but straightforward calculations then show that Y has third and fourth cumulants of orders $O(n^{-3/2})$ and $O(n^{-2})$ instead of the usual $O(n^{-1/2})$ and

$O(n^{-1})$. Higher-order cumulants are $O(n^{-3/2})$ or smaller. To this order of approximation, therefore, Y has a normal distribution and $2nK^*(\bar{X}_n)$ has a non-central χ_p^2 distribution for which the rth cumulant is

$$\kappa_r = \{1 + b/n\}^r 2^{r-1} p(r-1)! + O(n^{-2}).$$

Thus the correction factor $1 + b/n$, derived as a correction for the mean, corrects all the cumulants simultaneously to the same order of approximation. See also Exercises 6.16 and 6.17.

Details of the proof are not very interesting but can be found in McCullagh (1984b, Section 7.1).

6.2.6 *Tail probabilities in the univariate case*

All of the calculations of the previous section apply equally to the univariate case, particularly (6.12) and (6.13). In order to find a suitably accurate approximation to the distribution function $\text{pr}(X \le x)$, or to the tail probability $\text{pr}(X \ge x)$, there are several possible lines of attack. The first and most obvious is to attempt to integrate the saddlepoint approximation directly. This method has the advantage of preserving the excellent properties of the saddlepoint approximation but it is cumbersome because the integration must be carried out numerically. An alternative method that is less cumbersome but retains high accuracy is to transform from X to a new scale defined by

$$T(x) = \pm\{2K^*(x)\}^{1/2}$$

where the sign of $T(x)$ is the same as that of $x - \mu$. The random variable $T(X)$ is sometimes called the *signed likelihood ratio statistic* although the term is appropriate only in the context of the exponential family of densities (6.7).

From the discussion in the previous section and from Exercises 6.16, 6.17, it may be seen that $T(X)$ is nearly normally distributed with mean and variance

$$E(T(X)) \simeq -\rho_3/6$$
$$\text{var}(T(X)) \simeq 1 + (14\rho_3^2 - 9\rho_4)/36,$$

where $\rho_3 = \kappa_3/\kappa_2^{3/2}$ is the usual univariate standardized measure of skewness of X. In fact, the cumulants of $T(X)$ differ from those of the normal distribution having the above mean and variance, by $O(n^{-3/2})$ when X is a mean or total of n independent observations. Since $T(x)$ is increasing in x, it follows that

$$\operatorname{pr}(X \geq x) = \operatorname{pr}\{T(X) \geq T(x)\}$$
$$\simeq 1 - \Phi\left(\frac{T(x) + \rho_3/6}{1 + (14\rho_3^2 - 9\rho_4)/72}\right). \qquad (6.14)$$

A similar, but not identical, formula was previously given by Lugannani & Rice (1980). In fact, Daniels's (1987) version of the Lugannani-Rice formula may be written

$$\operatorname{pr}(X \geq x) \simeq 1 - \Phi(T(x)) + \phi(T(x))\left(\frac{1}{S(x)} - \frac{1}{T(x)}\right), \qquad (6.15)$$

where $K'(\hat{\xi}) = x$ defines the saddlepoint and $S(x) = \hat{\xi}\{K''(\hat{\xi})\}^{1/2}$ is a kind of Wald statistic. Note that the standard error is calculated under the supposition that the mean of X is at x rather than at μ.

Admittedly, approximation (6.14) has been derived in a rather dubious fashion. Neither the nature of the approximation nor the magnitude of the error have been indicated. In fact, the approximation may be justified as an asymptotic expansion if X is a sum or average of n independent random variables, in which case ρ_3 is $O(n^{-1/2})$ and ρ_4 is $O(n^{-1})$. The error incurred in using (6.14) for normal deviations is typically $O(n^{-3/2})$. By normal deviations, we mean values of x for which $K^*(x)$ is $O(1)$, or equivalently, values of x that deviate from $E(X)$ by a bounded multiple of the standard deviation. This range includes the bulk of the probability. Further adjustments to (6.14) are required in the case of discrete random variables: it often helps, for example, to make a correction for continuity.

For large deviations, the approximation of Lugannani & Rice has relative error of order $O(n^{-1})$. On the other hand, the relative error of (6.14) is $O(1)$, but despite this substantial asymptotic inferiority, (6.14) is surprisingly accurate even for moderately extreme tail probability calculations of the kind that occur in significance testing.

For further discussion of the above and related approximations, the reader is referred to Daniels (1987).

6.3   Derivation of the saddlepoint approximation

The most direct derivation of the saddlepoint approximation, by inversion of the cumulant generating function, was described in Section 6.2.2, if only briefly. A simpler method of derivation is to apply the Edgeworth approximation not to the density $f_X(x)$ directly, but to an appropriately chosen member of the conjugate family or exponential family

$$f_X(x;\theta) = \exp\{\theta_i x^i - K(\theta)\} f_X(x). \qquad (6.16)$$

We aim then, for each value of x, to chose the most advantageous value of θ in order to make the Edgeworth approximation to $f_X(x;\theta)$ as accurate as possible. This is achieved by choosing $\theta = \hat{\theta}(x)$ in such a way that x is at the mean of the conjugate density under $\hat{\theta}$. In other words, we chose $\hat{\theta}$ such that

$$K^r(\hat{\theta}) = x^r \quad \text{or} \quad \hat{\theta}_r = K_r^*(x).$$

As the notation suggests, $\hat{\theta}$ is the maximum likelihood estimate of θ based on x in the family (6.16).

From (5.5) or (5.10), the Edgeworth approximation for the log density at the mean may be written

$$-\tfrac{1}{2}p\log(2\pi) - \tfrac{1}{2}\log|\kappa^{r,s}| + (3\rho_4 - 3\rho_{13}^2 - 2\rho_{23}^2)/4! + \ldots .$$

Taking logs in (6.16) gives

$$\log f_X(x) = -K^*(x) + \log f_X(x;\hat{\theta}).$$

Applying the Edgeworth expansion to the second term on the right gives

$$\log f_X(x) = -\tfrac{1}{2}p\log(2\pi) - \tfrac{1}{2}\log|K^{rs}(\hat{\theta})|$$
$$- K^*(x) + (3\rho_4(\hat{\theta}) - 3\rho_{13}^2(\hat{\theta}) - 2\rho_{23}^2(\hat{\theta}))/4! + \ldots$$
$$= -\tfrac{1}{2}p\log(2\pi) + \tfrac{1}{2}\log|K_{rs}^*(x)|$$
$$- K^*(x) - (3\rho_4^*(x) - 4\rho_{23}^{*2}(x))/4! + \ldots .$$

The first three terms above constitute the saddlepoint approximation for the log density: the fourth term is a correction term that is $O(n^{-1})$ in large samples. Terms that are ignored are $O(n^{-2})$.

Alternatively, and sometimes preferably, we may write the approximate density function as

$$(2\pi c)^{-p/2}|K_{rs}^*|\exp\{-K^*(x)\} \qquad (6.17)$$

where c is a constant chosen to make the integral equal to one. To a first order of approximation, we may write

$$\log(c) = (3\bar{\rho}_{13}^2 + 2\bar{\rho}_{23}^2 - 3\bar{\rho}_4)/12.$$

Thus, $\log(c)$ is just the Bartlett adjustment term defined in (6.13). The error of approximation is $O(n^{-3/2})$ when the above approximation is used for the constant of integration.

The advantages of the saddlepoint approximation over the Edgeworth series are mainly connected with accuracy. Although both approximations are asymptotic, the saddlepoint approximation is often sufficiently accurate for statistical purposes even when n is small, less than 10, say. In addition, the saddlepoint approximation retains high relative accuracy over the whole range of possible values of x. The Edgeworth approximation, on the other hand, is valid only for values of $\bar{X}$ that deviate from $E(\bar{X})$ by $O(n^{-1/2})$. The implication of this restriction is that the Edgeworth series may not be of adequate accuracy to judge the probability of unusual events.

On the negative side, the saddlepoint approximation applies to the density, and if tail probability calculations are required, integration is necessary. Unfortunately, the saddlepoint approximation, unlike the Edgeworth series, cannot usually be integrated analytically. Numerical integration is one answer, but this is often cumbersome. An alternative and more convenient solution is described in Section 6.2.6. A second argument against the saddlepoint approximation in favour of the Edgeworth series is that in order to compute the saddlepoint, it is necessary to have an explicit formula for the cumulant generating function. To use the Edgeworth series, on the other hand, it is necessary only to know the first few cumulants, and these can often be computed without knowing the generating function. There may, in fact, be no closed form expression for the generating function: see, for example, Exercise 2.30. In short, the Edgeworth series is often easier to use in practice but is usually inferior in terms of accuracy, particularly in the far tails of the distribution.

6.4   Approximation to conditional distributions

6.4.1   *Conditional density*

Suppose that we require an approximation for the conditional distribution of a statistic X_2 given that $X_1 = x_1$. Both components may be vector valued. Calculations of this kind arise in a number of important areas of application. The following are a few examples.

(i) Elimination of nuisance parameters by conditioning, particularly where matched retrospective designs are used to study factors that influence the incidence of rare diseases (Breslow & Day, 1980, Chapter 7).

(ii) Conditioning to take account of the observed value of an ancillary statistic (Cox, 1958).

(iii) Testing for goodness of fit when the model to be tested contains unknown parameters (McCullagh, 1985).

The simplest and most natural way to proceed from the cumulant generating functions $K_{X_1 X_2}(.)$ and $K_{X_1}(.)$ is to compute the corresponding Legendre transformations, $K^*_{X_1 X_2}(x_1, x_2)$ and $K^*_{X_1}(x_1)$. The saddlepoint approximation is then used twice, once for the joint density and once for the marginal density of X_1. Thus,

$$f_{X_1 X_2}(x_1, x_2) \simeq c_{12} |K^*_{X_1 X_2; rs}|^{1/2} \exp\{-K^*_{X_1 X_2}(x_1, x_2)\}$$
$$f_{X_1}(x_1) \simeq c_1 |K^*_{X_1; rs}|^{1/2} \exp\{-K^*_{X_1}(x_1)\},$$

where c_{12} and c_1 are normalizing constants. On subtracting the approximate log densities, we find

$$\begin{aligned}
\log f_{X_2|X_1}(x_2|x_1) \simeq &\log c_{12} - \log c_1 \\
&+ \tfrac{1}{2} \log |K^*_{X_1 X_2; rs}| - \tfrac{1}{2} \log |K^*_{X_1; rs}| \\
&- K^*_{X_1 X_2}(x_1, x_2) + K^*_{X_1}(x_1).
\end{aligned} \tag{6.18}$$

In large samples, the error of approximation is $O(n^{-3/2})$ provided that the constants of integration c_1 and c_{12} are appropriately chosen.

Approximation (6.18) is sometimes called the *double saddlepoint approximation*. It is not the same as applying the saddlepoint approximation directly to the conditional cumulant generating function of X_2 given $X_1 = x_1$. For an example illustrating the differences, see Exercises 6.18–6.20.

6.4.2 Conditional tail probability

Suppose now that X_2 is a scalar and that we require an approximation to the conditional tail probability

$$\text{pr}(X_2 \geq x_2 | X_1 = x_1).$$

Expression (6.15) gives the required unconditional tail probability: the surprising fact is that the same expression, suitably reinterpreted, applies equally to conditional tail probabilities.

In the double saddlepoint approximation, there are, of course, two saddlepoints, one for the joint distribution of (X_1, X_2) and one for the marginal distribution of X_1. These are defined by

$$K^r(\hat{\xi}_1, \hat{\xi}_2) = x_1^r, \quad r = 1, ..., p-1; \quad K^p(\hat{\xi}_1, \hat{\xi}_2) = x_2$$

for the joint distribution, and

$$K^r(\tilde{\xi}_1, 0) = x_1^r, \quad r = 1, ..., p-1$$

for the marginal distribution of the $p - 1$ components of X_1. In the above expressions, ξ_1 has $p - 1$ components and ξ_2 is a scalar, corresponding to the partition of X.

The signed likelihood ratio statistic $T = T(x_2 | x_1)$ is most conveniently expressed in terms of the two Legendre transformations, giving

$$T = \text{sign}(\hat{\xi}_2)\{K^*_{X_1 X_2}(x_1, x_2) - K^*_{X_1}(x_1)\}.$$

Further, define the generalized conditional variance, $V = V(x_2 | x_1)$, by the determinant ratio

$$V = \frac{|K^{rs}(\hat{\xi}_1, \hat{\xi}_2)|}{|K^{rs}(\tilde{\xi}_1, 0)|} = \frac{|K^*_{X_1; rs}|}{|K^*_{X_1 X_2; rs}|}.$$

Using these expressions, the double saddlepoint approximation becomes

$$cV^{-1/2} \exp(-T^2/2).$$

In the conditional sample space, $V^{-1/2}$ plays the role of carrier measure and the exponent is invariant. The conditional tail probability is given by (6.15) using the Wald statistic

$$S(x) = \hat{\xi}_2 V^{1/2}.$$

This important result is due to I. Skovgaard (1986, personal communication) and is given here without proof.

6.5 Bibliographic notes

The following is a very brief description of a few key references. Further references can be found cited in these papers.

The Cramér-Chernoff large deviation result is discussed in greater detail by Bahadur (1971) and by Ellis (1985).

The derivation of the saddlepoint approximation by using an Edgeworth expansion for the exponentially tilted density goes back to the work of Esscher (1932), Cramér (1938), Chernoff (1952) and Bahadur & Ranga-Rao (1960). In a series of important papers, Daniels (1954, 1980, 1983) develops the saddlepoint method, derives the conditions under which the re-normalized approximation is exact and finds approximations for the density of a ratio and the solution of an estimating equation. Barndorff-Nielsen & Cox (1979) discuss double saddlepoint approximation as a device for approximating to conditional likelihoods.

The Legendre transformation plays an important role in the literature on large deviations, and is emphasized by Ellis (1985), Bahadur & Zabell (1979) and also to an extent, Daniels (1960).

The relationship between the Bartlett adjustment factor and the normalization factor in the saddlepoint formula is discussed by Barndorff-Nielsen & Cox (1984).

Tail probability calculations are discussed by Lugannani & Rice (1980), Robinson (1982) and by Daniels (1987).

6.6 Further results and exercises 6

6.1 Show that the array

$$M^{ij}(\xi) = E\{X^i X^j \exp(\xi_r x^r)\}$$

is positive definite for each ξ. Hence deduce that the function $M(\xi)$ is convex. Under what conditions is the inequality strict?

6.2 By using Hölder's inequality, show for any $0 \le \lambda \le 1$, that

$$K(\lambda \xi_1 + (1 - \lambda)\xi_2) \le \lambda K(\xi_1) + (1 - \lambda)K(\xi_2)$$

proving that $K(\xi)$ is a convex function.

6.3 Prove directly that $K^{rs}(\xi)$ is positive definite for each ξ in Ξ. Hence deduce that $K^*(x)$ is a convex function on $\mathcal{X}$.

6.4 Prove that $K^*(x) \geq 0$, with equality only if $x^i = \kappa^i$.

6.5 Prove the following extension of inequality (6.4) for vector-valued X

$$\text{pr}(X \in A) \leq \exp\{I(A)\}$$

where $I(A) = \sup_{x \in A}\{-K^*(x)\}$.

6.6 From the entropy limit (6.6), deduce the law of large numbers.

6.7 Show that the Legendre transformation of $Y = X_1 + \ldots + X_n$ is $nK^*(y/n)$, where the Xs are $i.i.d.$ with Legendre transformation $K^*(x)$.

6.8 Show that, for each θ in Ξ,

$$\exp(\theta_i x^i - K(\theta)) f_X(x)$$

is a distribution on $\mathcal{X}$. Find its cumulant generating function and the Legendre transformation.

6.9 From the definition

$$K_Y^*(y) = \sup_{\xi}\{\xi_i y^i - K_Y(\xi)\}$$

show that the Legendre transformation is invariant under affine transformation of coordinates on $\mathcal{X}$.

6.10 By writing ξ_i as a polynomial in z

$$\xi_i = a_{ir}z^r + a_{irs}z^r z^s/2! + a_{irst}z^r z^s z^t/3! + \ldots,$$

solve the equation

$$\kappa^{r,i}\xi_i + \kappa^{r,i,j}\xi_i\xi_j/2! + \kappa^{r,i,j,k}\xi_i\xi_j\xi_k/3! + \ldots \quad = z^r$$

by series reversal. Hence derive expansion (6.12) for $K^*(x)$.

6.11 Using expansion (6.12), find the mean of $2nK^*(\bar{X}_n)$ up to and including terms that are of order $O(n^{-1})$.

6.12 Show that the matrix inverse of $K^{rs}(\xi)$ is $K_{rs}^*(x)$, where $x^r = K^r(\xi)$ corresponds to the saddlepoint.

6.13 Using (6.12) or otherwise, show that, for each x in $\mathcal{X}$,

$$K_{ijk}^*(x) = -K^{rst}K_{ri}K_{sj}K_{tk}$$
$$K_{ijkl}^*(x) = -\{K^{rstu} - K^{rsv}K^{tuw}K_{vw}[3]\}K_{ri}K_{sj}K_{tk}K_{ul}$$

where all functions on the right are evaluated at $\xi_r = K_r^*(x)$, the saddlepoint image of x.

6.14 Using the results given in the previous exercise, show, using the notation of Section 6.3, that

$$\rho_{13}^2(\hat{\theta}) = \rho_{13}^{*2}(x) \qquad \rho_{23}^2(\hat{\theta}) = \rho_{23}^{*2}(x)$$
$$\rho_4(\hat{\theta}) = -\rho_4^*(x) + \rho_{13}^{*2}(x) + 2\rho_{23}^{*2}(x).$$

6.15 By using the expansion for ξ_i given in Section 6.2.4, show that the maximum likelihood estimate of θ based on $\bar{X}_n$ in the exponential family (6.14) has bias

$$E(n^{1/2}\hat{\theta}_r) = -\tfrac{1}{2}n^{-1/2}\kappa^{i,j,k}\kappa_{i,r}\kappa_{j,k} + O(n^{-3/2}).$$

6.16 Show that if $Z^r = \bar{X}_n^r - \kappa^r$ and

$$n^{-1/2}Y^r = Z^r - \kappa^{r,s,t}\kappa_{s,i}\kappa_{t,j}Z^iZ^j/6$$
$$+ \{8\kappa^{r,s,t}\kappa^{u,v,w}\kappa_{s,i}\kappa_{t,u}\kappa_{v,j}\kappa_{w,k} - 3\kappa^{r,s,t,u}\kappa_{s,i}\kappa_{t,j}\kappa_{u,k}\}Z^iZ^jZ^k/72$$

then $Y = O_p(1)$ and

$$2nK^*(\bar{X}_n) = Y^rY^s\kappa_{r,s} + O(n^{-2}).$$

6.17 Show that Y^r defined in the previous exercise has third cumulant of order $O(n^{-3/2})$ and fourth cumulant of order $O(n^{-1})$. Hence show that $2nK^*(\bar{X}_n)$ has a non-central χ_p^2 distribution for which the rth cumulant is

$$\{1 + b/n\}^r 2^{r-1}(r-1)!p + O(n^{-2}).$$

Find an expression for b in terms of the invariant cumulants of X.

6.18   Show that, in the case of the binomial distribution with index m and parameter π, the Legendre transformation is

$$y \log \left(\frac{y}{\mu}\right) + (m - y) \log \left(\frac{m - y}{m - \mu}\right)$$

where $\mu = m\pi$. Hence show that the saddlepoint approximation is

$$\frac{\pi^y (1 - \pi)^{m-y} m^{m+1/2}}{(2\pi)^{1/2} y^{y+1/2} (m - y)^{m-y+1/2}}.$$

In what circumstances is the saddlepoint approximation accurate? Derive the above as a double saddlepoint approximation to the conditional distribution of Y_1 given $Y_1 + Y_2 = m$, where the Ys are independent Poisson random variables.

6.19   Let X_1, X_2 be independent exponential random variables with common mean μ. Show that the Legendre transformation of the joint cumulant generating function is

$$K^*(x_1, x_2; \mu) = \frac{x_1 + x_2 - 2\mu}{\mu} - \log \left(\frac{x_1}{\mu}\right) - \log \left(\frac{x_2}{\mu}\right).$$

Show also that the Legendre transformation of the cumulant generating transformation of $\bar{X}$ is

$$K^*(\bar{x}; \mu) = 2 \left(\frac{\bar{x} - \mu}{\mu}\right) - 2 \log \left(\frac{\bar{x}}{\mu}\right).$$

Hence derive the double saddlepoint approximation for the conditional distribution of X_1 given that $X_1 + X_2 = 1$. Show that the re-normalized double saddlepoint approximation is exact.

6.20   Extend the results described in the previous exercise to gamma random variables having mean μ and indices ν_1, ν_2. Replace $\bar{X}$ by an appropriately weighted mean.

6.21   In the notation of Exercise 6.18, show that the second derivative of $K^*(x, 1 - x; 1/2)$ at $x = 1/2$ is 8, whereas the conditional variance of X_1 given that $X_1 + X_2 = 1$ is $1/12$. Hence deduce that the double saddlepoint approximation to the conditional density of X_1 is not the same as applying the ordinary saddlepoint approximation directly to the conditional cumulant generating function of X_1.

6.22 Using the asymptotic expansion for the normal tail probability

$$1 - \Phi(x) \simeq \frac{\phi(x)}{x} \qquad x \to \infty$$

and taking $x > E(X)$, show, using (6.14), that

$$n^{-1} \log \mathrm{pr}\{\bar{X}_n > x\} \to -K^*(x)$$

as $n \to \infty$, where $\bar{X}_n$ is the average of n independent and identically distributed random variables. By retaining further terms in the expansion, find the rate of convergence to the entropy limit (6.5).

6.23 Using (6.11), show that the Legendre transformation $K^*(x;\theta)$ of the exponentially tilted density satisfies the partial differential equations

$$\frac{\partial K^*(x;\theta)}{\partial x^r} = \hat{\theta}_r(x) - \theta_r$$

$$-\frac{\partial K^*(x;\theta)}{\partial \theta_i} = x^i - K^i(\theta).$$

Hence show that in the univariate case,

$$K^*(x;\theta) = \int_\mu^x \frac{x - t}{v(t)}\, dt,$$

where $\mu = K'(\theta)$ and $v(\mu) = K''(\theta)$, (Wedderburn, 1974; Nelder & Pregibon, 1986).

6.24 By using Taylor expansions for $S(x)$ and $T(x)$ in (6.15), show that, for normal deviations, the tail probability (6.15) reduces to

$$1 - \Phi(T) + \phi(T)\left(-\frac{\rho_3}{6} + \frac{5\rho_3^2 - 3\rho_4}{24}T\right) + O(n^{-3/2}).$$

Hence deduce (6.14).

6.25 Let X be a random variable with density function

$$f_X(x;\theta) = \exp\{\theta_i x^i - K(\theta)\} f_0(x).$$

depending on the unknown parameter θ. Let θ have Jeffreys's prior density

$$\pi(\theta) = |K^{rs}(\theta)|^{1/2}.$$

Using Bayes's theorem, show that the posterior density for θ given x is approximately

$$\pi(\theta|x) \simeq c \exp\{\theta_i x^i - K(\theta) - K^*(x)\}|K^{rs}(\theta)|^{1/2},$$

whereas the density of the random variable $\hat{\theta}$ is approximately

$$p(\hat{\theta}|\theta) \simeq c \exp\{\theta_i x^i - K(\theta) - K^*(x)\}|K^{rs}(\hat{\theta})|^{1/2}.$$

In the latter expression x is considered to be a function of $\hat{\theta}$.
Find expressions for the constants in both cases.

6.26   Consider the conjugate density $f_X(x;\theta)$ as given in the previous exercise, where $K(\theta)$ is the cumulant generating function for $f_0(x)$ and $K^*(x)$ is its Legendre transform. Show that

$$E_\theta \left\{ \log \left(\frac{f_X(X;\theta)}{f_0(X)} \right) \right\} = K^*(E_\theta(X))$$

where $E_\theta(.)$ denotes expectation under the conjugate density. [In this context, $K^*(E_\theta(X))$ is sometimes called the *Kullback-Leibler distance* between the conjugate density and the original density.]

CHAPTER 7

Likelihood functions

7.1 Introduction

Let Y be a random variable whose density or distribution function $f_Y(y; \theta)$ depends on the p-dimensional parameter vector θ. Usually, we think of Y as vector valued with n independent components, but this consideration is important only in large sample approximations. For the most part, since it is unnecessary to refer to the individual components, we write Y without indices. A realization of Y is called an observation and, when we wish to make a clear distinction between the observation and the random variable, we write y for the observation. Of course, y is just a number or ordered set of numbers but, implicit in Y is the sample space or set of possible observations, one of which is y. This distinction is made at the outset because of its importance in the remainder of this chapter.

The parameter vector θ with components $\theta^1, ..., \theta^p$ is assumed to lie in some subset, Θ, of R^p. Often, in fact, $\Theta = R^p$, but this assumption is not necessary in the discussion that follows. For technical reasons, it helps to assume that Θ is an open set in R^p: this condition ensures, for example, that there are no equality constraints among the components and that the parameter space is genuinely p-dimensional.

Associated with any observed value y on Y, there is a particular parameter value $\theta_T \in \Theta$, usually unknown, such that Y has density function $f_Y(y; \theta_T)$. We refer to θ_T as the 'true' value. Often, however, when we wish to test a null hypothesis value, we write θ_0 and in subsequent calculations, θ_0 is treated as if it were the true value. Ideally, we would like to know the value of θ_T, but apart from exceptional cases, the observed data do not determine θ_T uniquely or precisely. Inference, then, is concerned with probability statements concerning those values in Θ that are consistent with the observed y. Usually, it is both unreasonable and undesirable to quote a single 'most consistent' parameter value: interval estimates, either in the form of confidence sets or Bayes intervals, are preferred.

Since the eventual goal is to make probabilistic statements concerning those parameter values that are consistent in some sense with the observed data, the conclusions must be unaffected by two kinds of transformation:

(i) invertible transformation of Y
(ii) invertible transformation of θ.

Invariance under the first of these groups is guaranteed if we work with the log likelihood function, defined up to an arbitrary additive function of y. Invariance under re-parameterization is a main concern of this chapter. For that reason, we are interested in quantities that transform as tensors under change of coordinates on Θ. Thus Θ, and not the sample space, is here regarded as the space of primary interest. By convention, therefore, we use superscripts to represent the coordinates of an arbitrary point in Θ. In this respect, the notation differs from that used in the previous chapter, where transformations of the sample space were considered.

In almost all schools of statistical inference, the log likelihood function for the observed data

$$l(\theta; y) = \log f_Y(y; \theta)$$

plays a key role. One extreme view, (Edwards, 1972), is that nothing else matters. In the Bayesian framework on the other hand, it is necessary to acquire a prior distribution, $\pi(\theta)$ that describes 'degree of belief' or personal conviction prior to making the observation. Bayes's theorem then gives

$$\pi(\theta|y) = \pi(\theta) f_Y(y; \theta) / c(y)$$

as the posterior distribution for θ given y, where $c(y)$ is the normalization factor, $\int \pi(\theta) f_Y(y; \theta) d\theta$. All probability statements are then based on the posterior distribution of θ given y. Other schools of inference hold that, in order to conceive of probabilities as relative frequencies rather than as degrees of belief, it is necessary to take account of the sample space of possible observations. In other words, $l(\theta; y)$ must be regarded as the observed value of the random variable $l(\theta; Y)$. One difficulty with this viewpoint is that there is often some leeway in the choice of sample space.

A thorough discussion of the various schools of statistical inference is beyond the scope of this book, but can be found, for example, in Cox & Hinkley (1974) or Berger & Wolpert (1984). In the discussion that follows, our choice is to regard $l(\theta; Y)$ and its derivatives with respect to θ as random variables. The sample space is rarely mentioned explicitly, but it is implicit when we talk of moments or cumulants, which involve integration over the sample space.

Tensor methods are particularly appropriate and powerful in this context because of the requirement that any inferential statement should be materially unaffected by the parameterization chosen. The parameterization is simply a convenient but arbitrary way of specifying the various probability models under consideration. An inferential statement identifies a subset of these distributions and that subset should, in principle at least, be unaffected by the particular parameterization chosen. Unless otherwise stated, therefore, when we talk of tensors in this chapter, we refer implicitly to arbitrary invertible transformations of the parameter vector. Particular emphasis is placed on invariants, which may be used to make inferential statements independent of the coordinate system. The most important invariant is the log likelihood function itself. Other invariants are connected with the likelihood ratio statistic and its distribution.

7.2 Log likelihood derivatives

7.2.1 Null cumulants

In what follows, it is assumed that the log likelihood function has continuous partial derivatives up to the required order and that these derivatives have finite moments, again up to the required order, which is obvious from the context. These derivatives at an arbitrary point θ, are written as

$$U_r = u_r(\theta; Y) = \partial l(\theta; Y)/\partial \theta^r$$
$$U_{rs} = u_{rs}(\theta; Y) = \partial^2 l(\theta; Y)/\partial \theta^r \partial \theta^s$$
$$U_{rst} = u_{rst}(\theta; Y) = \partial^3 l(\theta; Y)/\partial \theta^r \partial \theta^s \partial \theta^t$$

and so on. Our use of subscripts here is not intended to imply that the log likelihood derivatives are tensors. In fact, the derivatives with respect to an alternative parameterization $\phi = \phi^1, ..., \phi^p$, are given by

$$\bar{U}_r = \theta_r^i U_i$$
$$\bar{U}_{rs} = \theta_r^i \theta_s^j U_{ij} + \theta_{rs}^i U_i \qquad (7.1)$$
$$\bar{U}_{rst} = \theta_r^i \theta_s^j \theta_t^k U_{ijk} + \theta_r^i \theta_{st}^j U_{ij}[3] + \theta_{rst}^i U_i$$

and so on, where $\theta_r^i = \partial\theta^i/\partial\phi^r$ is assumed to have full rank, $\theta_{rs}^i = \partial^2\theta^i/\partial\phi^r\partial\phi^s$ and so on. Thus U_r is a tensor but subsequent higher-order derivatives are not, on account of the higher derivatives that appear in the transformation formulae. The log likelihood derivatives are tensors under the smaller group of linear or affine transformations, but this is of no substantial importance in the present context.

For reasons that will become clear shortly, it is desirable to depart to some extent from the notation used in Chapters 2 and 3 for moments and cumulants. The null moments of U_r, U_{rs}, U_{rst},... are written as

$$\mu_r = E(U_r; \theta), \qquad \mu_{r,s} = E(U_r U_s; \theta),$$
$$\mu_{rs} = E(U_{rs}; \theta), \qquad \mu_{r,st} = E(U_r U_{st}; \theta),$$
$$\mu_{rst} = E(U_{rst}; \theta), \qquad \mu_{r,st,uvw} = E(U_r U_{st} U_{uvw}; \theta)$$

and so on. The word 'null' here refers to the fact that the twin processes of differentiation and averaging both take place at the same value of θ. The null cumulants are defined by

$$\kappa_r = \mu_r, \qquad \kappa_{r,s} = \mu_{r,s} - \mu_r\mu_s \qquad \kappa_{rs,tu} = \mu_{rs,tu} - \mu_{rs}\mu_{tu}$$

and so on.

Neither the set of moments nor the set of cumulants is linearly independent. To see how the linear dependencies arise, we note that for all θ, integration over the sample space gives

$$\int f_Y(y; \theta)dy = 1.$$

Differentiation with respect to θ and reversing the order of differentiation and integration gives

$$\mu_r = \kappa_r = \int u_r(\theta; y) f_Y(y; \theta) dy = 0.$$

Further differentiation gives

$$\mu_{[rs]} = \mu_{rs} + \mu_{r,s} = 0$$
$$\mu_{[rst]} = \mu_{rst} + \mu_{r,st}[3] + \mu_{r,s,t} = 0$$
$$\mu_{[rstu]} = \mu_{rstu} + \mu_{r,stu}[4] + \mu_{rs,tu}[3] + \mu_{r,s,tu}[6] + \mu_{r,s,t,u} = 0.$$

In terms of the null cumulants, we have

$$\kappa_{[rs]} = \kappa_{rs} + \kappa_{r,s} = 0$$
$$\kappa_{[rst]} = \kappa_{rst} + \kappa_{r,st}[3] + \kappa_{r,s,t} = 0 \qquad (7.2)$$
$$\kappa_{[rstu]} = \kappa_{rstu} + \kappa_{r,stu}[4] + \kappa_{rs,tu}[3] + \kappa_{r,s,tu}[6] + \kappa_{r,s,t,u} = 0,$$

and so on, with summation over all partitions of the indices. In the remainder of this chapter, the enclosure within square brackets of a set of indices implies summation over all partitions of that set, as in the expressions listed above. In addition, $\kappa_{r,[st]}$ is synonomous with the combination $\kappa_{r,s,t} + \kappa_{r,st}$, the rule in this case applying to a subset of the indices. Details of the argument leading to (7.2) are given in Exercise 7.1. In particular, to reverse the order of differentiation and integration, it is necessary to assume that the sample space does not depend on θ.

In the univariate case, power notation is often employed in the form

$$i_{rst} = E\left\{ \left(\frac{\partial l}{\partial \theta}\right)^r \left(\frac{\partial^2 l}{\partial \theta^2}\right)^s \left(\frac{\partial^3 l}{\partial \theta^3}\right)^t ; \theta \right\}.$$

The moment identities then become $i_{10} = 0$,

$$i_{01} + i_{20} = 0,$$
$$i_{001} + 3i_{11} + i_{30} = 0,$$
$$i_{0001} + 4i_{101} + 3i_{02} + 6i_{21} + i_{40} = 0.$$

Similar identities apply to the cumulants, but we refrain from writing these down, in order to avoid further conflict of notation.

7.2.2 Non-null cumulants

Given a test statistic based on the log likelihood derivatives at the hypothesized value, only the null distribution, or the null cumulants, are required in order to compute the significance level. However, in order to assess the suitability of a proposed test statistic, it is necessary to examine the sensitivity of the statistic to changes in the parameter value. Suppose then that U_r, U_{rs},... are the log likelihood derivatives at an arbitrary point θ and that the 'true' parameter point is θ_T. We may then examine how the cumulants of U_r, U_{rs},... depend on the value of $\delta = \theta_T - \theta$. Thus, we write in an obvious notation,

$$\mu_r(\theta; \theta_T) = E\{U_r; \theta_T\} = \int \frac{\partial \log f_Y(y; \theta)}{\partial \theta} f_Y(y; \theta_T) dy$$

and similarly for $\mu_{r,s}(\theta; \theta_T)$, $\mu_{rs}(\theta; \theta_T)$ and so on. The null values are $\mu_r(\theta; \theta) = \mu_r$, $\mu_{r,s}(\theta; \theta) = \mu_{r,s}$ and so on. The non-null cumulants are written $\kappa_r(\theta; \theta_T)$, $\kappa_{rs}(\theta; \theta_T)$, $\kappa_{r,s}(\theta; \theta_T)$ and so on, where, for example,

$$\kappa_r(\theta; \theta_T) = \mu_r(\theta; \theta_T)$$
$$\kappa_{r,s}(\theta; \theta_T) = \mu_{r,s}(\theta; \theta_T) - \mu_r(\theta; \theta_T)\mu_s(\theta; \theta_T).$$

For small values of δ, it is possible to express the non-null cumulants as a power series in δ, with coefficients that involve the null cumulants alone. In fact, from the Taylor expansion

$$\frac{f_Y(Y; \theta_T)}{f_Y(Y; \theta)} = 1 + U_r\delta^r + (U_{rs} + U_rU_s)\delta^r\delta^s/2!$$
$$+ (U_{rst} + U_rU_{st}[3] + U_rU_sU_t)\delta^r\delta^s\delta^t/3! + ...,$$

we find the following expansions.

$$\mu_r(\theta; \theta_T) = \mu_r + \mu_{r,s}\delta^s + \mu_{r,[st]}\delta^s\delta^t/2! + \mu_{r,[stu]}\delta^s\delta^t\delta^u/3! + ...$$
$$\mu_{r,s}(\theta; \theta_T) = \mu_{r,s} + \mu_{r,s,t}\delta^t + \mu_{r,s,[tu]}\delta^t\delta^u/2!$$
$$+ \mu_{r,s,[tuv]}\delta^t\delta^u\delta^v/3! + ... \qquad (7.3)$$
$$\mu_{r,st}(\theta; \theta_T) = \mu_{r,st} + \mu_{r,st,u}\delta^u + \mu_{r,st,[uv]}\delta^u\delta^v/2! + ...$$

where, for example,

$$\mu_{r,[stu]} = \mu_{r,stu} + \mu_{r,s,tu}[3] + \mu_{r,s,t,u}$$

involves summation over all partitions of the bracketed indices. Identical expansions hold for the cumulants. For example,

$$\kappa_{r,s}(\theta; \theta_T) = \kappa_{r,s} + \kappa_{r,s,t}\delta^t + \kappa_{r,s,[tu]}\delta^t\delta^u/2!$$
$$+ \kappa_{r,s,[tuv]}\delta^t\delta^u\delta^v/3! + \dots \quad (7.4)$$

where

$$\kappa_{r,s,[tuv]} = \kappa_{r,s,tuv} + \kappa_{r,s,t,uv}[3] + \kappa_{r,s,t,u,v}.$$

Note that $\kappa_{r,s,[t]}$, $\kappa_{r,s,[tu]}$, $\dots$ are the vector of first derivatives and the array of second derivatives with respect to the second argument only, of the null cumulant, $\kappa_{r,s}(\theta; \theta)$. There is therefore, a close formal similarity between these expansions and those derived by Skovgaard (1986a) for the derivatives of $\kappa_{r,s}(\theta) = \kappa_{r,s}(\theta; \theta)$ and similar null cumulants. Skovgaard's derivatives involve additional terms that arise from considering variations in the two arguments simultaneously. The derivation of (7.4) can be accomplished along the lines of Section 3 of Skovgaard's paper.

7.2.3 Tensor derivatives

One peculiar aspect of the identities (7.2) and also of expansions (7.3) and (7.4) is that, although the individual terms, in general, are not tensors, nevertheless the identities and expansions are valid in all coordinate systems. Consider, for example, the identity $\kappa_{rs} + \kappa_{r,s} = 0$ in (7.2). Now, U_r is a tensor and hence all its cumulants are tensors. Thus $\kappa_{r,s}$, $\kappa_{r,s,t}$ and so on are tensors and hence $\kappa_{rs} = -\kappa_{r,s}$ must also be a tensor even though U_{rs} is not a tensor. This claim is easily verified directly: see Exercise 7.2. Similarly, from the identity $\kappa_{rst} + \kappa_{r,st}[3] + \kappa_{r,s,t} = 0$, it follows that $\kappa_{rst} + \kappa_{r,st}[3]$ must be a tensor. However, neither κ_{rst} nor $\kappa_{r,st}$ are tensors. In fact, the transformation laws are

$$\bar{\kappa}_{r,st} = \theta^i_r\theta^j_s\theta^k_t\kappa_{i,jk} + \theta^i_r\theta^j_{st}\kappa_{i,j}$$
$$\ddot{\kappa}_{rst} = \theta^i_r\theta^j_s\theta^k_t\kappa_{ijk} + \theta^i_r\theta^j_{st}\kappa_{ij}[3].$$

From these, it can be seen that $\kappa_{rst} + \kappa_{r,st}[3]$ is indeed a tensor.

The principal objection to working with arrays that are not tensors is that it is difficult to recognize and to construct invariants. For example, the log likelihood ratio statistic is invariant and, when we expand it in terms of log likelihood derivatives, it is helpful if the individual terms in the expansion are themselves invariants. For that reason, we seek to construct arrays V_r, V_{rs}, V_{rst},... related to the log likelihood derivatives, such that the Vs are tensors. In addition, in order to make use of Taylor expansions such as (7.3) and (7.4), we require that the Vs be ordinary log likelihood derivatives in *some* coordinate system. This criterion excludes covariant derivatives as normally defined in differential geometry: it also excludes least squares residual derivatives, whose cumulants do not obey (7.2). See Exercise 7.3.

Let θ_0 be an arbitrary parameter value and define

$$\beta^r_{st} = \kappa^{r,i}\kappa_{i,st}. \quad \beta^r_{stu} = \kappa^{r,i}\kappa_{i,stu}, \quad \beta^r_{stuv} = \kappa^{r,i}\kappa_{i,stuv}.$$

where $\kappa^{r,s}$ is the matrix inverse of $\kappa_{r,s}$. In these definitions, only null cumulants at θ_0 are involved. This means that the β-arrays can be computed at each point in Θ without knowing the 'true' value, θ_T. The βs are the regression coefficients of U_{st}, U_{stu},... on U_r where the process of averaging is carried out under θ_0, the same point at which derivatives were computed. Now consider the parameter transformation defined in a neighbourhood of θ_0 by

$$\phi^r - \phi^r_0 = \theta^r - \theta^r_0 + \beta^r_{st}(\theta^s - \theta^s_0)(\theta^t - \theta^t_0)/2!$$
$$+ \beta^r_{stu}(\theta^s - \theta^s_0)(\theta^t - \theta^t_0)(\theta^u - \theta^u_0)/3! + \dots$$
$$(7.5)$$

Evidently, θ_0 transforms to ϕ_0. From (7.1), the derivatives at ϕ_0 with respect to ϕ, here denoted by V_i, V_{ij}, V_{ijk},..., satisfy

$$U_r = V_r$$
$$U_{rs} = V_{rs} + \beta^i_{rs}V_i$$
$$U_{rst} = V_{rst} + \beta^i_{rs}V_{it}[3] + \beta^i_{rst}V_i \qquad (7.6)$$
$$U_{rstu} = V_{rstu} + \beta^i_{rs}V_{itu}[6] + \beta^i_{rs}\beta^j_{tu}V_{ij}[3] + \beta^i_{rst}V_{iu}[4] + \beta^i_{rstu}V_i$$

with summation over all partitions of the free indices. Evidently, the Vs are genuine log likelihood derivatives having the property that the null covariances

$$\nu_{r,st} = \text{cov}(V_r, V_{st}), \quad \nu_{r,stu} = \text{cov}(V_r, V_{stu}), \dots$$

of V_r with the higher-order derivatives, are all zero. Note, however, that

$$\nu_{rs,tu} = \operatorname{cov}(V_{rs}, V_{tu}) = \kappa_{rs,tu} - \kappa_{rs,i}\kappa_{tu,j}\kappa^{i,j}$$
$$\nu_{r,s,tu} = \operatorname{cum}(V_r, V_s, V_{tu}) = \kappa_{r,s,tu} - \kappa_{r,s,i}\kappa_{tu,j}\kappa^{i,j}$$

are non-zero in general. It follows that identities (7.2) apply to the Vs in the form

$$\nu_{rs} + \nu_{r,s} = 0,$$
$$\nu_{rst} + \nu_{r,s,t} = 0,$$
$$\nu_{rstu} + \nu_{rs,tu}[3] + \nu_{r,s,tu}[6] + \nu_{r,s,t,u} = 0.$$

It remains to show that the Vs are tensors. To do so, we need to examine the effect on the Vs of applying a non-linear transformation to θ. One method of proof proceeds as follows. Take $\theta_0 = \phi_0 = 0$ for simplicity and consider the transformation

$$\bar{\theta}^r = a_i^r \theta^i + a_{ij}^r \theta^i \theta^j / 2! + a_{ijk}^r \theta^i \theta^j \theta^k / 3! + \dots$$

so that the log likelihood derivatives transform to $\bar{U}_r$, $\bar{U}_{rs}$, $\bar{U}_{rst}$ satisfying

$$
\begin{aligned}
U_r &= a_r^i \bar{U}_i \\
U_{rs} &= a_r^i a_s^j \bar{U}_{ij} + a_{rs}^i \bar{U}_i \\
U_{rst} &= a_r^i a_s^j a_t^k \bar{U}_{ijk} + a_r^i a_{st}^j \bar{U}_{ij}[3] + a_{rst}^i \bar{U}_i
\end{aligned}
\tag{7.7}
$$

and so on. These are the inverse equations to (7.1). Proceeding quite formally now, equations (7.6) may be written using matrix notation in the form

$$\mathbf{U} = \mathbf{B}\mathbf{V}. \tag{7.8}$$

Similarly, the relation between U and $\bar{U}$ in (7.8) may be written

$$\mathbf{U} = \mathbf{A}\bar{\mathbf{U}}.$$

These matrix equations encompass all of the derivatives up to whatever order is specified. In the particular case where $a_i^r = \delta_i^r$, but not otherwise, the array of coefficients $\mathbf{B}$ transforms to $\bar{\mathbf{B}}$, where

$$\mathbf{B} = \mathbf{A}\bar{\mathbf{B}},$$

as can be seen by examining equations (7.7) above. More generally, if $a_r^i \neq \delta_r^i$, we may write $\mathbf{B} = \mathbf{A}\bar{\mathbf{B}}\mathbf{A}^{*-1}$, where $\mathbf{A}^*$ is a direct product matrix involving a_r^i alone (Exercise 7.8). Premultiplication of (7.8) by $\mathbf{A}^{-1}$ gives

$$\bar{\mathbf{U}} = \mathbf{A}^{-1}\mathbf{U} = \mathbf{A}^{-1}\mathbf{B}\mathbf{V} = \bar{\mathbf{B}}\mathbf{V}.$$

Since, by definition, the transformed Vs satisfy $\bar{\mathbf{U}} = \bar{\mathbf{B}}\bar{\mathbf{V}}$, it follows that if $a_i^r = \delta_i^r$, then $\bar{\mathbf{V}} = \mathbf{V}$. Thus, the Vs are unaffected by non-linear transformation in which the leading coefficient is δ_i^r. Since all quantities involved are tensors under the general linear group, it follows that the Vs must be tensors under arbitrary smooth invertible parameter transformation.

The Vs defined by (7.6) are in no sense unique. In fact any sequence of coefficients τ_{rs}^i, $\tau_{rst}^i, \ldots$ that transforms like the βs will generate a sequence of symmetric tensors when inserted into (7.6). One possibility is to define

$$\tau_{st}^r = \kappa^{r,i}\kappa_{i,[st]}, \qquad \tau_{stu}^r = \kappa^{r,i}\kappa_{i,[stu]}$$

and so on. These arrays transform in the same way as the βs: any linear combination of the two has the same property. In fact, the Vs defined by (7.6) can be thought of as derivatives in the 'canonical' coordinate system: the corresponding tensors obtained by replacing β by τ are derivatives in the 'mean-value' coordinate system. This terminology is taken from the theory of exponential family models.

Barndorff-Nielsen (1986) refers to the process of obtaining tensors via (7.6) as the *intertwining* of *strings*, although *unravelling of strings* might be a better description of the process. In this terminology, the sequence of coefficients β_{st}^r, $\beta_{stu}^r, \ldots$ or the alternative sequence τ_{st}^r, $\tau_{stu}^r, \ldots$ is known as a *connection* string. The log likelihood derivatives themselves, and any sequence that transforms like (7.1) under re-parameterization, forms an infinite *co-string*. *Contra-strings* are defined by analogy. The main advantage of this perspective is that it forces one to think of the *sequence* of derivatives as an indivisible object: the higher-order derivatives have no invariant interpretation in the absence of the lower-order derivatives. See Foster (1986) for a light-hearted but enlightened discussion on this point.

7.3   Large sample approximation

7.3.1   *Log likelihood derivatives*

Suppose now that Y has n independent and identically distributed components so that the log likelihood for the full data may be written as the sum

$$l(\theta; Y) = \sum_i l(\theta; Y_i).$$

The derivatives U_r, U_{rs},... are then expressible as sums of n independent and identically distributed random variables. Under mild regularity conditions, therefore, the joint distribution of U_r, U_{rs},... may be approximated for large n by the normal distribution, augmented, if necessary by Edgeworth corrections.

It is convenient in the calculations that follow to make the dependence on n explicit by writing

$$U_r = n^{1/2} Z_r$$
$$U_{rs} = n\kappa_{rs} + n^{1/2} Z_{rs} \qquad (7.9)$$
$$U_{rst} = n\kappa_{rst} + n^{1/2} Z_{rst}$$

and so on for the higher-order derivatives. Thus,

$$\kappa_{r,s} = -\kappa_{rs} = -E\{\partial^2 l(\theta; Y_i)/\partial\theta^r \partial\theta^s; \theta\}$$

is the Fisher information per observation and κ_{rst}, κ_{rstu},... are higher-order information measures per observation. Moreover, assuming θ to be the true value, it follows that Z_r, Z_{rs}, Z_{rst},... are $O_p(1)$ for large n.

More generally, if, as would normally be the case, the components of Y are not identically distributed but still independent, $\kappa_{r,s}$ is the average Fisher information per observation and κ_{rst}, κ_{rstu},... are higher-order average information measures per observation. Additional fairly mild assumptions, in the spirit of the Lindeberg-Feller condition, are required to ensure that $\kappa_{r,s} = O(1)$, $Z_r = O_p(1)$, $Z_{rs} = O_p(1)$ and so on. Such conditions are taken for granted in the expansions that follow.

7.3.2 *Maximum likelihood estimation*

The likelihood equations $u_r(\hat{\theta}; Y) = 0$ may be expanded in a Taylor series in $\hat{\delta} = n^{1/2}(\hat{\theta} - \theta)$ to give

$$0 = n^{1/2} Z_r + (n\kappa_{rs} + n^{1/2} Z_{rs})\hat{\delta}^s/n^{1/2} + (n\kappa_{rst} + n^{1/2} Z_{rst})\hat{\delta}^s \hat{\delta}^t/(2n)$$
$$+ n\kappa_{rstu}\hat{\delta}^s \hat{\delta}^t \hat{\delta}^u/(6n^{3/2}) + O_p(n^{-1}).$$

For future convenience, we write

$$\kappa^{rs} = \kappa^{r,i}\kappa^{s,j}\kappa_{ij}, \quad \kappa^{rst} = \kappa^{r,i}\kappa^{s,j}\kappa^{t,k}\kappa_{ijk},$$

and so on, using the tensor $\kappa_{i,j}$ and its matrix inverse $\kappa^{i,j}$ to lower and raise indices. We may now solve for $\hat{\delta}$ in terms of the Zs, whose joint cumulants are known, giving

$$\hat{\delta}^r = \kappa^{r,s} Z_s + n^{-1/2}(\kappa^{r,s}\kappa^{t,u} Z_{st} Z_u + \kappa^{rst} Z_s Z_t/2)$$
$$+ n^{-1}\left(\kappa^{r,s}\kappa^{t,u}\kappa^{v,w} Z_{st} Z_{uv} Z_w + \kappa^{rst}\kappa^{u,v} Z_s Z_{tu} Z_v\right.$$
$$+ \kappa^{r,s}\kappa^{tuv} Z_{st} Z_u Z_v/2 + \kappa^{rst}\kappa^{uvw}\kappa_{t,w} Z_s Z_u Z_w/2 \qquad (7.10)$$
$$\left. + \kappa^{r,s}\kappa^{t,u}\kappa^{v,w} Z_{suw} Z_t Z_v/2 + \kappa^{rstu} Z_s Z_t Z_u/6\right) + O_p(n^{-3/2}).$$

Terms have been grouped here in powers of $n^{1/2}$. It is worth pointing out at this stage that $\hat{\delta}^r$ is not a tensor. Hence the expression on the right of the above equation is also not a tensor. However, the first-order approximation, namely $\kappa^{r,s} Z_s$, is a tensor.

From the above equation, or at least from the first two terms of the equation, some useful properties of maximum likelihood estimates may be derived. For example, we find

$$E(\hat{\delta}^r) = n^{-1/2}(\kappa^{r,s}\kappa^{t,u}\kappa_{st,u} + \kappa^{rst}\kappa_{s,t}/2) + O(n^{-3/2})$$
$$= -n^{-1/2}\kappa^{r,s}\kappa^{t,u}(\kappa_{s,t,u} + \kappa_{s,tu})/2 + O(n^{-3/2}).$$

This is $n^{1/2}$ times the bias of $\hat{\theta}^r$. In addition, straightforward calculations give

$$\text{cov}(\hat{\delta}^r, \hat{\delta}^s) = \kappa^{r,s} + O(n^{-1})$$
$$\text{cum}(\hat{\delta}^r, \hat{\delta}^s, \hat{\delta}^t) = n^{-1/2}\kappa^{r,i}\kappa^{s,j}\kappa^{t,k}(\kappa_{ijk} - \kappa_{i,j,k}) + O(n^{-3/2}).$$

Higher-order cumulants are $O(n^{-1})$ or smaller.

In the univariate case we may write

$$E(\hat{\theta} - \theta) = -n^{-1}(i_{30} + i_{11})/(2i_{20}^2) + O(n^{-2})$$
$$\text{var}(\hat{\theta}) = i_{20}/n + O(n^{-2})$$
$$\kappa_3(\hat{\theta}) = n^{-2}(i_{001} - i_{30})/i_{20}^3 + O(n^{-3}),$$

where i_{rst} is the generalized information measure per observation.
More extensive formulae of this type are given by Shenton & Bowman (1977, Chapter 3): their notation, particularly their version of the summation convention, differs from that used here. See also Peers & Iqbal (1985), who give the cumulants of the maximum likelihood estimate up to and including terms of order $O(n^{-1})$. In making comparisons, note that $\kappa^{ij} = -\kappa^{i,j}$ is used by Peers & Iqbal to raise indices.

7.4   Maximum likelihood ratio statistic

7.4.1   *Invariance properties*

In the previous section, the approximate distribution of the maximum likelihood estimate was derived through an asymptotic expansion in the log likelihood derivatives at the true parameter point. Neither $\hat{\theta}$ nor $\hat{\delta}$ are tensors and consequently the asymptotic expansion is not a tensor expansion. For that reason, the algebra tends to be a little complicated: it is not evident how the arrays involved should transform under a change of coordinates. In this section, we work with the maximized log likelihood ratio statistic defined by

$$W(\theta) = 2l(\hat{\theta}; Y) - 2l(\theta; Y)$$

where $l(\theta; Y)$ is the log likelihood for the full data comprising n independent observations. Since W is invariant under re-parameterization, it may be expressed in terms of other simpler invariants. The distributional calculations can be rendered tolerably simple if we express W as an asymptotic expansion involving invariants derived from the tensors $V_r, V_{rs}, V_{rst}, \ldots$ and their joint cumulants. The known joint cumulants of the Vs can then be used to determine the approximate distribution of W to any required order of approximation.

First, however, we derive the required expansion in arbitrary coordinates.

7.4.2 Expansion in arbitrary coordinates

Taylor expansion of $l(\hat{\theta}; Y) - l(\theta; Y)$ about θ gives

$$\tfrac{1}{2}W(\theta) = l(\hat{\theta}; Y) - l(\theta; Y)$$
$$= U_r \hat{\delta}^r / n^{1/2} + U_{rs} \hat{\delta}^r \hat{\delta}^s / (2n) + U_{rst} \hat{\delta}^r \hat{\delta}^s \hat{\delta}^t / (6n^{3/2}) + \dots$$

Note that U_r, U_{rs},... are the log likelihood derivatives at θ, here assumed to be the true parameter point. If we now write

$$\hat{\delta}^r = Z^r + c^r / n^{1/2} + d^r / n + O_p(n^{-3/2}),$$

where $Z^r = \kappa^{r,s} Z_s$, c^r and d^r are given by (7.10), we find that $\tfrac{1}{2}W(\theta)$ has the following expansion:

$$n^{1/2} Z_r (Z^r + c^r / n^{1/2} + d^r / n + \dots) / n^{1/2}$$
$$+ (n\kappa_{rs} + n^{1/2} Z_{rs}) \{ Z^r Z^s + 2 Z^r c^s / n^{1/2} + (c^r c^s + 2 Z^r d^s) / n + \dots \} / (2n)$$
$$+ (n\kappa_{rst} + n^{1/2} Z_{rst}) (Z^r Z^s Z^t + 3 Z^r Z^s c^t / n^{1/2} + \dots) / (6n^{3/2})$$
$$+ (n\kappa_{rstu} + n^{1/2} Z_{rstu}) (Z^r Z^s Z^t Z^u + \dots) / (24n^2) + O_p(n^{-3/2}).$$

This expansion includes all terms up to order $O_p(n^{-1})$ in the null case and involves quartic terms in the expansion of $l(\hat{\theta}; Y)$. On collecting together terms that are of equal order in n, much cancellation occurs. For example, in the $O_p(n^{-1/2})$ term, the two expressions involving c^r cancel, and likewise for the two expressions involving d^r in the $O_p(n^{-1})$ term. For this reason, the expansion to order $O_p(n^{-1})$ of $W(\theta)$ does not involve d^r, and c^r occurs only in the $O_p(n^{-1})$ term.

Further simplification using (7.10) gives

$$\tfrac{1}{2}W(\theta) = \tfrac{1}{2} Z_r Z_s \kappa^{r,s} + n^{-1/2} \{ \kappa_{rst} Z^r Z^s Z^t / 3! + Z_{rs} Z^r Z^s / 2! \}$$
$$+ n^{-1} \{ (Z_{ri} Z^i + \tfrac{1}{2} \kappa_{rij} Z^i Z^j) \kappa^{r,s} (Z_{si} Z^i + \tfrac{1}{2} \kappa_{sij} Z^i Z^j) / 2$$
$$+ \kappa_{rstu} Z^r Z^s Z^t Z^u / 4! + Z_{rst} Z^r Z^s Z^t / 3! \} \qquad (7.11)$$

when terms that are $O_p(n^{-3/2})$ are ignored. Note that the $O_p(1)$, $O_p(n^{-1/2})$ and $O_p(n^{-1})$ terms are each invariant. This is not immediately obvious because Z_{rs}, Z_{rst}, κ_{rst} and κ_{rstu} are not tensors. In particular, the individual terms in the above expansion are not invariant.

7.4.3 *Invariant expansion*

The simplest way to obtain an invariant expansion is to use (7.11) in the coordinate system defined by (7.5). We simply replace all κs by νs and re-define the Zs to be

$$
\begin{aligned}
Z_r &= n^{-1/2} V_r \\
Z_{rs} &= n^{-1/2} (V_{rs} - n\nu_{rs}) \\
Z_{rst} &= n^{-1/2} (V_{rst} - n\nu_{rst}).
\end{aligned}
\tag{7.12}
$$

The leading term in the expansion for $W(\theta)$ is $Z_r Z_s \nu^{r,s}$, also known as the score statistic or the quadratic score statistic. The $O_p(n^{-1/2})$ term is

$$
n^{-1/2}(Z_{rs} Z^r Z^s - \nu_{r,s,t} Z^r Z^s Z^t / 3),
$$

which involves the skewness tensor of the first derivatives as well as a 'curvature' correction involving the residual second derivative, Z_{rs}. Note that Z_{rs} is zero for full exponential family models in which the dimension of the sufficient statistic is the same as the dimension of the parameter. See Section 6.2.2, especially Equation (6.7).

7.4.4 *Bartlett factor*

From (7.11) and (7.12), the mean of $W(\theta)$ can be obtained in the form

$$
\begin{aligned}
p + n^{-1}\{&\nu_{rst}\nu^{r,s,t}/3 + \nu_{r,s,tu}\nu^{r,t}\nu^{s,u} + \nu_{rij}\nu_{skl}\nu^{r,s}\nu^{i,j}\nu^{k,l}/4 \\
&+ \nu_{rij}\nu_{skl}\nu^{r,s}\nu^{i,k}\nu^{j,l}/2 + \nu_{rs,tu}\nu^{r,t}\nu^{s,u} + \nu_{rstu}\nu^{r,s}\nu^{t,u}/4\} \\
&+ O(n^{-3/2}).
\end{aligned}
$$

Often the mean is written in the form

$$
E(W(\theta); \theta) = p\{1 + b(\theta)/n + O(n^{-3/2})\}
$$

where $b(\theta)$, known as the Bartlett correction factor, is given by

$$
\begin{aligned}
pb(\theta) = {}&\rho_{13}^2/4 + \rho_{23}^2/6 - (\nu_{r,s,t,u} - \nu_{rs,tu}[3])\nu^{r,s}\nu^{t,u}/4 \\
&- (\nu_{r,s,tu} + \nu_{rs,tu})\nu^{r,s}\nu^{t,u}/2.
\end{aligned}
\tag{7.13}
$$

In deriving the above, we have made use of the identities

$$\nu_{rst} = -\nu_{r,s,t}$$
$$\nu_{rstu} = -\nu_{r,s,t,u} - \nu_{r,s,tu}[6] - \nu_{rs,tu}[3]$$

derived in Sections 7.2.1 and 7.2.3, and also,

$$\rho_{13}^2 = \nu_{i,j,k}\nu_{l,m,n}\nu^{i,j}\nu^{k,l}\nu^{m,n},$$
$$\rho_{23}^2 = \nu_{i,j,k}\nu_{l,m,n}\nu^{i,l}\nu^{j,m}\nu^{k,n},$$
$$\rho_4 = \nu_{i,j,k,l}\nu^{i,j}\nu^{k,l},$$

which are the invariant standardized cumulants of V_r.

The reason for the unusual grouping of terms in (7.13) is that, not only are the individual terms invariant under re-parameterization, but with this particular grouping they are nearly invariant under the operation of conditioning on ancillary statistics. For example, ρ_4 defined above is not invariant under conditioning. This point is examined further in the following chapter: it is an important point because the expression for $b(\theta)$ demonstrates that the conditional mean of $W(\theta)$ is independent of all ancillary statistics, at least to the present order of approximation. In fact, subsequent calculations in the following chapter show that $W(\theta)$ is statistically independent of *all* ancillary statistics to a high order of approximation.

Since Z_r is asymptotically normal with covariance matrix $\kappa_{r,s}$, it follows from (7.11) that the likelihood ratio statistic is asymptotically χ_p^2. This is a first-order approximation based on the leading term in (7.11). The error term in the distributional approximation appears to be $O(n^{-1/2})$, but as we shall see, it is actually $O(n^{-1})$. In fact, it will be shown that the distribution of

$$W' = W/\{1 + b(\theta)/n\}, \tag{7.14}$$

the Bartlett corrected statistic, is $\chi_p^2 + O(n^{-3/2})$. Thus, not only does the Bartlett factor correct the mean of W to this order of approximation, but it also corrects all of the higher-order cumulants of W to the same order of approximation. This is an unusual and surprising result. There is no similar correction for the quadratic score statistic, $U_r U_s \kappa^{r,s}$, which is also asymptotically χ_p^2.

7.4.5 *Tensor decomposition of W*

In order to decompose the likelihood ratio statistic into single degree of freedom contrasts, we define the vector with components

$$
\begin{aligned}
W_r &= Z_r + n^{-1/2}\{Z_{rs}Z^s/2 + \nu_{rst}Z^sZ^t/3!\} \\
&\quad + n^{-1}\{Z_{rst}Z^sZ^t/3! + \nu_{rstu}Z^sZ^tZ^u/4! + 3Z_{rs}Z^{st}Z_t/8 \\
&\quad + 5Z_{rs}Z_tZ_u\nu^{stu}/12 + \nu_{rst}\nu_{uvw}\nu^{t,u}Z^sZ^vZ^w/9\}. \qquad (7.15)
\end{aligned}
$$

It is then easily shown that

$$
W = W_rW_s\nu^{r,s} + O_p(n^{-3/2}).
$$

Further, taking the Zs as defined in (7.12), W_r is a tensor.

The idea in transforming from W to W_r is that the components of W_r may be interpreted as single degree of freedom contrasts, though they are not independent. In addition, as we now show, the joint distribution of W_r is very nearly normal, a fact that enables us to derive the distribution of W.

A straightforward but rather lengthy calculation shows that the joint cumulants of the W_r are

$$
\begin{aligned}
E(W_r;\theta) &= n^{-1/2}\nu_{rst}\nu^{s,t}/3! + O(n^{-3/2}) \\
&= -n^{-1/2}\nu_{r,s,t}\nu^{s,t}/3! + O(n^{-3/2}) \\
\mathrm{cov}(W_r,W_s;\theta) &= \nu_{r,s} + n^{-1}\left(\nu_{rstu}\nu^{t,u}/4 + \nu_{rt,su}\nu^{t,u} + \nu_{r,t,su}\nu^{t,u}\right. \\
&\quad \left. + \nu_{r,i,j}\nu_{s,k,l}\nu^{i,k}\nu^{j,l}/6 + 2\nu_{r,s,i}\nu_{j,k,l}\nu^{i,j}\nu^{k,l}/9\right) + O(n^{-2})
\end{aligned}
$$

$$
\begin{aligned}
\mathrm{cum}(W_r,W_s,W_t;\theta) &= O(n^{-3/2}) \\
\mathrm{cum}(W_r,W_s,W_t,W_u;\theta) &= O(n^{-2}).
\end{aligned}
$$

Higher-order joint cumulants are of order $O(n^{-3/2})$ or smaller. In other words, ignoring terms that are of order $O(n^{-3/2})$, the components of W_r are jointly normally distributed with the above mean and covariance matrix. To the same order of approximation, it follows that W has a scaled non-central χ_p^2 distribution with non-centrality parameter

$$
n^{-1}\nu_{r,s,t}\nu_{u,v,w}\nu^{r,s}\nu^{t,u}\nu^{v,w} = n^{-1}p\bar{\rho}_{12}^3 = a/n,
$$

which is a quadratic form in $E(W_r; \theta)$. The scale factor in this distribution is a scalar formed from the covariance matrix of W_r, namely

$$1 + \{\nu_{rstu}\nu^{r,s}\nu^{t,u}/4! + \nu_{rt,su}\nu^{r,s}\nu^{t,u} + \rho_{23}^2/6 + 2\rho_{13}^2/9\}/(np)$$
$$= 1 + c/n.$$

The rth cumulant of W (Johnson & Kotz, 1970, p. 134) is

$$2^{r-1}(r-1)!(1 + c/n)^r\{1 + ar/(np)\} + O(n^{-3/2})$$
$$= 2^{r-1}(r-1)!p\{1 + b/n\}^r + O(n^{-3/2})$$

where $b = b(\theta)$ is given by (7.13). Thus the rth cumulant of $W' = W/(1 + b/n)$ is $2^{r-1}(r-1)!p + O(n^{-3/2})$, to this order of approximation, the same as the rth cumulant of a χ_p^2 random variable. We conclude from this that the corrected statistic (7.14) has the χ_p^2 distribution to an unusually high order of approximation.

The argument just given is based entirely on formal calculations involving moments and cumulants. While it is true quite generally, for discrete as well as continuous random variables, that the cumulants of W' differ from those of χ_p^2 by $O(n^{-3/2})$, additional regularity conditions are required in order to justify the 'obvious' conclusion that $W' \sim \chi_p^2 + O(n^{-3/2})$. Discreteness has an effect that is of order $O(n^{-1/2})$, although the error term can often be reduced to $O(n^{-1})$ if a continuity correction is made. Despite these caveats, the correction is often beneficial even for discrete random variables for which the 'obvious' step cannot readily be justified. The argument is formally correct provided only that the joint distribution of W_r has a valid Edgeworth expansion up to and including the $O(n^{-1})$ term.

7.5 Some examples

7.5.1 *Exponential regression model*

Suppose, independently for each i, that Y_i has the exponential distribution with mean μ_i satisfying the log-linear model

$$\eta^i = \log(\mu_i) = x_r^i\beta^r. \tag{7.16}$$

The notation used here is close to that used in the literature on generalized linear models where η is known as the *linear predictor*, $\mathbf{X} = \{x_r^i\}$ is called the model matrix and β is the vector of unknown parameters. If we let $Z_i = (Y_i - \mu_i)/\mu_i$, then the first two derivatives of the log likelihood may be written in the form

$$U_r = x_r^i Z_i \quad \text{and.} \quad U_{rs} = -\sum_i x_r^i x_s^i (Y_i/\mu_i).$$

The joint cumulants are as follows

$$n\kappa_{r,s} = x_r^i x_s^j \delta_{ij} = \mathbf{X}^T\mathbf{X}, \qquad \kappa^{r,s} = n(\mathbf{X}^T\mathbf{X})^{-1}$$

$$n\kappa_{r,s,t} = 2x_r^i x_s^j x_t^k \delta_{ijk} \qquad n\kappa_{r,st} = -x_r^i x_s^j x_t^k \delta_{ijk}$$

$$n\kappa_{r,s,t,u} = 6x_r^i x_s^j x_t^k x_u^l \delta_{ijkl} \qquad n\kappa_{r,s,tu} = -2x_r^i x_s^j x_t^k x_u^l \delta_{ijkl}$$

$$n\kappa_{rs,tu} = x_r^i x_s^j x_t^k x_u^l \delta_{ijkl}.$$

In addition, we have the following tensorial cumulants

$$\nu_{rs,tu} = \kappa_{rs,tu} - \kappa_{rs,i}\kappa_{tu,j}\kappa^{i,j}$$

$$\nu_{r,s,tu} = \kappa_{r,s,tu} - \kappa_{r,s,i}\kappa_{tu,j}\kappa^{i,j}.$$

In order to express the Bartlett adjustment factor using matrix notation, it is helpful to define the following matrix and vector, both of order n.

$$\mathbf{P} = \mathbf{X}(\mathbf{X}^T\mathbf{X})^{-1}\mathbf{X}^T, \quad \mathbf{V} = \operatorname{diag}(\mathbf{P}).$$

Note that $\mathbf{P}$ is the usual projection matrix that projects on to the column space of $\mathbf{X}$: it is also the asymptotic covariance matrix of $\hat{\eta}$, the maximum likelihood estimate of η. Thus, the components of $\mathbf{P}$ and $\mathbf{V}$ are $O(n^{-1})$. Straightforward substitution now reveals that the invariant constants in the Bartlett correction term may be written as follows

$$n^{-1}\rho_{13}^2 = 4\mathbf{V}^T\mathbf{P}\mathbf{V}, \quad n^{-1}\rho_{23}^2 = 4\sum_{ij} P_{ij}^3, \quad n^{-1}\rho_4 = 6\mathbf{V}^T\mathbf{V},$$

$$n^{-1}\nu^{r,s}\nu^{t,u}\nu_{rt,su} = \mathbf{V}^T\mathbf{V} - \sum_{ij} P_{ij}^3,$$

$$n^{-1}\nu^{r,s}\nu^{t,u}\nu_{rs,tu} = \mathbf{V}^T\mathbf{V} - \mathbf{V}^T\mathbf{P}\mathbf{V},$$

$$n^{-1}\nu^{r,s}\nu^{t,u}\nu_{r,s,tu} = -2\mathbf{V}^T\mathbf{V} + 2\mathbf{V}^T\mathbf{P}\mathbf{V}.$$

After collecting terms, we find that

$$\epsilon_p = n^{-1}pb(\theta) = \sum_{ij} P_{ij}^3/6 - \mathbf{V}^T(\mathbf{I} - \mathbf{P})\mathbf{V}/4 \qquad (7.17)$$

which is independent of the value of the parameter, as is to be expected from considerations of invariance.

In the particular case where $p = 1$ and $\mathbf{X} = \mathbf{1}$, a vector of 1s, we have $\mathbf{V}^T(\mathbf{I} - \mathbf{P})\mathbf{V} = 0$ and the adjustment reduces to $b = 1/6$. This is the adjustment required in testing the hypothesis $H_0 : \mu_i = 1$ (or any other specified value) against the alternative $H_1 : \mu_i = \mu$, where the value μ is left unspecified. More generally, if the observations are divided into k sets each of size m, so that $n = km$, we may wish to test for homogeneity of the means over the k sets. For this purpose, it is convenient to introduce two indices, i indicating the set and j identifying the observation within a set. In other words, we wish to test H_1 against the alternative $H_2 : \mu_{ij} = \mu_i$, where the k means, $\mu_1, ..., \mu_k$, are left unspecified. In this case, $\mathbf{X}$ is the incidence matrix for a balanced one-way or completely randomized design. Again, we have $\mathbf{V}^T(\mathbf{I} - \mathbf{P})\mathbf{V} = 0$ and the value of the adjustment is given by

$$\epsilon_k = \sum_{ij} P_{ij}^3/6 = k/(6m).$$

This is the adjustment appropriate for testing H_0 against H_2. To find the adjustment appropriate for the test of H_1 against H_2, we subtract, giving

$$\epsilon_k - \epsilon_1 = k/(6m) - 1/(6n).$$

More generally, if the k sets are of unequal sizes, $m_1, ..., m_k$, it is a straightforward exercise to show that

$$\epsilon_k - \epsilon_1 = \sum m_i^{-1}/6 - n^{-1}/6.$$

The test statistic in this case,

$$T = -2 \sum m_i \log(\bar{y}_i/\bar{y})$$

in an obvious notation, is formally identical to Bartlett's (1937) test for homogeneity of variances. It is not difficult to verify directly that the first few cumulants of $(k-1)T/\{k-1+\epsilon_k-\epsilon_1\}$ are the same as those of χ^2_{k-1} when terms of order $O(m_i^{-2})$ are ignored (Bartlett, 1937).

The claim just made does not follow from the results derived in Section 7.4.5, which is concerned only with simple null hypotheses. In the case just described, H_1 is composite because the hypothesis does not determine the distribution of the data. Nevertheless, the adjustment still corrects all the cumulants.

7.5.2 Poisson regression model

Following closely the notation of the previous section, we assume that Y_i has the Poisson distribution with mean value μ_i satisfying the log-linear model (7.16). The first two derivatives of the log likelihood are

$$U_r = x_r^i(Y_i - \mu_i) \quad \text{and} \quad U_{rs} = -\sum_i x_r^i x_s^i \mu_i = -\mathbf{X}^T\mathbf{W}\mathbf{X},$$

where $\mathbf{W} = \mathrm{diag}(\mu_i)$. In this case, U_{rs} is a constant and all cumulants involving U_{rs} vanish. Since all cumulants of Y_i are equal to μ_i, the cumulants of U_r are

$$n\kappa_{r,s} = \sum_i x_r^i x_s^i \mu_i = \mathbf{X}^T\mathbf{W}\mathbf{X}, \quad \kappa^{r,s} = n(\mathbf{X}^T\mathbf{W}\mathbf{X})^{-1}$$

$$n\kappa_{r,s,t} = \sum_i x_r^i x_s^i x_t^i \mu_i, \qquad n\kappa_{r,s,t,u} = \sum_i x_r^i x_s^i x_t^i x_u^i \mu_i.$$

Now define the matrix $\mathbf{P}$ and the vector $\mathbf{V}$ by

$$\mathbf{P} = \mathbf{X}(\mathbf{X}^T\mathbf{W}\mathbf{X})^{-1}\mathbf{X}^T \quad \text{and} \quad \mathbf{V} = \mathrm{diag}(\mathbf{P})$$

so that $\mathbf{P}$ is the asymptotic covariance matrix of $\hat{\eta}$ and $\mathbf{P}\mathbf{W}$ projects on to the column space of $\mathbf{X}$. It follows then that the invariant standardized cumulants of U_r are

$$n^{-1}\rho^2_{13} = \mathbf{V}^T\mathbf{W}\mathbf{P}\mathbf{W}\mathbf{V}$$

$$n^{-1}\rho^2_{23} = \sum_{ij} \mu_i\mu_j P^3_{ij}$$

$$n^{-1}\rho_4 = \mathbf{V}^T\mathbf{W}\mathbf{V}.$$

Thus, the Bartlett adjustment is given by

$$\epsilon_p = n^{-1}pb(\theta) = \sum_{ij} \mu_i\mu_j P_{ij}^3/6 - \mathbf{V}^T\mathbf{W}(\mathbf{I} - \mathbf{PW})\mathbf{V}/4. \quad (7.18)$$

This expression can be simplified to some extent in those cases where $\mathbf{X}$ is the incidence matrix for a decomposable log-linear model (Williams, 1976). See also Cordeiro (1983) who points out that, for decomposable models, $\mathbf{V}$ lies in the column space of $\mathbf{X}$ implying that the second term on the right of (7.18) vanishes for such models. In general, however, the second term in (7.18) is not identically zero: see Exercise 7.15.

7.5.3 Inverse Gaussian regression model

The inverse Gaussian density function, which arises as the density of the first passage time of Brownian motion with positive drift, may be written

$$f_Y(y; \mu, \nu) = \left(\frac{\nu}{2\pi y^3}\right)^{1/2} \exp\left(\frac{-\nu(y-\mu)^2}{2\mu^2 y}\right) \qquad y, \mu > 0. \quad (7.19)$$

For a derivation, see Moran (1968, Section 7.23). This density is a member of the two-parameter exponential family. The first four cumulants of Y are

$$\kappa_1 = \mu, \quad \kappa_3 = 3\mu^5/\nu^2, \quad \kappa_2 = \mu^3/\nu, \quad \kappa_4 = 15\mu^7/\nu^3.$$

These can be obtained directly from the generating function

$$K_Y(\xi) = \nu\{b(\theta + \xi/\nu) - b(\theta)\}$$

where $\theta = -(2\mu^2)^{-1}$, $b(\theta) = -(-2\theta)^{1/2}$ and θ is called the canonical parameter if ν is known. See, for example, Tweedie (1957a,b).

Evidently, ν is a precision parameter or 'effective sample size' and plays much the same role as σ^{-2} in normal-theory models. To construct a linear regression model, we suppose for simplicity that ν is given and constant over all observations. The means of the independent random variables $Y_1, ..., Y_n$ are assumed to satisfy the inverse linear regression model

$$\eta^i = x_r^i\beta^r, \qquad \eta^i = 1/\mu_i,$$

where $\beta^1, ..., \beta^p$ are unknown parameters. This is a particular instance of a generalized linear model in which the variance function is cubic and the link function is the reciprocal. The 'canonical' link function in this instance is $\theta = \mu^{-2}$. In applications, other link functions, particularly the log, might well be found to give a better fit: it is essential, therefore, to check for model adequacy but this aspect will not be explored here.

Using matrix notation, the first two derivatives of the log likelihood may be written

$$U_r = -\nu\{\mathbf{X}^T\mathbf{Y}\mathbf{X}\beta - \mathbf{X}^T\mathbf{1}\},$$
$$U_{rs} = -\nu\mathbf{X}^T\mathbf{Y}\mathbf{X},$$

where $\mathbf{Y} = \text{diag}\{y_1, ..., y_n\}$ is a diagonal matrix of observed random variables.

The above derivatives are formally identical to the derivatives of the usual normal-theory log likelihood with $\mathbf{Y}$ and $\mathbf{1}$ taking the place of the weight matrix and response vector respectively. This analogy is obvious from the interpretation of Brownian motion and from the fact that the likelihood does not depend on the choice of stopping rule. For further discussion of this and related points, see Folks & Chhikara (1978) and the ensuing discussion of that paper.

It follows that the maximum likelihood estimate of β is

$$\hat{\beta} = (\mathbf{X}^T\mathbf{Y}\mathbf{X})^{-1}\mathbf{X}^T\mathbf{1}.$$

This is one of the very rare instances of a generalized linear model for which closed form estimates of the regression parameters exist whatever the model matrix.

The joint cumulants of the log likelihood derivatives are

$$n\kappa_{r,s} = \nu\sum_i x_r^i x_s^i \mu_i = \nu\mathbf{X}^T\mathbf{W}\mathbf{X}, \quad \kappa^{r,s} = n(\mathbf{X}^T\mathbf{W}\mathbf{X})^{-1}/\nu$$

$$n\kappa_{r,s,t} = -3\nu\sum_i x_r^i x_s^i x_t^i \mu_i^2, \qquad n\kappa_{r,s,t,u} = 15\nu\sum_i x_r^i x_s^i x_t^i x_u^i \mu_i^3,$$

$$n\kappa_{r,st} = \nu\sum_i x_r^i x_s^i x_t^i \mu_i^2, \qquad n\kappa_{rs,tu} = \nu\sum_i x_r^i x_s^i x_t^i x_u^i \mu_i^3,$$

$$n\kappa_{r,s,tu} = -3\nu\sum_i x_r^i x_s^i x_t^i x_u^i \mu_i^3.$$

With $\mathbf{P}$, $\mathbf{V}$ and $\mathbf{W}$ as defined in the previous section, it follows after the usual routine calculations that the $O(n^{-1})$ bias of $\hat{\beta}$ is

$$\text{bias}(\hat{\beta}) = (\mathbf{X}^T \mathbf{W} \mathbf{X})^{-1} \mathbf{X}^T \mathbf{W}^2 \mathbf{V}/\nu.$$

In addition, the Bartlett adjustment factor may be simplified to

$$\epsilon_p = \left(\sum_{ij} P_{ij}^3 \mu_i^2 \mu_j^2 + \mathbf{V}^T \mathbf{W}^2 \mathbf{P} \mathbf{W}^2 \mathbf{V} - 2\mathbf{V}^T \mathbf{W}^3 \mathbf{V} \right) /\nu. \quad (7.20)$$

This expression is suitable for routine computation using simple matrix manipulations. It may be verified that if $\mathbf{X}$ is the incidence matrix for a one-way layout, then $\epsilon_p = 0$, in agreement with the known result that the likelihood ratio statistic in this case has an exact χ_p^2 distribution. More generally, for designs that have sufficient structure to discriminate between different link functions, the Bartlett adjustment is non-zero, showing that the exact results do not extend beyond the one-way layout.

7.6 Bibliographic notes

The emphasis in this chapter has been on likelihood ratio statistics and other invariants derived from the likelihood function. In this respect, the development follows closely to that of McCullagh & Cox (1986). Univariate invariants derived from the likelihood are discussed by Peers (1978).

The validity of adjusting the likelihood ratio statistic by means of a straightforward multiplicative correction was first demonstrated by Lawley (1956). His derivation is more comprehensive than ours in that it also covers the important case where nuisance parameters are present.

Bartlett factors and their role as normalization factors are discussed by Barndorff-Nielsen & Cox (1984).

Geometrical aspects of normal-theory non-linear models have been discussed by a number of authors from a slightly different perspective: see, for example, Beale (1960), Bates & Watts (1980) or Johansen (1983).

Amari (1985) deals with arbitrary likelihoods for regular problems, and emphasises the geometrical interpretation of various invariants. These 'interpretations' involve various kinds of *curvatures* and *connection coefficients.* It is not clear that such notions, borrowed from the differential geometry of esoteric spaces, necessarily shed much light on the interpretation of statistical invariants. The subject, however, is still young!

For a slightly more positive appraisal of the role of differential geometry in statistical theory, see the review paper by Barndorff-Nielsen, Cox & Reid (1986).

Further aspects of differential geometry in statistical theory are discussed in a forthcoming IMS monograph by Amari, Barndorff-Nielsen, Kass, Lauritzen and Rao.

7.7   Further results and exercises 7

7.1   Let X_r, X_{rs}, $X_{rst},...$ be a sequence of arrays of arbitrary random variables. Such a sequence will be called *triangular.* Let the joint moments and cumulants be denoted as in Section 7.2.1 by

$$\mu_{r,st,uvw} = E\{X_r X_{st} X_{uvw}\}$$
$$\kappa_{r,st,uvw} = \text{cum}\{X_r, X_{st}, X_{uvw}\}$$

and so on. Now write $\mu_{[...]}$ and $\kappa_{[...]}$ for the sum over all partitions of the subscripts as follows

$$\mu_{[rs]} = \mu_{rs} + \mu_{r,s}$$
$$\mu_{[rst]} = \mu_{rst} + \mu_{r,st}[3] + \mu_{r,s,t}$$

and so on, with identical definitions for $\kappa_{[rs]}$, $\kappa_{[rst]}$ and so on. Show that $\kappa_{[r]} = \mu_{[r]}$,

$$\kappa_{[rs]} = \mu_{[rs]} - \mu_{[r]}\mu_{[s]}$$
$$\kappa_{[rst]} = \mu_{[rst]} - \mu_{[r]}\mu_{[st]}[3] + 2\mu_{[r]}\mu_{[s]}\mu_{[t]}$$
$$\kappa_{[rstu]} = \mu_{[rstu]} - \mu_{[r]}\mu_{[stu]}[4] - \mu_{[rs]}\mu_{[tu]}[3] + 2\mu_{[r]}\mu_{[s]}\mu_{[tu]}[6]$$
$$- 6\mu_{[r]}\mu_{[s]}\mu_{[t]}\mu_{[u]}.$$

Hence show that the cumulants of the log likelihood derivatives satisfy $\kappa_{[...]} = 0$, whatever the indices.

7.2   Give a probabilistic interpretation of $\mu_{[...]}$ and $\kappa_{[...]}$ as defined in the previous exercise.

7.3 Give the inverse formulae for $\mu_{[\ldots]}$ in terms of $\kappa_{[\ldots]}$.

7.4 By first examining the derivatives of the null moments of log likelihood derivatives, show that the derivatives of the null cumulants satisfy

$$\frac{\partial \kappa_{r,s}(\theta)}{\partial \theta^t} = \kappa_{rt,s} + \kappa_{r,st} + \kappa_{r,s,t}$$

$$\frac{\partial \kappa_{rs}(\theta)}{\partial \theta^t} = \kappa_{rst} + \kappa_{rs,t}$$

$$\frac{\partial \kappa_{r,st}(\theta)}{\partial \theta^u} = \kappa_{ru,st} + \kappa_{r,stu} + \kappa_{r,st,u}$$

$$\frac{\partial \kappa_{r,s,t}(\theta)}{\partial \theta^u} = \kappa_{ru,s,t} + \kappa_{r,su,t} + \kappa_{r,s,tu} + \kappa_{r,s,t,u}.$$

State the generalization of this result that applies to

(i) cumulants of arbitrary order (Skovgaard, 1986a)
(ii) derivatives of arbitrary order.

Hence derive identities (7.2) by repeated differentiation of $\kappa_r(\theta)$.

7.5 Using expansions (7.3) for the non-null moments, derive expansions (7.4) for the non-null cumulants.

7.6 Express the four equations (7.6) simultaneously using matrix notation in the form

$$\mathbf{U} = \mathbf{B}\mathbf{V}$$

and give a description of the matrix $\mathbf{B}$. It may be helpful to define $\beta_r^i = \delta_r^i$.

7.7 Show that equations (7.7) may be written in the form

$$\mathbf{U} = \mathbf{A}\bar{\mathbf{U}}$$

where $\mathbf{A}$ has the same structure as $\mathbf{B}$ above.

7.8 Show that under transformation of coordinates on Θ, the coefficient matrix $\mathbf{B}$ transforms to $\bar{\mathbf{B}}$, where

$$\mathbf{B} = \mathbf{A}\bar{\mathbf{B}}\mathbf{A}^{*-1}$$

and give a description of the matrix $\mathbf{A}^*$.

7.9 Using the results of the previous three exercises, show that the arrays V_r, V_{rs},..., defined at (7.6), behave as tensors under change of coordinates on Θ.

7.10 Let s_i^2, $i = 1, ..., k$ be k independent mean squares calculated from independent normal random variables. Suppose

$$E(s_i^2) = \sigma_i^2, \quad \mathrm{var}(s_i^2) = 2\sigma_i^4/m_i$$

where m_i is the number of degrees of freedom for s_i^2. Derive the likelihood ratio statistic, W, for testing the hypothesis $H_0 : \sigma_i^2 = \sigma^2$ against the alternative that leaves the variances unspecified. Using the results of Section 7.5.1, show that under H_0,

$$E(W) = k - 1 + \tfrac{1}{3} \sum m_i^{-1} - \tfrac{1}{3} m_\bullet^{-1}$$

where $m_\bullet$ is the total degrees of freedom. Hence derive Bartlett's test for homogeneity of variances (Bartlett, 1937).

7.11   Suppose that $Y_1, ..., Y_n$ are independent Poisson random variables with mean μ. Show that the likelihood ratio statistic for testing $H_0 : \mu = \mu_0$ against an unspecified alternative is

$$W = 2n\{\bar{Y} \log(\bar{Y}/\mu_0) - (\bar{Y} - \mu_0)\}.$$

By expanding in a Taylor series about $\bar{Y} = \mu_0$ as far as the quartic term, show that

$$E(W; \mu_0) = 1 + \frac{1}{6n\mu_0} + O(n^{-2}).$$

7.12 Derive the result stated in the previous exercise directly from (7.18). Check the result numerically for $\mu_0 = 1$ and for $n = 1, 5, 10$. Also, check the variance of W numerically. You will need either a computer or a programmable calculator.

7.13 Using the notation of the previous two exercises, let $\pm W^{1/2}$ be the signed square root of W, where the sign is that of $\bar{Y} - \mu_0$. Using the results given in Section 7.4.5, or otherwise, show that

$$E(\pm W^{1/2}) = -\frac{1}{6(n\mu_0)^{1/2}} + O(n^{-3/2}),$$

$$\mathrm{var}(\pm W^{1/2}) = 1 + \frac{1}{8n\mu_0} + O(n^{-2}),$$

$$\kappa_3(\pm W^{1/2}) = O(n^{-3/2}), \qquad \kappa_4(\pm W^{1/2}) = O(n^{-2}).$$

Hence show that under $H_0 : \mu = \mu_0$,

$$S = \frac{\pm W^{1/2} + (n\mu_0)^{-1/2}/6}{1 + (16n\mu_0)^{-1}}$$

has the same moments as those of $N(0,1)$ when terms of order $O(n^{-3/2})$ are ignored. Why is it technically wrong in this case to say that

$$S \sim N(0,1) + O(n^{-3/2})?$$

7.14 Repeat the calculations of the previous exercise, this time for the exponential distribution in place of the Poisson. Compare numerically the transformation $\pm W^{1/2}$ with the Wilson-Hilferty cube root transformation

$$3n^{1/2}\left\{\left(\frac{\bar{Y}}{\mu_0}\right)^{1/3} + \frac{1}{9n} - 1\right\},$$

which is also normally distributed to a high order of approximation. Show that the cube root transformation has kurtosis of order $O(n^{-1})$, whereas $\pm W^{1/2}$ has kurtosis of order $O(n^{-2})$. For a derivation of the above result, see Kendall & Stuart (1977, Section 16.7).

7.15 Show that the second term on the right in (7.17) is zero if **X** is the incidence matrix for

(i)   an unbalanced one-way layout

(ii) a randomized blocks design (two-way design) with equal numbers of replications per cell.

(iii) a Latin square design.

Show that the second term is not zero if **X** is the model matrix for an ordinary linear regression model in which more than two x-values are observed.

7.16 Find expressions for the first term in (7.17) for the four designs mentioned in the previous exercise.

7.17 Simplify expression (7.18) in the case of a two-way contingency table and a model that includes no interaction term. Show that the second term is zero.

7.18 Using expression (7.13) for $b(\theta)$ together with the expressions given for the cumulants in Section 7.5.1, derive (7.17) as the Bartlett correction applicable to the exponential regression model (7.16).

7.19   Comment on the similarity between the correction terms, (7.17) and (7.18).

7.20   Using expression (7.15) for W_r, show that the joint third and fourth cumulants are $O(n^{-3/2})$ and $O(n^{-2})$ respectively. Derive the mean vector and covariance matrix. Hence justify the use of the Bartlett adjustment as a multiplicative factor.

7.21   *Normal circle model*: Suppose Y is a bivariate normal random vector with mean $(\rho\cos(\theta), \rho\sin(\theta))$ and covariance matrix $n^{-1}I_2$. Let $\rho = \rho_0$ be given.

(i) Find the maximum likelihood estimate of θ.

(ii) Derive the likelihood ratio statistic for testing the hypothesis $H_0 : \theta = 0$.

(iii) Interpret the first two log likelihood derivatives and the likelihood ratio statistic geometrically.

(iv) Show that the Bartlett correction for testing the hypothesis in (ii) is $b(\theta) = 1/(4\rho_0^2)$.

(v) Show that the first derivative of the log likelihood function is normally distributed: find its variance.

7.22   Using the results derived in the previous exercise for $n = 4$ and $\rho_0 = 1$, construct 95% confidence intervals for θ based on (a) the score statistic and (b) the likelihood ratio statistic. For numerical purposes, consider the two data values $(y_1, y_2) = (0.5, 0.0)$ and $(1.5, 0.0)$. Is the value $\theta = \pi$ consistent with either observation? Plot the log likelihood function. Plot the first derivative against θ. Comment on the differences between the confidence intervals.

7.23   *Normal spherical model*: Suppose Y is a trivariate normal random vector with mean $(\rho\cos(\theta)\cos(\phi), \rho\cos(\theta)\sin(\phi), \rho\sin(\theta))$ and covariance matrix $n^{-1}I_3$. Let $\rho = \rho_0$ be given.

(i) Find the maximum likelihood estimate of (θ, ϕ).

(ii) Derive the likelihood ratio statistic for testing the hypothesis $H_0 : \theta = 0, \phi = 0$.

(iii) Show that the Bartlett correction for testing the hypothesis in (ii) is identically zero regardless of the value of ρ_0.

7.24   *Normal hypersphere model*: Repeat the calculations of the previous exercise, replacing the spherical surface in 3-space by a p-dimensional spherical surface in R^{p+1}. Show that the Bartlett adjustment reduces to

$$b(\theta) = \frac{-(p-2)}{4\rho_0^2},$$

which is negative for $p \geq 3$. (McCullagh & Cox, 1986).

7.25 By considering the sum of two independent inverse Gaussian random variables, justify the interpretation of ν in (7.19) as an 'effective sample size'.

7.26 Show that (7.20) vanishes if $\mathbf{X}$ is the incidence matrix for an unbalanced one-way layout.

7.27 Derive the expression analogous to (7.20) for the log link function replacing the reciprocal function. Simplify in the case $p = 1$.

7.28 For the exponential regression model of Section 7.5.1, show that the $O(n^{-1})$ bias of $\hat{\beta}$ is

$$\text{bias}(\hat{\beta}) = -(\mathbf{X}^T\mathbf{X})^{-1}\mathbf{X}^T\mathbf{V}/2.$$

Show that, for a simple random sample of size 1, the bias is exactly $-\gamma$, where $\gamma = 0.57721$ is Euler's constant. Find the exact bias for a simple random sample of size n and compare with the approximate formula.

7.29 Repeat the calculations of the previous exercise for the Poisson log-linear model of Section 7.5.2.

CHAPTER 8

Ancillary statistics

8.1 Introduction

In problems of parametric inference, an ancillary statistic is one whose distribution is the same for all parameter values. This is the simplest definition and the one that will be used here. Certain other authors, e.g. Cox & Hinkley (1974), require that the statistic in question be a component of the minimal sufficient statistic, and this is often a reasonable requirement if only to ensure uniqueness in certain special cases. More complicated definitions are required to deal satisfactorily with the case where nuisance parameters are present. For simplicity, nuisance parameters are not considered here.

Ancillary statistics usually arise in one of two ways. One simple example of each will suffice to show that no frequency theory of inference can be considered entirely satisfactory or complete if ancillary statistics are ignored. Even an approximate theory must attempt to come to grips with conditional inference if approximations beyond the first order are contemplated.

The first and most common way in which an ancillary statistic occurs is as a random sample size or other random index identifying the experiment actually performed. Suppose that N is a random variable whose probability distribution is

$$\mathrm{pr}(N = 1) = 0.5, \quad \mathrm{pr}(N = 100) = 0.5.$$

Conditionally on $N = n$, the random variables $Y_1, ..., Y_n$ are normally distributed with mean θ and unit variance. On the basis of the observed values n and $y_1, ..., y_n$, probabilistic statements concerning the value of θ are required.

The same problem can be posed in a number of ways. For instance, either of two machines can be used to measure a certain quantity θ. Both machines are unbiased, the first with variance 1.0, the second with variance 1/100. A machine is selected at random; a measurement is taken and the machine used is recorded.

In the first case N is ancillary: in the second case the indicator variable for the machine used is an ancillary statistic. In either case, the ancillary statistic serves as an index or indicator for the experiment actually performed.

Whatever the distribution of N, few statisticians would quarrel with the statement that $\hat{\theta} = \bar{Y}_N$ is normally distributed with mean θ and variance $1/N$, even though N is a random variable. Implicitly, we invoke the conditionality principle, which states that values of the ancillary other than that observed are irrelevant for the purposes of inference. The unconditional variance of $\hat{\theta}$, namely $E(1/N) = 1/2 + 1/200 = 0.505$, is relevant only if N is not observed and if its distribution as stated above is correct.

In fact, ancillary statistics of this kind arise frequently in scientific work where observations are often censored or unavailable for reasons not connected with the phenomenon under investigation. Accidents will happen, so the saying goes.

The second type of ancillary arises in the context of a translation model or, more generally, a group transformation model. To take a very simple example, suppose that $Y_1, ..., Y_n$ are independent and identically distributed uniformly on the interval $(\theta, \theta + 1)$. This is an instance of a non-regular problem in which the sample space is parameter dependent. The sufficient statistic is the pair of extreme values, $(Y_{(1)}, Y_{(n)})$, or equivalently, the minimum $Y_{(1)}$ and the range $R = Y_{(n)} - Y_{(1)}$. Clearly, by invariance, the distribution of R is unaffected by the value of θ and hence R is ancillary. Given that $R = r$, the conditional distribution of $Y_{(1)}$ is uniform on the interval $(\theta, \theta + 1 - r)$. Equivalently, the conditional distribution of $Y_{(n)} = Y_{(1)} + r$ is uniform over the interval $(\theta + r, \theta + 1)$. Either conditional distribution leads to conditional equi-tailed confidence intervals

$$\{y_{(1)} - (1 - r)(1 - \alpha/2), \quad y_{(1)} - (1 - r)\alpha/2\} \qquad (8.1)$$

at level $1 - \alpha$. Although these intervals also have the required coverage probability unconditionally as well as conditionally, their unconditional properties seem irrelevant if the aim is to summarise the information in the data concerning the value of θ.

On the other hand, if the ancillary is ignored, we may base our inferences on the pivot $Y_{(1)} - \theta$ whose density is $n(1 - y)^{n-1}$ over the interval $(0, 1)$. If $n = 2$,

$$(Y_{(1)} - 0.7764, \quad Y_{(1)} - 0.0253) \tag{8.2}$$

is an exact equi-tailed 90% confidence interval for θ. Even though it is technically correct and exact, the problem with this statement is that it is inadequate as a summary of the data actually observed. In fact, if $R = 0.77$, not a particularly extreme value, the conditional coverage of (8.2) is approximately 96% rather than the 90% quoted. At the other extreme, if $R \geq 0.975$, the conditional coverage of (8.2) is exactly zero so that the unconditional 90% coverage is quite misleading.

These examples demonstrate the poverty of any frequency theory of inference that ignores ancillary statistics. Having made that point, it must be stated that the difficulties involved in establishing a coherent frequency theory of conditional inference are formidable. Current procedures, even for an approximate conditional theory, are incomplete although much progress has been made in the past decade. Two difficulties are connected with the existence and uniqueness of ancillary statistics. In general, exactly ancillary statistics need not exist. Even where they do exist they need not be unique. If two ancillary statistics exist, they are not, in general, jointly ancillary and an additional criterion may be required to choose between them (Cox, 1971). One device that is quite sensible is to require ancillaries to be functions of the minimal sufficient statistic. Even this does not guarantee uniqueness. Exercises 8.2–8.4 discuss one example where the maximal ancillary is not unique but where the conclusions given such an ancillary are unaffected by the choice of ancillary.

Given the overwhelming magnitude of these difficulties, it appears impossible to devise an exact frequency theory of conditional inference to cover the most general cases. For these reasons, we concentrate here on approximate theories based on statistics that are ancillary in a suitably defined approximate sense. It is tempting here to make a virtue out of necessity, but the approximate theory does have the advantage of greater 'continuity' between those models for which exact analysis is possible and the majority of models for which no exact theory exists.

8.2   Joint cumulants

8.2.1   *Joint cumulants of A and U*

Suppose that A has a distribution not depending on the parameter
θ. The moments and cumulants do not depend on θ and hence their
derivatives with respect to θ must vanish. Thus, differentiation of

$$E(A(Y)) = \int A(y) f_Y(y;\theta) dy = \text{const}$$

with respect to θ gives

$$\int A(y) u_r(\theta;y) f_Y(y;\theta) dy = \text{cov}(A, U_r) = 0.$$

This final step assumes the usual regularity condition that it is le-
gitimate to interchange the order of differentiation and integration.
In particular, the sample space must be the same for all parameter
values.

Further differentiation gives

$$\text{cov}(A, U_{[rs]}) = 0; \quad \text{cov}(A, U_{[rst]}) = 0, \dots$$

and so on.

If A is ancillary, all functions of A are also ancillary. Hence it
follows that the joint cumulants

$$\kappa_3(A, A, U_r) = 0, \qquad \kappa_4(A, A, A, U_r) = 0, \dots$$
$$\kappa_3(A, A, U_{[rs]}) = 0, \quad \kappa_4(A, A, A, U_{[rs]}) = 0, \dots$$

also vanish. These results extend readily to higher-order joint
cumulants provided that these cumulants involve one of U_r, $U_{[rs]}$,
$U_{[rst]}, \dots$ exactly once.

In general, although U_r is uncorrelated with A and with all
functions of A, the conditional variance of U_r given A may be
heavily dependent on the value of A. For instance, if A is a random
sample size, the conditional covariance matrix of U_r given $A = a$
is directly proportional to a.

These results apply to *any* exact ancillary whatsoever: they
also apply approximately to any approximate ancillary suitably
defined.

8.2.2 Conditional cumulants given A

The conditional distribution of the data given $A = a$ is

$$f_{Y|A}(y|a;\theta) = f_Y(y;\theta)/f_A(a).$$

Hence, the conditional log likelihood differs from the unconditional log likelihood by a function not involving θ. Derivatives with respect to θ of the conditional log likelihood are therefore identical to derivatives of the unconditional log likelihood. With some exceptions as noted below, the conditional cumulants are different from the unconditional cumulants. In particular, it follows from differentiating the identity

$$\int f_{Y|A}(y|a;\theta) = 1$$

that the following identities hold

$$E(U_r|A) = 0$$
$$E(U_{[rs]}|A) = 0$$
$$E(U_{[rst]}|A) = 0.$$

Thus, the Bartlett identities (7.2) for the null cumulants of log likelihood derivatives are true conditionally on the value any ancillary statistic as well as unconditionally. Other cumulants or combinations of cumulants are affected by conditioning in a way that depends on the particular ancillary selected.

8.3   Local ancillarity

Exactly ancillary statistics arise usually in one of two ways, either as a random sample size or other index of the experiment actually performed, or as as a configuration statistic in location problems. These instances of exactly ancillary statistics are really rather special and in the majority of problems, no exactly ancillary statistic exists. In some cases it is known that no exact ancillary exists: in other cases no exact ancillary is known and it is suspected that none exists. For that reason, we concentrate our attention here on statistics that are approximately ancillary in some suitable sense.

Let θ_0 be an arbitrary but specified parameter value and let $A = A(Y;\theta_0)$ be a candidate ancillary statistic. If the distribution of A is approximately the same for all parameter values in some suitably chosen neighbourhood of θ_0, we say that A is approximately ancillary or locally ancillary in the vicinity of θ_0. More precisely, if the distribution of A under the density evaluated at $\theta_0 + n^{-1/2}\delta$ satisfies

$$f_A(a; \theta_0 + n^{-1/2}\delta) = f_A(a; \theta_0)\{1 + O(n^{-q/2})\}, \qquad (8.3)$$

we say that A is qth-order locally ancillary in the vicinity of θ_0 (Cox, 1980). Two approximations are involved in this definition of approximate ancillarity. First, there is the $O(n^{-q/2})$ tolerance in (8.3), where n is the sample size or other measure of the extent of the observational record. Second, the definition is local, applying only to parameter values in an $O(n^{-1/2})$ neighbourhood of the target value, θ_0. The latter restriction is reasonable, at least in large samples, because if θ_0 is the true parameter point, the likelihood function eventually becomes negligible outside this $O(n^{-1/2})$ neighbourhood.

From (8.3) it can be seen that the log likelihood based on A satisfies

$$l_A(\theta_0 + n^{-1/2}\delta) = l_A(\theta_0) + O(n^{-q/2}). \qquad (8.4)$$

Since $\partial l/\partial\theta = n^{1/2}\partial l/\partial\delta$, it follows that the Fisher information based on A for θ at θ_0 must be $O(n^{1-q/2})$ at most. This criterion is also used as a definition of approximate ancillarity.

If the distribution of A can be approximated by an Edgeworth series in which the cumulants are κ_1, κ_2, κ_3,..., condition (8.3) is equivalent to

$$\nabla\kappa_r \equiv \kappa_r(\theta_0 + n^{-1/2}\delta) - \kappa_r(\theta_0) = O(n^{-q/2}), \quad r = 1, 2, \dots \quad (8.5)$$

provided that A has been standardized so that $A - \kappa_1(\theta_0)$ is $O_p(1)$. Since κ_r is $O(n^{1-r/2})$, it is necessary only to check the first q cumulants of A in order to verify the property of local ancillarity. Usually (8.5) is much easier to verify than (8.3). Note, however, that A is usually a vector or array of random variables so that the rth cumulant array has a large number of components, each of which must satisfy (8.5).

We now proceed to show that approximately ancillary statistics exist and that they may be constructed from log likelihood derivatives. Since there is little merit in considering derivatives that are not tensors, we work with the tensorial derivatives V_r, $V_{rs}, \ldots$ at θ_0, as defined in (7.6). These derivatives are assumed to be based on n independent observations, to be asymptotically normal and to have all cumulants of order $O(n)$.

Under θ_0, the expectation of V_{rs} is $n\nu_{rs}$. By (7.4), the expectation of V_{rs} under the density at $\theta_0 + n^{-1/2}\delta$ is

$$n\{\nu_{rs} + \nu_{rs,i}\delta^i/n^{1/2} + \nu_{rs,[ij]}\delta^i\delta^j/(2n) + \ldots\}.$$

In terms of the standardized random variables $Z_{rs} = n^{-1/2}(V_{rs} - n\nu_{rs})$, which are $O_p(1)$ both under θ_0 and under the density at $\theta_0 + \delta/n^{1/2}$, we have

$$E(Z_{rs}; \theta_0 + \delta/n^{1/2}) = \nu_{rs,i}\delta^i + n^{-1/2}\nu_{rs,[ij]}\delta^i\delta^j/2 + O(n^{-1}).$$

Thus, since $\nu_{rs,i} = 0$, it follows from (8.5) that Z_{rs} is first-order locally ancillary in the vicinity of θ_0. Hence V_{rs}, the tensorial array of second derivatives, is first-order locally ancillary.

A similar argument shows that V_{rst} is also first-order locally ancillary but not ancillary to second order.

To construct a statistic that is second-order locally ancillary, we begin with a first-order ancillary statistic, Z_{rs}, that is $O_p(1)$ and aim to make a suitable adjustment of order $O_p(n^{-1/2})$. Thus, we seek coefficients $\beta_{rs}^{i,j}$, $\beta_{rs}^{i,jk}$ such that the cumulants of

$$A_{rs} = Z_{rs} - n^{-1/2}\beta_{rs}^{i,j}Z_iZ_j - n^{-1/2}\beta_{rs}^{i,jk}Z_iZ_{jk} \qquad (8.6)$$

satisfy the ancillarity conditions up to second order. Since ancillarity is preserved under transformation, it is unnecessary in (8.6) to include quadratic terms in Z_{rs}. The differences between the cumulants of Z_r, Z_{rs} under θ_0 and under $\theta_0 + n^{-1/2}\delta$ are given by

$$\nabla E(Z_r) = \nu_{r,i}\delta^i + n^{-1/2}\nu_{r,[ij]}\delta^i\delta^j/2 + O(n^{-1})$$
$$\nabla E(Z_{rs}) = n^{-1/2}\nu_{rs,[ij]}\delta^i\delta^j/2 + O(n^{-1})$$
$$\nabla \mathrm{cov}(Z_r, Z_s) = n^{-1/2}\nu_{r,s,i}\delta^i + O(n^{-1})$$
$$\nabla \mathrm{cov}(Z_r, Z_{st}) = n^{-1/2}\nu_{r,st,i}\delta^i + O(n^{-1}).$$

It follows that the mean of A_{rs} changes by an amount

$$\nabla E(A_{rs}) = n^{-1/2}\{\nu_{rs,[ij]} - 2\beta_{rs}^{k,l}\nu_{i,k}\nu_{j,l}\}\delta^i\delta^j + O(n^{-1}).$$

Thus, if we choose the coefficients

$$\beta_{rs}^{i,j} = \nu_{rs,[kl]} \nu^{i,k} \nu^{j,l}/2, \tag{8.7}$$

it follows that $\nabla E(A_{rs}) = O(n^{-1})$ as required. The coefficients $\beta_{rs}^{i,j}$ are uniquely determined by second-order ancillarity.

To find the remaining coefficients, it is necessary to compute $\nabla \mathrm{cov}(A_{rs}, A_{tu})$ and to ensure that this difference is $O(n^{-1})$. Calculations similar to those given above show that

$$\nabla \mathrm{cov}(A_{rs}, A_{tu}) = n^{-1/2}\{\nu_{rs,tu,i}\delta^i - \beta_{rs}^{i,jk}\nu_{jk,tu}\nu_{i,l}\delta^l[2]\} + O(n^{-1}).$$

The coefficients $\beta_{rs}^{i,jk}$ are not uniquely determined by the requirement of second-order ancillarity unless the initial first-order statistic is a scalar. Any set of coefficients that satisfies

$$\{\beta_{rs}^{i,jk}\nu_{jk,tu} + \beta_{tu}^{i,jk}\nu_{jk,rs}\}\nu_{i,v} = \nu_{rs,tu,v} \tag{8.8}$$

gives rise to a statistic that is locally ancillary to second order. If, to the order considered, all such ancillaries gave rise to the same sample space conditionally, non-uniqueness would not be a problem. In fact, however, two second-order ancillaries constructed in the above manner need not be jointly ancillary to the same order.

So far, we have not imposed the obvious requirement that the coefficients $\beta_{rs}^{i,jk}$ should satisfy the transformation laws of a tensor. Certainly, (8.8) permits non-tensorial solutions. With the restriction to tensorial solutions, it might appear that the only solution to (8.8) is

$$\beta_{rs}^{i,jk} = \nu_{rs,tu,v} \nu^{tu,jk} \nu^{i,v}/2,$$

where $\nu^{rs,tu}$ is a generalized inverse of $\nu_{rs,tu}$. Without doubt, this is the most obvious and most 'reasonable' solution, but it is readily demonstrated that it is not unique. For instance, if we define the tensor $\epsilon_{rs}^{i,jk}$ by

$$\epsilon_{rs}^{i,jk}\nu_{jk,tu} = \{\nu_{rs,t,u,v} - \nu_{r,s,tu,v}\}\nu^{i,v},$$

it is easily seen that $\beta_{rs}^{i,jk} + \epsilon_{rs}^{i,jk}$ is also a solution to (8.8). In fact, any scalar multiple of ϵ can be used here.

It is certainly possible to devise further conditions that would guarantee uniqueness or, alternatively, to devise criteria in order to select the most 'relevant' of the possible approximate ancillaries. In the remainder of this section, however, our choice is to tolerate the non-uniqueness and to explore its consequences for conditional inference.

The construction used here for improving the order of ancillarity is taken from Cox (1980), who considered the case of one-dimensional ancillary statistics, and from McCullagh (1984a), who dealt with the more general case. Skovgaard (1986c) shows that, under suitable regularity conditions, the order of ancillarity may be improved indefinitely by successive adjustments of decreasing orders. Whether it is desirable in practice to go much beyond second or third order is quite another matter.

8.4 Stable combinations of cumulants

In Section 8.2.2, it was shown that, conditionally on any ancillary however selected, certain combinations of cumulants of log likelihood derivatives are identically zero. Such combinations whose value is unaffected by conditioning may be said to be *stable*. Thus, for instance, ν_i and $\nu_{ij} + \nu_{i,j}$ are identically zero whether we condition on A or not. It is important to recognize stable combinations in order to determine the effect, if any, of the choice of ancillary on the conclusions reached.

In this section, we demonstrate that, for the type of ancillary considered in the previous section, certain cumulants and cumulant combinations, while not exactly stable, are at least stable to first order in n. By way of example, it will be shown that $\nu_{i,j,k,l}$ and $\nu_{ij,kl}$ are both unstable but that the combination

$$\nu_{i,j,k,l} - \nu_{ij,kl}[3] \tag{8.9}$$

is stable to first order.

Suppose then that A, with components A_r, is ancillary, either exactly, or approximately to some suitable order. It is assumed that the joint distribution of (A, V_i, V_{ij}) may be approximated by an Edgeworth series and that all joint cumulants are $O(n)$. The dimension of A need not equal p but must be fixed as $n \to \infty$. The mean and variance of A are taken to be 0 and $n\lambda_{r,s}$ respectively. Thus, $A_r = O_p(n^{1/2})$ and $A^r = n^{-1}\lambda^{r,s}A_s$ is $O_p(n^{-1/2})$.

From the results given in Section 5.6, the conditional covariance of V_i and V_j is

$$\kappa_2(V_i, V_j | A) = n\{\nu_{i,j} + \nu_{r;i,j} A^r + O(n^{-1})\}$$
$$= n\nu_{i,j} + O(n^{1/2}),$$

where $\nu_{r;i,j}$ is the third-order joint cumulant of A_r, V_i, V_j. Note that the ancillarity property

$$\mathrm{cov}(A_r, V_i) = \nu_{r;i} = 0$$

greatly simplifies these calculations. Similarly, in the case of the third cumulant, we have

$$\kappa_3(V_i, V_j, V_k | A) = n\{\nu_{i,j,k} + \nu_{r;i,j,k} A^r + O(n^{-1})\}$$
$$= n\nu_{i,j,k} + O(n^{1/2}),$$

Thus, to first order at least, $\nu_{i,j}$ and $\nu_{i,j,k}$ are unaffected by conditioning. We say that these cumulants are *stable to first order*.

On the other hand, from the final equation in Section 5.6.2, we find that the conditional fourth cumulant of V_i, V_j, V_k, V_l given A is

$$\kappa_4(V_i, V_j, V_k, V_l | A) = n\{\nu_{i,j,k,l} - \nu_{r;i,j}\nu_{s;k,l}\lambda^{r,s}[3] + O(n^{-1/2})\}.$$

In this case, unlike the previous two calculations, conditioning has a substantial effect on the leading term. Thus, $\nu_{i,j,k,l}$ is unstable to first order: its interpretation is heavily dependent on the conditioning event.

Continuing in this way, it may be seen that the conditional covariance matrix of V_{ij} and V_{kl} is

$$\kappa_2(V_{ij}, V_{kl} | A) = n\{\nu_{ij,kl} - \nu_{r;ij}\nu_{s;kl}\lambda^{r,s}[3] + O(n^{-1/2})\}.$$

Again, this is an unstable cumulant. However, from the identity $\nu_{r;ij} = -\nu_{r,i,j}$ it follows that the combination

$$\kappa_4(V_i, V_j, V_k, V_l | A) - \kappa_2(V_{ij}, V_{kl} | A)[3] = n\{\nu_{i,j,k,l} - \nu_{ij,kl}[3] + O(n^{-1/2})\}$$

is stable to first order. Similarly, the conditional third cumulant

$$\kappa_3(V_i, V_j, V_{kl} | A) = n\{\nu_{i,j,kl} - \nu_{r;i,j}\nu_{s;kl}\lambda^{r,s} + O(n^{-1/2})\},$$

is unstable, whereas the combination

$$\nu_{i,j,kl} + \nu_{ij,kl} \qquad (8.10)$$

is stable to first order.

These calculations are entirely independent of the choice of ancillary. They do not apply to the random sample size example discussed in Section 8.1 unless $\mu = E(N) \to \infty$, $\mathrm{var}(N) = O(\mu)$ and certain other conditions are satisfied. However, the calculations do apply to the approximate ancillary constructed in the previous section. Note that, conditionally on the ancillary (8.6), the conditional covariance of V_{ij} and V_{kl} is reduced from $O(n)$ to $O(1)$, whereas the third-order joint cumulant of V_i, V_j and V_{kl} remains $O(n)$. This is a consequence of the stability of the combination (8.10).

In this context, it is interesting to note that the Bartlett adjustment factor (7.11) is a combination of the four stable combinations derived here. It follows that, up to and including terms of order $O(n^{-1})$, the likelihood ratio statistic is independent of all ancillary and approximately ancillary statistics.

8.5   Orthogonal statistic

Numerical computation of conditional distributions and conditional tail areas is often a complicated unappealing task. In many simple problems, particularly those involving the normal-theory linear model, the problem can be simplified to a great extent by 'regressing out' the conditioning statistic and forming a 'pivot' that is independent of the conditioning statistic. In normal-theory and other regression problems, the reasons for conditioning are usually connected with the elimination of nuisance parameters, but the same device of constructing a pivotal statistic can also be used to help cope with ancillary statistics. The idea is to start with an arbitrary statistic, V_i say, and by making a suitable minor adjustment, ensure that the adjusted statistic, S_i, is independent of the ancillary to the order required. Inference can then be based on the marginal distribution of S_i: this procedure is sometimes labelled 'conditional inference without tears' or 'conditional inference without conditioning'.

In those special cases where there is a complete sufficient statistic, S, for the parameters, Basu's theorem (Basu, 1955, 1958) tells us that all ancillaries are independent of S. This happy state of affairs means that inferences based on the marginal distribution of S are automatically conditional and are safe from conditionality criticisms of the type levelled in Section 8.1. By the same token, the result given at the end of the previous section shows that the maximized likelihood ratio statistic is similarly independent of all ancillaries and of approximate ancillaries to third order in n. Inferences based on the marginal distribution of the likelihood ratio statistic are, in large measure, protected against criticism on grounds of conditionality.

The score statistic, V_i, is asymptotically independent of all ancillaries, but only to first order in n. In other words, $\text{cov}(V_i, A) = 0$ for all ancillaries. First-order independence is a very weak requirement and is occasionally unsatisfactory if n is not very large or if the ancillary takes on an unusual or extreme value. For this reason we seek an adjustment to V_i to make the adjusted statistic independent of all ancillaries to a higher order in n. For the resulting statistic to be useful, it is helpful to insist that it have a simple null distribution, usually normal, again to the same high order of approximation.

Thus, we seek coefficients $\gamma_r^{i,j}$, $\gamma_r^{i,jk}$ such that the adjusted statistic

$$S_r = Z_r + n^{-1/2}\{\gamma_r^{i,j}(Z_i Z_j - \nu_{i,j}) + \gamma_r^{i,jk} Z_i Z_{jk}\}$$

is independent of A to second order and also normally distributed to second order. Both of these calculations are made under the null density at θ_0.

For the class of ancillaries (8.6) based on the second derivatives of the log likelihood, we find

$$\text{cov}(S_r, A_{st}) = O(n^{-1})$$
$$\kappa_3(S_r, S_s, A_{tu}) = n^{-1/2}\{\nu_{r,s,tu} - 2\beta_{tu}^{i,j}\nu_{i,r}\nu_{j,s} + \gamma_r^{i,jk}\nu_{jk,tu}\nu_{i,s}[2]\} + O(n^{-1})$$

On using (8.7), we find

$$(\gamma_r^{i,jk}\nu_{i,s} + \gamma_s^{i,jk}\nu_{i,r})\nu_{jk,tu} = \nu_{rs,tu} \tag{8.11}$$

as a condition for orthogonality to second order. One solution, but by no means the only one, unless $p = 1$, is given by

$$\gamma_r^{i,jk}\nu_{i,s} = \delta_{rs}^{jk}/2. \tag{8.12}$$

The remaining third-order joint cumulant

$$\kappa_3(S_r, A_{st}, A_{uv}) = n^{-1/2}\{\nu_{r,st,uv} - \beta_{st}^{i,jk}\nu_{i,r}\nu_{jk,uv}[2]\} + O(n^{-1})$$

is guaranteed to be $O(n^{-1})$ on account of (8.8). The choice of ancillary among the coefficients satisfying (8.8) is immaterial.

Finally, in order to achieve approximate normality to second order, we require that the third-order cumulant of S_r, S_s and S_t be $O(n^{-1})$. This condition gives

$$\nu_{r,s,t} + \gamma_r^{i,j}\nu_{i,s}\nu_{j,t}[6] = 0.$$

Again, the solution is not unique unless $p = 1$, but it is natural to consider the 'symmetric' solution

$$\gamma_r^{i,j} = -\nu_{r,s,t}\nu^{i,s}\nu^{j,t}/6. \qquad (8.13)$$

For the particular choice of coefficients (8.12) and (8.13), comparison with (7.13) shows that

$$S_r = W_r + \nu_{r,s,t}\nu^{s,t}/(6n^{1/2}) + O_p(n^{-1}),$$

where W_r are the components in the tensor decomposition of the likelihood ratio statistic given in Section 7.4.5.

In the case of scalar parameters, and using the coefficients (8.12) and (8.13), $W_r/i_{20}^{1/2}$ is equal to the signed square root of the likelihood ratio statistic. If we denote by ρ_3 the third standardized cumulant of $\partial l/\partial\theta$, then the orthogonal statistic may be written in the form

$$S = \pm\{l(\hat{\theta}) - l(\theta_0)\}^{1/2} + \rho_3/6. \qquad (8.14)$$

This statistic is distributed as $N(0,1) + O(n^{-1})$ independently of all ancillaries. The sign is chosen according to the sign of $\partial l/\partial\theta$.

8.6   Conditional distribution given A

In the previous sections it was shown that ancillary statistics, whether approximate or exact, are, in general, not unique, but yet certain useful formulae can be derived that are valid conditionally on any ancillary, however chosen. Because of the non-uniqueness of ancillaries, the most useful results must apply to as wide a class of *relevant* ancillaries as possible. This section is devoted to finding convenient expressions for the distributions of certain statistics such as the score statistic U_r or the maximum likelihood estimate $\hat{\theta}^r$, given the value of A. At no stage in the development is the ancillary specified. The only condition required is one of relevance, namely that the statistic of interest together with A should be sufficient to high enough order. Minimal sufficiency is not required.

It turns out that there is a particularly simple expression for the conditional distribution of $\hat{\theta}$ given A and that this expression is either exact or, if not exact, accurate to a very high order of asymptotic approximation. This conditional distribution, which may be written in the form

$$p(\hat{\theta}; \theta | A) = (2\pi c)^{-p/2} |\hat{j}_{rs}|^{1/2} \exp\{l(\theta) - l(\hat{\theta})\}\{1 + O(n^{-3/2})\}$$

is known as Barndorff-Nielsen's formula (Barndorff-Nielsen, 1980, 1983). One peculiar aspect of the formula is that the ancillary is not specified and the formula appears to be correct for a wide range of ancillary statistics. For this reason the description 'magical mystery formula' is sometimes used. In this expression $\log c = b$, the Bartlett adjustment factor, $\hat{j}$ is the *observed* information matrix regarded as a function of $\hat{\theta}$ and A, and the formula is correct for any relevant ancillary.

It is more convenient at the outset to work with the score statistic with components $U_r = \partial l / \partial \theta$ evaluated at $\theta = 0$, a value chosen here for later convenience of notation. All conditional cumulants of U are assumed to be $O(n)$ as usual. In what follows, it will be assumed that A is locally ancillary to third order and that the pair (U, A) is jointly sufficient to the same order. In other words, for all θ in some neighbourhood of the origin, the conditional log likelihood based on U satisfies

$$l_{U|A}(\theta) - l_{U|A}(0) = l_Y(\theta) - l_Y(0) + O(n^{-3/2}). \qquad (8.15)$$

Ancillary statistics satisfying these conditions, at least for θ in an $O(n^{-1/2})$ neighbourhood of the origin, can be constructed along the lines described in Section 8.3, but starting with the second and third derivatives jointly. In the case of location models, or more generally for group transformation models, exactly ancillary statistics exist that satisfy the above property for all θ. Such ancillaries typically have dimension of order $O(n)$. Since no approximation will be used here for the marginal distribution of A, it is not necessary to impose restrictions on its dimension. Such restrictions would be necessary if Edgeworth or saddlepoint approximations were used for the distribution of A.

The first step in the derivation is to find an approximation to the log likelihood function in terms of the conditional cumulants of U given $A = a$. Accordingly, let $K(\xi)$ be the conditional cumulant generating function of U given A at $\theta = 0$. Thus, the conditional cumulants $\kappa_{r,s}$, $\kappa_{r,s,t},\ldots$ are functions of a. By the saddlepoint approximation, the conditional log density of U given A at $\theta = 0$ is

$$-K^*(u) + \log|K^{*rs}(u)|/2 - p\log(2\pi)/2 - (3\rho_4^* - 4\rho_{23}^{*2})/4! \\ + O(n^{-3/2}), \qquad (8.16)$$

where $K^*(u)$ is the Legendre transformation of $K(\xi)$. This is the approximate log likelihood at $\theta = 0$. To obtain the value of the log likelihood at an arbitrary point, θ, we require the conditional cumulant generating function $K(\xi; \theta)$ of U given A at θ.

For small values of θ, the conditional cumulants of U have their usual expansions about $\theta = 0$ as follows.

$$E(U_r; \theta) = \kappa_{r,s}\theta^s + \kappa_{r,[st]}\theta^s\theta^t/2! + \kappa_{r,[stu]}\theta^s\theta^t\theta^u/3! + \ldots$$
$$\mathrm{cov}(U_r, U_s; \theta) = \kappa_{r,s} + \kappa_{r,s,t}\theta^t + \kappa_{r,s,[tu]}\theta^t\theta^u/2! + \ldots$$
$$\kappa_3(U_r, U_s, U_t; \theta) = \kappa_{r,s,t} + \kappa_{r,s,t,u}\theta^u + \ldots .$$

These expansions can be simplified by suitable choice of coordinate system. For any given value of A, we may choose a coordinate system in the neighbourhood of the origin satisfying $\kappa_{r,st} = 0$, $\kappa_{r,stu} = 0$, so that all higher-order derivatives are conditionally uncorrelated with U_r. This property is achieved using the transformation (7.5). Denoting the cumulants in this coordinate system by ν with appropriate indices, we find after collecting certain terms that

$$E(U_r; \theta) = K_r(\theta) + \nu_{r,s,tu}\theta^s\theta^t\theta^u[3]/3! + \ldots$$

$$\text{cov}(U_r, U_s; \theta) = K_{rs}(\theta) + \nu_{r,s,tu}\theta^t\theta^u/2! + \dots \qquad (8.17)$$
$$\kappa_3(U_r, U_s, U_t; \theta) = K_{rst}(\theta) + \dots .$$

Thus, the conditional cumulant generating function of U under θ is

$$K(\xi; \theta) = K(\theta + \xi) - K(\theta) + \nu_{r,s,tu}\xi^r\theta^s\theta^t\theta^u[3]/3!$$
$$+ \nu_{r,s,tu}\xi^r\xi^s\theta^t\theta^u/4 + \dots .$$

The final two terms above measure a kind of departure from simple exponential family form even after conditioning. To this order of approximation, an arbitrary model cannot be reduced to a full exponential family model by conditioning.

To find the Legendre transformation of $K(\xi; \theta)$, we note first that

$$K^*(u) - \theta^r u_r + K(\theta)$$

is the Legendre transformation of $K(\theta + \xi) - K(\theta)$. To this must be added a correction term of order $O(n^{-1})$ involving $\nu_{r,s,tu}$. Note that $\nu_{rs,tu}$, the so-called 'curvature' tensor or covariance matrix of the residual second derivatives, does not enter into these calculations. A straightforward calculation using Taylor expansion shows that the Legendre transformation of $K(\xi; \theta)$ is

$$K^*(u; \theta) = K^*(u) - \theta^r u_r + K(\theta) - \nu_{r,s,tu}u^r u^s\theta^t\theta^u/4$$
$$+ \nu_{r,s,tu}\theta^r\theta^s\theta^t\theta^u/4 + O(n^{-3/2}).$$

It may be checked that $K^*(u; \theta)$, evaluated at the mean of U under θ given by (8.17), is zero as it ought to be.

We are now in a position to use the saddlepoint approximation for a second time, but first we require the log determinant of the array of second derivatives. In subsequent calculations, terms that have an effect of order $O(n^{-3/2})$ on probability calculations are ignored without comment. Hence, the required log determinant is

$$\log \det K^{*rs}(u; \theta) = \log \det K^{*rs}(u; 0) - \nu_{r,s,tu}\nu^{r,s}\theta^t\theta^u/2.$$

Exercise 1.16 is useful for calculating log determinants.

The log likelihood function or the log density function may be written in terms of the Legendre transformation as

$$l(\theta) - l(0) = -K^*(u; \theta) + K^*(u)$$
$$+ \tfrac{1}{2}\log \det K^{*rs}(u; \theta) - \tfrac{1}{2}\log \det K^{*rs}(u)$$
$$= \theta^r u_r - K(\theta) + \nu_{r,s,tu}u^r u^s\theta^t\theta^u/4$$
$$- \nu_{r,s,tu}\theta^r\theta^s\theta^t\theta^u/4 - \nu_{r,s,tu}\nu^{r,s}\theta^t\theta^u/4 \qquad (8.18)$$

from which it can be seen that the conditional third cumulant $\nu_{r,s,tu}$ governs the departure from simple exponential form. This completes the first step in our derivation.

It is of interest here to note that the change in the derivative of $l(\theta)$, at least in its stochastic aspects, is governed primarily by the third term on the right of (8.18). It may be possible to interpret $\nu_{r,s,tu}$ as a curvature or torsion tensor, though its effect is qualitatively quite different from Efron's (1975) curvature, which is concerned mainly with the variance of the residual second derivatives. The latter notion of curvature is sensitive to conditioning.

In many instances, the likelihood function and the maximized likelihood ratio statistic can readily be computed either analytically or numerically whereas the conditional cumulant generating function, $K(\xi)$ and the conditional Legendre transformation, $K^*(u)$ cannot, for the simple reason that A is not specified explicitly. Thus, the second step in our derivation is to express $K^*(u)$ and related quantities directly in terms of the likelihood function and its derivatives.

The likelihood function given above has its maximum at the point

$$\hat{\theta}^r = K^{*r}(u) + \epsilon^r(u) + O(n^{-2}) \qquad (8.19)$$

where ϵ^r is $O(n^{-3/2})$ given by

$$\epsilon^r = -\nu_{st,u,v}\nu^{r,s}\nu^{u,v}u^t/2 - \nu_{s,t,uv}\nu^{r,s}u^tu^uu^v/2.$$

On substituting $\hat{\theta}$ into (8.18), further simplification shows that the log likelihood ratio statistic is

$$l(\hat{\theta}) - l(0) = K^*(u) - \nu_{r,s,tu}\nu^{r,s}u^tu^u/4 + O(n^{-3/2}),$$

which differs from the conditional Legendre transformation of $K(\xi)$ by a term of order $O(n^{-1})$. Each of these calculations involves a small amount of elementary but tedious algebra that is hardly worth reproducing.

We now observe that

$$-\{l(\hat{\theta}) - l(0)\} + \tfrac{1}{2}\log\det K^{*rs}(u;\hat{\theta}) = -K^*(u) + \tfrac{1}{2}\log\det K^{*rs}(u),$$
$$(8.20)$$

which is the dominant term in the saddlepoint approximation for the conditional log density of U given A. It now remains to express the left member of the above equation solely in terms of the log likelihood function and its derivatives.

On differentiating (8.18), we find that

$$u_{rs}(\theta) = -K_{rs}(\theta) - \nu_{rs,t,u}\nu^{t,u}/2 - \nu_{r,s,tu}\theta^t\theta^u/2 - \nu_{r,t,su}[2]\theta^t\theta^u.$$
$$+ \nu_{rs,t,u}u^t u^u/2 - \nu_{rs,t,u}\theta^t\theta^u/2.$$

Hence, the observed Fisher information at $\hat{\theta}$ is

$$\hat{j}_{rs} = -u_{rs}(\hat{\theta}) = K_{rs}(\hat{\theta}) + \nu_{r,s,tu}\hat{\theta}^t\hat{\theta}^u/2 + \nu_{rs,t,u}\nu^{t,u}/2 + \nu_{r,t,su}[2]\hat{\theta}^t\hat{\theta}^u$$

and the observed information determinant is given by

$$\log\det\hat{j}_{rs} = -\log\det K^{*rs}(u;\hat{\theta}) + 2\nu_{r,t,su}\nu^{r,s}\hat{\theta}^t\hat{\theta}^u + \nu_{r,s,tu}\nu^{r,s}\nu^{t,u}/2.$$

On substituting into (8.20), it is seen that the saddlepoint approximation with one correction term for the conditional log density of U given A is

$$-\{l(\hat{\theta}) - l(\theta)\} - \tfrac{1}{2}\log\det\hat{j}_{rs} - \tfrac{1}{2}p\log(2\pi) + \nu_{r,t,su}\nu^{r,s}\hat{\theta}^t\hat{\theta}^u - \tfrac{1}{2}pb(\theta),$$
$$(8.21)$$

where $b(\theta)$ is the Bartlett adjustment, given in this instance by

$$pb(\theta) = (3\rho_{13}^2 + 2\rho_{23}^2 - 3\rho_4)/12 - \nu_{r,s,tu}\nu^{r,s}\nu^{t,u}/2.$$

As pointed out in Section 8.4, $b(\theta)$ can be computed from the unconditional cumulants using the formula (7.11). Similarly, $\nu_{r,t,su}$ in (8.21) can be computed from the unconditional cumulants using (8.10).

Expression (8.21) can be simplified even further. The derivative of the transformation (8.19) from u to $\hat{\theta}$ is

$$\hat{\theta}^{rs} = K^{*rs}(u) - \nu^{r,t}\nu^{s,u}\nu^{v,w}\nu_{tu,v,w}/2 - \nu^{r,t}\nu^{s,u}u^v u^w\nu_{t,u,vw}/2$$
$$- \nu^{r,t}\nu^{v,s}u^u u^w\nu_{tu,v,w}[2]/2$$
$$= K^{*rs}(u;\hat{\theta}) - \nu^{r,t}\nu^{s,u}\nu^{v,w}\nu_{tu,v,w}/2 - \nu^{r,t}\nu^{v,s}u^u u^w\nu_{tu,v,w}.$$

Hence, the log determinant of the transformation is

$$\log\det\hat{\theta}^{rs} = \log\det K^{*rs}(u;\hat{\theta}) - \nu^{r,s}\nu^{t,u}\nu_{rs,t,u}/2 - \nu^{r,s}u^t u^u\nu_{rt,s,u}$$
$$= -\log\det\hat{j}_{rs} + \nu_{r,t,su}\nu^{r,s}u^t u^u. \qquad (8.22)$$

Hence, under the assumption that $\theta = 0$ is the true parameter point, the conditional log density of $\hat{\theta}$ given A is

$$-\{l(\hat{\theta}) - l(\theta)\} + \tfrac{1}{2}\log\det\hat{j}_{rs} - p\log(2\pi) - pb(\hat{\theta})/2.$$

More generally, if the true parameter point is θ, the conditional density of $\hat{\theta}$ as a function of θ and the conditioning variable may be written

$$p(\hat{\theta}; \theta | A) = (2\pi\hat{c})^{-p/2}\exp\{l(\theta) - l(\hat{\theta})\}|\hat{j}|^{1/2} \qquad (8.23)$$

where $\hat{c} = \log b(\hat{\theta})$. The log likelihood function $l(\theta)$, in its dependence on the data, is to be regarded as a function of $(\hat{\theta}, a)$. Similarly for the observed information determinant. Thus, for example, $l(\theta) = l(\theta; \hat{\theta}, a)$ is the log likelihood function, $u_{rs}(\theta) = u_{rs}(\theta; \hat{\theta}, a)$ and $\hat{j}_{rs} = u_{rs}(\hat{\theta}; \hat{\theta}, a)$ is the observed information matrix.

Approximation (8.23) gives the conditional density of the maximum likelihood estimate of the canonical parameter corresponding to the transformation (7.5). On transforming to any other parameter, the form of the approximation remains the same. In fact, the approximation is an invariant of *weight* 1 under re-parameterization in the sense of Section 6.1. Thus, (8.23) is equally accurate or inaccurate in all parameterizations and the functional form of the approximation is the same whatever parameterization is chosen.

8.7 Bibliographic notes

It would be impossible in the limited space available here to discuss in detail the various articles that have been written on the subject of ancillary statistics and conditional inference. What follows is a minimal set of standard references.

Fisher (1925, 1934) seems to have been first to recognize the need for conditioning to ensure that probability calculations are relevant to the data observed. His criticism of Welch's test (Fisher, 1956) was based on its unsatisfactory conditional properties. Other important papers that discuss the need for conditioning and the difficulties that ensue are Cox (1958, 1971), Basu (1964), Pierce (1973), Robinson (1975, 1978), Kiefer (1977), Lehmann (1981) and Buehler (1982). The book by Fraser (1968) is exceptional for the emphasis placed on group structure as an integral part of the model specification.

The question of the existence or otherwise of ancillary statistics (or similar regions) was first posed by R.A. Fisher as the 'problem of the Nile' in his 1936 Harvard Tercentenary lecture. It is now known that, in the continuous case, exactly ancillary statistics always exist if the parameter space is finite or if attention is restricted to any finite set of parameter values, however numerous. This conclusion follows from Liapounoff's theorem (Halmos, 1948), which states that the range of a vector measure is closed and, in the non-atomic case, convex. On the other hand, it is also known that no such regions, satisfying reasonable continuity conditions, exist for the Behrens-Fisher problem (Linnik, 1968). It seems, then, that as the number of parameter points under consideration increases, regions whose probability content is exactly the same for all parameter values are liable to become increasingly 'irregular' in some sense. However, acceptable regions whose probability content is approximately the same for all parameter values do appear to exist in most instances.

Kalbfleisch (1975) makes a distinction, similar to that made in Section 8.1, between experimental ancillaries and mathematical ancillaries. Lloyd (1985ab), on the other hand, distinguishes between internal and external ancillaries. The former are functions of the minimal sufficient statistic.

Much of the recent work has concentrated on the notion of approximate ancillarity or asymptotic ancillarity. See, for example, Efron & Hinkley (1978), Cox (1980), Hinkley (1980), Barndorff-Nielsen (1980) for further details. The results given in Section 8.5 are taken from McCullagh (1984a).

Formula (8.23) was first given by Barndorff-Nielsen (1980) synthesizing known exact conditional results for translation models due to Fisher (1934), and approximate results for full exponential family models based on the saddlepoint approximation. In subsequent papers, (Barndorff-Nielsen, 1983, 1984, 1985), the formula has been developed and used to obtain conditional confidence intervals for one-dimensional parameters. More recent applications of the formula have been to problems involving nuisance parameters. The formulae in Section 6.4 are special cases of what is called the 'modified profile likelihood'.

The derivation of Barndorff-Nielsen's formula in Section 8.6 appears to be new.

8.8   Further results and exercises 8

8.1   Show that if (X_1, X_2) has the bivariate normal distribution with zero mean, variances σ_1^2, σ_2^2 and covariance $\rho\sigma_1\sigma_2$, then the ratio $U = X_1/X_2$ has the Cauchy distribution with median $\theta = \rho\sigma_1/\sigma_2$ and dispersion parameter $\tau^2 = \sigma_1^2(1 - \rho^2)/\sigma_2^2$. Explicitly,

$$f_U(u; \theta, \tau) = \tau^{-1}\pi^{-1}\{1 + (u - \theta)^2/\tau^2\}^{-1},$$

where $-\infty < \theta < \infty$ and $\tau > 0$. Deduce that $1/U$ also has the Cauchy distribution with median $\theta/(\tau^2 + \theta^2)$ and dispersion parameter $\tau^2/(\tau^2 + \theta^2)^2$. Interpret the conclusion that θ/τ is invariant.

8.2   Let $X_1, ..., X_n$ be independent and identically distributed Cauchy random variables with unknown parameters (θ, τ). Let $\bar{X}$ and s_X^2 be the sample mean and sample variance respectively. By writing $X_i = \theta + \tau\epsilon_i$, show that the joint distribution of the *configuration statistic* A with components $A_i = (X_i - \bar{X})/s_X$ is independent of the parameters and hence that A is ancillary. [This result applies equally to any location-scale family where the ϵ_i are *i.i.d.* with known distribution.]

8.3   Using the notation of the previous exercise, show that for any constants a, b, c, d satisfying $ad - bc \neq 0$,

$$Y_i = (a + bX_i)/(c + dX_i) \qquad i = 1, ..., n$$

are independent and identically distributed Cauchy random variables. Deduce that the derived statistic A^* with components $A_i^* = (Y_i - \bar{Y})/s_Y$ has a distribution not depending on (θ, τ). Hence conclude that the maximal ancillary for the problem described in Exercise 8.2 is not unique. Demonstrate explicitly that two such ancillaries are not jointly ancillary. [This construction is specific to the two-parameter Cauchy problem.]

8.4   Suppose, in the notation previously established, that $n = 3$. Write the ancillary in the form $\{\text{sign}(X_3 - X_2), \text{sign}(X_2 - X_1)\}$, together with an additional component

$$A_X = (X_{(3)} - X_{(2)})/(X_{(2)} - X_{(1)}),$$

where $X_{(j)}$ are the ordered values of X. Show that A_X is a function of the sufficient statistic, whereas the first two components are not. Let $Y_i(t) = 1/(X_i - t)$ and denote by $A(t)$ the corresponding ancillary computed as a function of the transformed values. Show that the function $A(t)$ is continuous except at the three points $t = X_i$ and hence deduce that the data values may be recovered from the set of ancillaries $\{A(t), -\infty < t < \infty\}$. [In fact, it is enough to know the values of $A(t)$ at three distinct points interlacing the observed values. However, these cannot be specified in advance.]

8.5 Suppose that (X_1, X_2) are bivariate normal variables with zero mean, unit variance and unknown correlation ρ. Show that $A_1 = X_1$ and $A_2 = X_2$ are each ancillary, though not jointly ancillary, and that neither is a component of the sufficient statistic. Let $T = X_1 X_2$. Show that

$$T|A_1 = a_1 \sim N\{\rho a_1^2, (1 - \rho^2) a_1^2\}$$
$$T|A_2 = a_2 \sim N\{\rho a_2^2, (1 - \rho^2) a_2^2\}.$$

8.6 In the notation of the previous exercise, suppose that it is required to test the hypothesis $H_0 : \rho = 0$, and that the observed values are $x_1 = 2$, $x_2 = 1$. Compute the conditional tail areas $\text{pr}(T \geq t|A_1 = a_1)$ and $\text{pr}(T \geq t|A_2 = a_2)$. Comment on the appropriateness of these tail areas as measures of evidence against H_0.

8.7 Suppose that (X_1, X_2, X_3) have the trivariate normal distribution with zero mean and intra-class covariance matrix with variances σ^2 and correlations ρ. Show that $-\frac{1}{2} \leq \rho \leq 1$. Prove that the moments of $X_1 + \omega X_2 + \omega^2 X_3$ and $X_1 + \omega X_3 + \omega^2 X_2$ are independent of both parameters, but that neither statistic is ancillary $[\omega = \exp(2\pi i/3)]$.

8.8 Suppose in the previous exercise that $\sigma^2 = 1$. Show that this information has no effect on the sufficient statistic but gives rise to ancillaries, namely X_1, X_2, X_3, no two of which are jointly ancillary.

8.9 Show that the tensorial decomposition of the likelihood ratio statistic in (7.13) is not unique but that all such decompositions are orthogonal statistics in the sense used in Section 8.5 above.

8.10 Show that the Legendre transformation of

$$K(\xi; \theta) = K(\theta + \xi) - K(\theta) + \nu_{r,s,tu} \xi^r \theta^s \theta^t \theta^u [3]/3!$$
$$+ \nu_{r,s,tu} \xi^r \xi^s \theta^t \theta^u /4 + \dots$$

with respect to the first argument is approximately

$$K^*(u;\theta) = K^*(u) - \theta^r u_r + K(\theta) - \nu_{r,s,tu} u^r u^s \theta^t \theta^u/4$$
$$+ \nu_{r,s,tu} \theta^r \theta^s \theta^t \theta^u/4.$$

8.11 Show that the Legendre transformation, $K^*(u;\theta)$, evaluated at

$$u_r = K_r(\theta) + \nu_{r,s,tu} \theta^s \theta^t \theta^u [3]/3!,$$

is zero to the same order of approximation.

8.12 Show that the maximum of the log likelihood function (8.18) is given by (8.19).

8.13 Beginning with the canonical coordinate system introduced at (8.17), transform from θ to ϕ with components

$$\phi_r = K_r(\theta) + \nu_{r,s,tu} \theta^s \theta^t \theta^u/2 + \nu_{r,s,tu} \theta^s \nu^{t,u}/2.$$

Show that, although $E(U_r;\theta) \neq \phi_r$, nevertheless $\hat{\phi}_r = U_r$. Show also that the observed information determinant with respect to the components of ϕ satisfies

$$\tfrac{1}{2} \log \det K^{*rs}(u;\hat{\theta}) = \tfrac{1}{2} \log \det \hat{j}_\phi^{rs} + \nu_{r,s,tu} \nu^{r,s} \nu^{t,u}$$

at the maximum likelihood estimate. Hence deduce (8.23) directly from (8.20).

8.14 Suppose that $Y_1, ..., Y_n$ are independent and identically distributed on the interval θ, $\theta+1$. Show that the likelihood function is constant in the interval $(y_{(n)} - 1, y_{(1)})$ and is zero otherwise. Hence, interpret $r = y_{(n)} - y_{(1)}$ as an indicator of the shape of the likelihood function.

Complementary set partitions

A catalogue of complementary set partitions is an indispensable tool in any serious work involving moments or cumulants of low-order polynomials. To be useful, items in such a catalogue need to be grouped in various ways so that, whatever calculations are contemplated, all partitions in the same class contribute equally to the final answer. If, as is often the case, the number of classes is small compared to the total number of partitions, the labour of calculation can be reduced to manageable proportions. Such a list, satisfying the most common needs in statistics, is provided in the tables that follow. First, however, it is necessary to describe the three nested groupings or equivalence relations that have been employed in compiling the present catalogue.

Suppose, by way of example, that we require a simplified expression for the covariance of two quadratic forms, $a_{ij}X^iX^j$ and $b_{ij}X^iX^j$. The answer, $a_{ij}b_{kl}\kappa^{ij,kl}$, needs to be expressed in terms of ordinary cumulants, for which we need a list of all partitions complementary to $ij|kl$. Consulting Table 1 in the column headed 4 under the entry $\Upsilon^* = 12|34$, we find the following list of 11 partitions grouped into five finer equivalence classes or four coarser classes formed by combining two of the five.

$$1234\ [\mathbf{1}]$$
$$123|4\ [\mathbf{2}][\mathbf{2}]$$
$$134|2$$
$$13|24\ [\mathbf{2}]$$
$$13|2|4\ [\mathbf{4}]$$

Two partitions Υ_a and Υ_b are considered to be in the same equivalence class if the intersection matrices $\Upsilon^* \cap \Upsilon_a$ and $\Upsilon^* \cap \Upsilon_b$ are equal, possibly after permuting columns. In particular, Υ_a and Υ_b must have the same number of blocks and the same block sizes, but this criterion is not sufficient to guarantee equivalence. By way of illustration, consider the four partitions 123|4, 124|3, 134|2 and 234|1, which give rise to the following intersection matrices with $\Upsilon^* = 12|34$:

$$
\begin{array}{cccccccc}
2 & 0 & & 2 & 0 & & 1 & 1 & & 1 & 1 \\
1 & 1 & & 1 & 1 & & 2 & 0 & & 2 & 0
\end{array}
$$

These four partitions are therefore properly catalogued as 123|4 [2] and 134|2 [2]. In other words, we list one member of the class followed by the class size in brackets. In Table 1, we have written 123|4 [2][2] followed by 134|2 in smaller type. The meaning is the same: the reason for arranging things this way is that the lines in smaller type can subsequently be omitted for brevity, for example in Table 3. No partitions in small type appear unless Υ^* has at least two blocks of equal size.

If row permutations of the intersection matrix are permitted, the two classes just described coalesce, implying that the lines in smaller type can be omitted. Both equivalence relationships are of interest in statistics and the catalogue has been compiled with both in mind.

To return now to the covariance of two quadratics, we find, assuming a_{ij} and b_{ij} to be symmetric, that

$$
\begin{aligned}
a_{ij}b_{kl}\kappa^{ij,kl} = a_{ij}b_{kl}\{ & \kappa^{i,j,k,l} + 2\kappa^i\kappa^{j,k,l} + 2\kappa^k\kappa^{i,j,l} \\
& + 2\kappa^{i,k}\kappa^{j,l} + 4\kappa^i\kappa^k\kappa^{j,l} \},
\end{aligned}
$$

so that the sizes of the equivalence classes become ordinary arithmetic factors. The two classes $i|jkl[2]$ and $k|ijl[2]$ do not contribute equally to this sum and must, therefore, be kept separated. To see that this is so, we need only consider the case where $a_{ij}X^iX^j$ is the residual sum of squares after linear regression and $b_{ij}X^iX^j$ is an arbitrary quadratic form. It follows then, under the usual assumptions, that $a_{ij}\kappa^i = a_{ij}\kappa^j = 0$. Thus the partitions $i|jkl$ and $j|ikl$ contribute zero whereas $k|ijl$ and $l|ijk$ make a non-zero contribution.

In the important special case where $a_{ij} = b_{ij}$, the two classes in question coalesce and the coarser grouping is then appropriate.

In the univariate case, only the block sizes are relevant. The corresponding grouping is indicated in the catalogue by means of extra space at the appropriate places. For example, if $\Upsilon^* = 123|45$, there are 10 equivalence classes determined by the intersection matrices. However, if X is a scalar, we have

$$\mathrm{cov}(X^2, X^3) = \kappa_5 + 5\kappa_1\kappa_4 + 9\kappa_2\kappa_3 + 9\kappa_1^2\kappa_3 + 9\kappa_1\kappa_2^2 + 6\kappa_1^3\kappa_2$$

using power notation. Thus, the original 10 classes are reduced to six if only block sizes are relevant.

For each partition Υ^* of $S_n = \{1, 2, ..., n\}$ and for $n \leq 6$, Table 1 gives a complete list of all complementary partitions, grouped in the way we have described. For $n = 7, 8$, and for each Υ^*, Table 2 gives lists of complementary partitions excluding those that have unit parts (a block of unit size). For statistical purposes, this omission is not serious because, if the random variables have zero mean, partitions having a unit part do not contribute. For $n = 9, 10, 12$ and for selected Υ^*, Table 3 gives the required lists with unit parts omitted and with the lines in small type also omitted. The missing partitions can be inferred from the figures in brackets on the preceding partition.

All three tables were computed using a 'C' program written by Allan Wilks of AT&T Bell Laboratories. Table 1 and parts of Table 2 were checked against lists previously compiled by hand. To compute Table 3, Wilks used a primitive form of parallel processor (up to 10 VAX and SUN computers running simultaneously but independently). Much of the computer time was occupied by the task of grouping partitions into equivalence classes as opposed to checking the connectivity criterion. Total computer time is difficult to determine precisely, but is certainly measured in CPU-months! The overwhelming majority of this time was lavished on Table 3. Evidently, the algorithm used by the computer is much less efficient than the algorithm used by the human brain: fortunately, the computer algorithm, despite its inefficiency, is much the more reliable!

Because of its excessive length, Table 3, which deals with sets of sizes 9,10,12, is not included in this Appendix.

For computational and algorithmic details of this operation, see McCullagh & Wilks (1985a,b).

Table 1: *Complementary set partitions*

1	2	3	4
1	**12**	**123**	**1234**
1 [1]	12 [1]	123 [1]	1234 [1]
	1|2 [1]	12|3 [3]	123|4 [4]
		1|2|3 [1]	12|34 [3]
	1|2		12|3|4 [6]
	12 [1]	**12|3**	1|2|3|4 [1]
		123 [1]	
		13|2 [2]	**123|4**
			1234 [1]
		1|2|3	124|3 [3]
		123 [1]	12|34 [3]
			14|2|3 [3]
			12|34
			1234 [1]
			123|4 [2][2]
			134|2
			13|24 [2]
			13|2|4 [4]
			12|3|4
			1234 [1]
			134|2 [2]
			13|24 [2]
			1|2|3|4
			1234 [1]

TABLE 1   255

Table 1: *Complementary set partitions* (*contd.*)

5

12345
12345 [1]
1234|5 [5]
123|45 [10]
123|4|5 [10]
12|34|5 [15]
12|3|4|5 [10]
1|2|3|4|5 [1]

1234|5
12345 [1]
1235|4 [4]
123|45 [4]
125|34 [6]
125|3|4 [6]
12|35|4 [12]
15|2|3|4 [4]

123|45
12345 [1]
1234|5 [2]
1245|3 [3]
124|35 [6]
145|23 [3]
124|3|5 [6]
145|2|3 [3]
12|34|5 [6]
14|25|3 [6]
14|2|3|5 [6]

123|4|5
12345 [1]
1245|3 [3]

124|35 [3][2]
125|34
145|23 [3]
145|2|3 [3]
14|25|3 [6]

12|34|5
12345 [1]
1235|4 [2][2]
1345|2
123|45 [2][2]
134|25
135|24 [4]
135|2|4 [4]
13|25|4 [4][2]
13|45|2

12|3|4|5
12345 [1]
1345|2 [2]
134|25 [2][3]
135|24
145|23

1|2|3|4|5
12345 [1]

6

123456
123456 [1]
12345|6 [6]
1234|56 [15]
1234|5|6 [15]
123|456 [10]
123|45|6 [60]
123|4|5|6 [20]
12|34|56 [15]
12|34|5|6 [45]
12|3|4|5|6 [15]
1|2|3|4|5|6 [1]

12345|6
123456 [1]
12346|5 [5]
1234|56 [5]
1236|45 [10]
1236|4|5 [10]
123|456 [10]
123|46|5 [20]
126|34|5 [30]
126|3|4|5 [10]
12|34|56 [15]
12|36|4|5 [30]
16|2|3|4|5 [5]

1234|56
123456 [1]
12345|6 [2]
12356|4 [4]
1235|46 [8]
1256|34 [6]

1235|4|6 [8]
1256|3|4 [6]
123|456 [4]
125|346 [6]
123|45|6 [8]
125|34|6 [12]
125|36|4 [24]
156|23|4 [12]
125|3|4|6 [12]
156|2|3|4 [4]
12|35|46 [12]
12|35|4|6 [24]
15|26|3|4 [12]
15|2|3|4|6 [8]

1234|5|6
123456 [1]
12356|4 [4]
1235|46 [4][2]
1236|45
1256|34 [6]
1256|3|4 [6]
123|456 [4]
125|346 [6]
125|36|4 [12][2]
126|35|4
156|23|4 [12]
156|2|3|4 [4]
12|35|46 [12]
15|26|3|4 [12]

123|456
123456 [1]
12345|6 [3][2]
12456|3
1234|56 [3][2]
1456|23

Table 1: *Complementary set partitions* (*contd.*)

1245|36 [9]
1234|5|6 [3][2]
1456|2|3
1245|3|6 [9]
124|356 [9]
124|35|6 [18][2]
145|26|3
124|56|3 [9][2]
145|23|6
124|3|5|6 [9][2]
145|2|3|6
12|34|56 [9]
14|25|36 [6]
12|34|5|6 [9][2]
14|56|2|3
14|25|3|6 [18]
14|2|3|5|6 [9]

123|45|6
123456 [1]
12346|5 [2]
12456|3 [3]
1234|56 [2]
1245|36 [3]
1246|35 [6]
1456|23 [3]
1246|3|5 [6]
1456|2|3 [3]
124|356 [6]
126|345 [3]
124|36|5 [6]
124|56|3 [6]
126|34|5 [6]
145|26|3 [6]
146|23|5 [6]
146|25|3 [12]
146|2|3|5 [6]
12|34|56 [6]

14|25|36 [6]
14|26|3|5 [12]
14|56|2|3 [6]

123|4|5|6
123456 [1]
12456|3 [3]
1245|36 [3][3]
1246|35
1256|34
1456|23 [3]
1456|2|3 [3]
124|356 [3][3]
125|346
126|345
145|26|3 [6][3]
146|25|3
156|24|3
14|25|36 [6]

12|34|56
123456 [1]
12345|6 [2][3]
12356|4
13456|2
1235|46 [4][3]
1345|26
1356|24
1235|4|6 [4][3]
1345|2|6
1356|2|4
123|456 [2][3]
125|346
134|256
135|246 [4]

123|45|6 [4][6]
125|36|4
134|25|6
156|23|4
356|14|2
345|16|2
135|46|2 [8][3]
135|26|4
135|24|6
135|2|4|6 [8]
13|25|46 [8]
13|25|4|6 [8][3]
13|45|2|6
15|36|2|4

12|34|5|6
123456 [1]
12356|4 [2][2]
13456|2
1235|46 [2][4]
1236|45
1345|26
1346|25
1356|24 [4]
1356|2|4 [4]
123|456 [2][2]
134|256
135|246 [4]
135|26|4 [4][4]
135|46|2
136|25|4
136|45|2
156|23|4 [4][2]
356|14|2
13|25|46 [4][2]
13|26|45

12|3|4|5|6
123456 [1]
13456|2 [2]
1345|26 [2][4]
1346|25
1356|24
1456|23
134|256 [2][3]
135|246
136|245

1|2|3|4|5|6
123456 [1]

TABLE 2   257

Table 2: *Complementary set partitions omitting unit parts*

7

1234567
1234567 [1]
12345|67 [21]
1234|567 [35]
123|45|67 [105]

123456|7
1234567 [1]
12345|67 [6]
12347|56 [15]
1234|567 [15]
1237|456 [20]
123|45|67 [60]
127|34|56 [45]

12345|67
1234567 [1]
12346|57 [10]
12367|45 [10]
1234|567 [5]
1236|457 [20]
1267|345 [10]
123|46|57 [20]
126|34|57 [60]
167|23|45 [15]

12345|6|7
1234567 [1]
12346|57 [5][2]
12347|56
12367|45 [10]
1234|567 [5]
1236|457 [10][2]
1237|456
1267|345 [10]

123|46|57 [20]
126|34|57 [30][2]
127|34|56
167|23|45 [15]

1234|567
1234567 [1]
12345|67 [3]
12356|47 [12]
12567|34 [6]
1235|467 [12]
1256|347 [18]
1567|234 [4]
123|45|67 [12]
125|34|67 [18]
125|36|47 [36]
156|23|47 [36]

1234|56|7
1234567 [1]
12345|67 [2]
12356|47 [4]
12357|46 [8]
12567|34 [6]
1235|467 [8]
1237|456 [4]
1256|347 [6]
1257|346 [12]
1567|234 [4]
123|45|67 [8]
125|34|67 [12]
125|36|47 [24]
127|35|46 [12]
156|23|47 [12]
157|23|46 [24]

1234|5|6|7
1234567 [1]
12356|47 [4][3]
12357|46
12367|45
12567|34 [6]
1235|467 [4][3]
1236|457
1237|456
1256|347 [6][3]
1257|346
1267|345
1567|234 [4]
125|36|47 [12][3]
126|35|47
127|35|46
156|23|47 [12][3]
157|23|46
167|23|45

123|456|7
1234567 [1]
12345|67 [3][2]
12456|37
12347|56 [3][2]
14567|23
12457|36 [9]
1234|567 [3][2]
1456|237
1245|367 [9]
1247|356 [9][2]
1457|236
124|35|67 [18][2]
145|26|37
124|37|56 [9][2]
145|23|67
127|34|56 [9][2]
457|12|36
147|23|56 [9]
147|25|36 [18]

123|45|67
1234567 [1]
12346|57 [4]
12456|37 [6][2]
12467|35
14567|23 [3]
1234|567 [2][2]
1236|457
1245|367 [3][2]
1267|345
1246|357 [12]
1456|237 [6][2]
1467|235
124|36|57 [12][2]
126|34|57
145|26|37 [6][2]
167|24|35
146|23|57 [12]
146|25|37 [24]
456|12|37 [6][2]
467|12|35

123|45|6|7
1234567 [1]
12346|57 [2][2]
12347|56
12456|37 [3][2]
12457|36
12467|35 [6]
14567|23 [3]
1234|567 [2]
1245|367 [3]
1246|357 [6][2]
1247|356
1267|345 [3]
1456|237 [3][2]
1457|236
1467|235 [6]
124|36|57 [6][2]
124|37|56

Table 2: *Complementary set partitions (contd.)*

126|34|57 [6][2]
127|34|56
145|26|37 [6]
146|23|57 [6][2]
147|23|56
146|25|37 [12][2]
147|25|36
167|24|35 [6]
467|12|35 [6]

123|4|5|6|7
1234567 [1]
12456|37 [3][4]
12457|36
12467|35
12567|34
14567|23 [3]
1245|367 [3][6]
1246|357
1247|356
1256|347
1257|346
1267|345
1456|237 [3][4]
1457|236
1467|235
1567|234
145|26|37 [6][6]
146|25|37
147|25|36
156|24|37
157|24|36
167|24|35

12|34|56|7
1234567 [1]
12345|67 [2][3]
12356|47
13456|27

12357|46 [4][3]
13457|26
13567|24
1235|467 [4][3]
1356|247
1345|267
1257|346 [2][6]
1347|256
1237|456
1567|234
3457|126
3567|124
1357|246 [8]
123|45|67 [4][6]
125|36|47
134|25|67
356|14|27
156|23|47
345|16|27
135|27|46 [8][3]
135|24|67
135|26|47
157|23|46 [8][3]
137|25|46
357|14|26

12|34|5|6|7
1234567 [1]
12356|47 [2][6]
12357|46
12367|45
13456|27
13457|26
13467|25
13567|24 [4]
1235|467 [2][6]
1236|457
1237|456
1345|267
1346|257
1347|256

1356|247 [4][3]
1357|246
1367|245
1567|234 [2][2]
3567|124
135|26|47 [4][6]
135|27|46
136|25|47
136|27|45
137|25|46
137|26|45
156|23|47 [4][6]
157|23|46
167|23|45
356|14|27
357|14|26
367|14|25

12|3|4|5|6|7
1234567 [1]
13456|27 [2][5]
13457|26
13467|25
13567|24
14567|23
1345|267 [2][10]
1346|257
1347|256
1356|247
1357|246
1367|245
1456|237
1457|236
1467|235
1567|234

1|2|3|4|5|6|7
1234567 [1]

8

12345678
12345678 [1]
123456|78 [28]
12345|678 [56]
1234|5678 [35]
1234|56|78 [210]
123|456|78 [280]
12|34|56|78 [105]

1234567|8
12345678 [1]
123456|78 [7]
123458|67 [21]
12345|678 [21]
12348|567 [35]
1234|5678 [35]
1234|56|78 [105]
1238|45|67 [105]
123|456|78 [70]
123|458|67 [210]
12|34|56|78 [105]

123456|78
12345678 [1]
123457|68 [12]
123478|56 [15]
12345|678 [6]
12347|568 [30]
12378|456 [20]
1234|5678 [15]
1237|4568 [20]
1234|57|68 [30]
1237|45|68 [120]
1278|34|56 [45]

TABLE 2   259

Table 2: *Complementary set partitions* (*contd.*)

123|457|68 [120]	1236|47|58 [60]	**12345|6|7|8**	12345|678 [4][2]
123|478|56 [60]	1267|34|58 [90]	12345678 [1]	15678|234
127|348|56 [90]	1678|23|45 [15]	123467|58 [5][3]	12356|478 [24][2]
12|34|57|68 [90]	123|456|78 [30]	123468|57	12567|348
	123|467|58 [60]	123478|56	
123456|7|8	126|347|58 [90]	123678|45 [10]	1235|4678 [16]
12345678 [1]	126|378|45 [90]	12346|578 [5][3]	1256|3478 [18]
123457|68 [6][2]	12|34|56|78 [45]	12347|568	1235|46|78 [48][2]
123458|67	12|36|47|58 [60]	12348|567	1567|23|48
123478|56 [15]		12367|458 [10][3]	1256|34|78 [36]
12345|678 [6]	**12345|67|8**	12368|457	1256|37|48 [72]
12347|568 [15][2]	12345678 [1]	12378|456	123|456|78 [24][2]
12348|567	123456|78 [2]	12678|345 [10]	125|678|34
12378|456 [20]	123467|58 [5]	1234|5678 [5]	123|567|48 [16]
1234|5678 [15]	123468|57 [10]	1236|4578 [10][3]	125|346|78 [36][2]
1237|4568 [20]	123678|45 [10]	1237|4568	156|278|34
1234|57|68 [30]	12346|578 [10]	1238|4567	125|367|48 [144]
1237|45|68 [60][2]	12348|567 [5]	1236|47|58 [20][3]	12|35|46|78 [72]
1238|45|67	12367|458 [10]	1237|46|58	15|26|37|48 [24]
1278|34|56 [45]	12368|457 [20]	1238|46|57	
123|457|68 [60][2]	12678|345 [10]	1267|34|58 [30][3]	**1234|567|8**
123|458|67	1234|5678 [5]	1268|34|57	12345678 [1]
123|478|56 [60]	1236|4578 [20]	1278|34|56	123456|78 [3]
127|348|56 [90]	1238|4567 [10]	1678|23|45 [15]	123458|67 [3]
12|34|57|68 [90]	1234|56|78 [10]	123|467|58 [20][3]	123567|48 [4]
	1236|45|78 [20]	123|468|57	123568|47 [12]
12345|678	1236|47|58 [40]	123|478|56	125678|34 [6]
12345678 [1]	1238|46|57 [20]	126|347|58 [30][3]	12345|678 [3]
123456|78 [3]	1267|34|58 [30]	126|348|57	12356|478 [12]
123467|58 [15]	1268|34|57 [60]	127|348|56	12358|467 [12]
123678|45 [10]	1678|23|45 [15]	126|378|45 [30][3]	12567|348 [6]
12346|578 [15]	123|456|78 [20]	127|368|45	12568|347 [18]
12367|458 [30]	123|467|58 [20]	128|367|45	15678|234 [4]
12678|345 [10]	123|468|57 [40]	12|36|47|58 [60]	1235|4678 [12]
1234|5678 [5]	126|347|58 [30]		1238|4567 [4]
1236|4578 [30]	126|348|57 [60]	**1234|5678**	1256|3478 [18]
1234|56|78 [15]	126|378|45 [60]	12345678 [1]	1235|46|78 [24]
1236|45|78 [30]	128|367|45 [30]	123456|78 [6][2]	1235|48|67 [12]
	12|34|56|78 [30]	125678|34	1238|45|67 [12]
	12|36|47|58 [60]	123567|48 [16]	1256|34|78 [18]
			1256|37|48 [36]

Table 2: *Complementary set partitions* (contd.)

```
1258|34|67 [18]      1257|34|68 [24]      1257|3468 [12]        12356|478 [4][6]
1258|36|47 [36]      1257|36|48 [48]      1235|47|68 [8][2]     12357|468
1567|23|48 [12]      1567|23|48 [24][2]   1235|48|67            12358|467
1568|23|47 [36]      1578|23|46           1237|45|68 [8][2]     12367|458
123|456|78 [12]      123|457|68 [16]      1238|45|67            12368|457
123|458|67 [12]      123|567|48 [8][2]    1256|37|48 [12]       12378|456
123|568|47 [12]      123|578|46           1257|34|68 [12][2]    12567|348 [6][4]
125|346|78 [18]      125|347|68 [24]      1258|34|67            12568|347
125|348|67 [18]      125|367|48 [48][2]   1257|36|48 [24][2]    12578|346
125|367|48 [36]      127|358|46           1258|36|47            12678|345
125|368|47 [72]      125|378|46 [24][2]   1278|35|46 [12]       15678|234 [4]
125|678|34 [18]      127|356|48           1567|23|48 [12][2]    1235|4678 [4][4]
128|356|47 [36]      125|678|34 [12][2]   1568|23|47            1236|4578
156|278|34 [36]      127|568|34           1578|23|46 [24]       1237|4568

12|35|46|78 [36]     156|278|34 [12]      123|457|68 [8][2]     1238|4567
12|35|48|67 [36]     157|268|34 [24]      123|458|67            1256|3478 [6][3]
15|26|37|48 [24]                          123|578|46 [8]        1257|3468
                     12|35|47|68 [48]     125|347|68 [12][2]    1258|3467
                     15|26|37|48 [24]     125|348|67            1256|37|48 [12][6]
1234|56|78                                125|367|48 [24][2]    1257|36|48
                     1234|56|7|8          125|368|47            1258|36|47
12345678 [1]                              125|378|46 [24]       1267|35|48
                     12345678 [1]         125|678|34 [12]       1268|35|47
123457|68 [4]                             127|356|48 [12][2]    1278|35|46
123567|48 [8][2]     123457|68 [2][2]     128|356|47            1567|23|48 [12][4]
123578|46            123458|67            127|358|46 [24][2]    1568|23|47
125678|34 [6]        123567|48 [4][2]     128|357|46            1578|23|46
                     123568|47            156|278|34 [12]       1678|23|45
12345|678 [2][2]     123578|46 [8]        157|268|34 [24]       125|367|48 [12][12]
12347|568            125678|34 [6]        12|35|47|68 [24][2]   125|368|47
12356|478 [4][2]     12345|678 [2]        12|35|48|67           125|378|46
12378|456            12356|478 [4]        15|26|37|48 [24]      126|357|48
12357|468 [16]       12357|468 [8][2]                           126|358|47
12567|348 [12][2]    12358|467            1234|5|6|7|8          126|378|45
12578|346            12378|456 [4]                              127|356|48
15678|234 [4]        12567|348 [6][2]     12345678 [1]          127|358|46
1235|4678 [8][2]     12568|347            123567|48 [4][4]      127|368|45
1237|4568            12578|346 [12]       123568|47             128|356|47
1256|3478 [6]        15678|234 [4]        123578|46             128|357|46
1257|3468 [12]       1235|4678 [8]        123678|45             128|367|45
1235|47|68 [16][2]   1237|4568 [4][2]     125678|34 [6]
1237|45|68           1238|4567
1256|37|48 [12][2]   1256|3478 [6]
1278|35|46
```

TABLE 2   261

Table 2: *Complementary set partitions (contd.)*

156|278|34 [12][3]
157|268|34
158|267|34
15|26|37|48 [24]

123|456|78
12345678 [1]
123457|68 [6][2]
124567|38
123478|56 [3][2]
145678|23
124578|36 [9]
12345|678 [3][2]
12456|378
12347|568 [6][2]
14567|238
12457|368 [18]
12478|356 [9][2]
14578|236
1234|5678 [3][2]
1278|3456
1237|4568 [2]
1245|3678 [9]
1247|3568 [18]
1234|57|68 [6][2]
1456|27|38
4567|12|38 [6][2]
1237|45|68
1245|37|68 [18]
1247|35|68 [36][2]
1457|26|38
1247|38|56 [18][2]
1457|23|68
1278|34|56 [9][2]
4578|12|36
1478|23|56 [9]
1478|25|36 [18]
124|357|68 [36][2]
145|267|38

124|378|56 [9][2]
145|678|23
124|567|38 [18][2]
127|345|68
124|578|36 [18][2]
145|278|36
127|348|56 [18][2]
147|568|23
127|458|36 [18]
147|258|36 [36]
12|34|57|68 [18][2]
14|27|38|56
12|37|45|68 [18]
14|25|37|68 [36]

123|456|7|8
12345678 [1]
123457|68 [3][4]
123458|67
124567|38
124568|37
123478|56 [3][2]
145678|23
124578|36 [9]
12345|678 [3][2]
12456|378
12348|567 [3][4]
12347|568
14567|238
14568|237
12457|368 [9][2]
12458|367
12478|356 [9][2]
14578|236
1234|5678 [3][2]
1278|3456
1245|3678 [9]
1248|3567 [9][2]
1247|3568

1234|57|68 [6][2]
1456|27|38
1245|38|67 [9][2]
1245|37|68
1247|35|68 [18][4]
1248|35|67
1457|26|38
1458|26|37
1247|38|56 [9][4]
1248|37|56
1457|23|68
1458|23|67
1278|34|56 [9][2]
4578|12|36
1478|23|56 [9]
1478|25|36 [18]
124|357|68 [18][4]
124|358|67
145|267|38
145|268|37
145|678|23 [9][2]
124|378|56
127|345|68 [9][4]
124|568|37
124|567|38
128|345|67
145|278|36 [18][2]
124|578|36
128|347|56 [9][4]
127|348|56
148|567|23
147|568|23
128|457|36 [9][2]
127|458|36
147|258|36 [36]
12|34|57|68 [18][2]
14|27|38|56
14|25|37|68 [18][2]
14|25|38|67

123|45|67|8
12345678 [1]
123456|78 [2][2]
123467|58
123468|57 [4]
124567|38 [3]
124568|37 [6][2]
124678|35
145678|23 [3]
12346|578 [4]
12348|567 [2][2]
12368|457
12456|378 [6][2]
12467|358
12458|367 [3][2]
12678|345
12468|357 [12]
14567|238 [3]
14568|237 [6][2]
14678|235
1234|5678 [2][2]
1236|4578
1245|3678 [3][2]
1267|3458
1246|3578 [12]
1248|3567 [6][2]
1268|3457
1234|56|78 [4][2]
1236|47|58
1245|36|78 [6][2]
1267|34|58
1246|35|78 [12][2]
1246|37|58
1246|38|57 [12]
1248|36|57 [12][2]
1268|34|57
1456|23|78 [6][2]
1467|23|58
1467|25|38 [12][2]
1456|27|38

Table 2: *Complementary set partitions (contd.)*

1458|26|37 [6][2]	124567|38 [3][3]	1467|25|38 [12][3]	124567|38 [3][5]
1678|24|35	124568|37	1468|25|37	124568|37
1468|23|57 [12]	124578|36	1478|25|36	124578|36
1468|25|37 [24]	124678|35 [6]	1678|24|35 [6]	124678|35
4568|12|37 [6][2]	145678|23 [3]	4678|12|35 [6]	125678|34
4678|12|35	12346|578 [2][3]	124|367|58 [6][3]	145678|23 [3]
124|356|78 [12][2]	12347|568	124|368|57	12456|378 [3][10]
126|347|58	12348|567	124|378|56	12457|368
124|367|58 [6][2]	12456|378 [3][3]	124|567|38 [6][3]	12458|367
126|345|78	12457|368	124|568|37	12467|358
126|348|57 [12][2]	12458|367	124|578|36	12468|357
124|368|57	12467|358 [6][3]	126|347|58 [6][6]	12478|356
124|567|38 [6][2]	12468|357	126|348|57	12567|348
126|457|38	12478|356	128|347|56	12568|347
124|568|37 [12][2]	12678|345 [3]	127|346|58	12578|346
126|478|35	14567|238 [3][3]	127|348|56	12678|345
128|346|57 [12]	14568|237	128|346|57	14567|238 [3][5]
128|456|37 [6][2]	14578|236	127|468|35 [6][3]	14568|237
128|467|35	14678|235 [6]	126|478|35	14578|236
145|267|38 [6]	1234|5678 [2]	128|467|35	14678|235
145|268|37 [12][2]	1245|3678 [3]	145|267|38 [6][3]	15678|234
148|267|35	1246|3578 [6][3]	145|268|37	1245|3678 [3][10]
146|257|38 [12]	1247|3568	145|278|36	1246|3578
146|258|37 [24][2]	1248|3567	146|257|38 [12][3]	1247|3568
146|278|35	1267|3458 [3][3]	146|258|37	1248|3567
146|578|23 [12]	1268|3457	147|258|36	1256|3478
148|567|23 [6][2]	1278|3456	146|278|35 [12][3]	1257|3468
168|457|23	1246|37|58 [6][6]	147|268|35	1258|3467
12|34|56|78 [12][2]	1246|38|57	148|267|35	1267|3458
12|36|47|58	1247|36|58	147|568|23 [6][3]	1268|3457
14|25|36|78 [12][2]	1247|38|56	146|578|23	1278|3456
14|26|37|58	1248|36|57	148|567|23	1456|27|38 [6][10]
14|26|38|57 [24]	1248|37|56	14|26|37|58 [12][3]	1457|26|38
	1267|34|58 [6][3]	14|26|38|57	1458|26|37
123|45|6|7|8	1268|34|57	14|27|38|56	1467|25|38
12345678 [1]	1278|34|56		1468|25|37
123467|58 [2][3]	1456|27|38 [6][3]	**123|4|5|6|7|8**	1478|25|36
123468|57	1457|26|38	12345678 [1]	1567|24|38
123478|56	1458|26|37		1568|24|37
	1467|23|58 [6][3]		1578|24|36
	1478|23|56		1678|24|35
	1468|23|57		

TABLE 2   263

Table 2: *Complementary set partitions* (contd.)

Col 1	Col 2	Col 3	Col 4
145|267|38 [6][15]	1235|4678 [4][6]	134|567|28 [4][12]	12358|467 [4][6]
145|268|37	1237|4568	127|356|48	12357|468
145|278|36	1257|3468	127|345|68	13458|267
146|257|38	1345|2678	134|578|26	13568|247
146|258|37	1347|2568	125|378|46	13457|268
146|278|35	1356|2478	125|347|68	13567|248
147|256|38	1357|2468 [8]	123|578|46	13478|256 [2][6]
147|258|36		123|567|48	12378|456
147|268|35	1235|47|68 [8][12]	178|356|24	12578|346
148|256|37	1237|45|68	156|378|24	34578|126
148|257|36	1257|36|48	156|347|28	15678|234
148|267|35	1345|27|68	178|345|26	35678|124
156|278|34	1347|25|68	137|258|46 [16][6]	13578|246 [8]
157|268|34	1356|27|48	137|458|26	
158|267|34	1378|25|46	135|247|68	1235|4678 [4][3]
	1578|23|46	157|368|24	1345|2678
12|34|56|78	3578|14|26	135|267|48	1356|2478
12345678 [1]	3567|14|28	135|467|28	1238|4567 [2][6]
123457|68 [4][6]	1567|23|48	13|25|47|68 [16][3]	1257|3468
123567|48	3457|16|28	13|27|45|68	1258|3467
123578|46	1357|24|68 [16][3]	15|27|36|48	1237|4568
134567|28	1357|26|48		1348|2567
134578|26	1357|28|46	**12|34|56|7|8**	1347|2568
135678|24		12345678 [1]	1357|2468 [8]
	123|457|68 [8][12]	123457|68 [2][6]	
12345|678 [2][12]	137|456|28	123458|67	1235|47|68 [4][6]
12347|568	137|256|48	123567|48	1235|48|67
12356|478	135|678|24	123568|47	1356|28|47
13478|256	125|367|48	134568|27	1345|28|67
12378|456	137|568|24	134567|28	1345|27|68
12567|348	135|278|46	123578|46 [4][3]	1356|27|48
12578|346	157|346|28	134578|26	1257|36|48 [4][12]
13456|278	127|358|46	135678|24	1258|36|47
15678|234	134|257|68	12345|678 [2][3]	1347|25|68
35678|124	157|348|26	12356|478	1348|25|67
34578|126	135|478|26	13456|278	1238|45|67
34567|128			1237|45|68
13578|246 [8][4]			3568|14|27
13567|248			3567|14|28
13457|268			3457|16|28
12357|468			3458|16|27
			1568|23|47
			1567|23|48

Table 2: *Complementary set partitions (contd.)*

1358|27|46 [8][6]	13|25|47|68 [8][6]	1235|4678 [2][8]	135|267|48 [4][24]
1358|26|47	13|25|48|67	1236|4578	135|268|47
1357|24|68	13|27|45|68	1237|4568	135|278|46
1358|24|67	13|28|45|67	1238|4567	135|467|28
1357|28|46	15|27|36|48	1345|2678	135|468|27
1357|26|48	15|28|36|47	1346|2578	135|478|26
1578|23|46 [8][3]		1347|2568	136|257|48
1378|25|46	**12|34|5|6|7|8**	1348|2567	136|258|47
3578|14|26	12345678 [1]	1356|2478 [4][3]	136|278|45
123|457|68 [4][12]	123567|48 [2][8]	1357|2468	136|457|28
123|458|67	123568|47	1358|2467	136|458|27
134|258|67	123578|46		136|478|25
125|367|48	123678|45	1356|27|48 [4][12]	137|256|48
125|368|47	134567|28	1356|28|47	137|258|46
158|346|27	134568|27	1357|26|48	137|268|45
134|257|68	134578|26	1357|28|46	137|456|28
157|346|28	134678|25	1358|26|47	137|458|26
138|456|27	135678|24 [4]	1358|27|46	137|468|25
138|256|47	12356|478 [2][12]	1367|25|48	138|256|47
137|256|48	12357|468	1367|28|45	138|257|46
137|456|28	12358|467	1368|25|47	138|267|45
134|578|26 [4][6]	12367|458	1368|27|45	138|456|27
123|578|46	12368|457	1378|25|46	138|457|26
125|378|46	12378|456	1378|26|45	138|467|25
178|356|24	13456|278	1567|23|48 [4][8]	156|378|24 [4][6]
173|345|26	13457|268	1568|23|47	157|368|24
156|378|24	13458|267	1578|23|46	158|367|24
135|468|27 [8][6]	13467|258	1678|23|45	167|358|24
135|467|28	13468|257	3567|14|28	168|357|24
135|267|48	13478|256	3568|14|27	178|356|24
135|248|67	13567|248 [4][4]	3578|14|26	
135|268|47	13568|247	3678|14|25	
135|247|68	13578|246		**12|3|4|5|6|7|8**
135|278|46 [8][3]	13678|245		12345678 [1]
135|478|26	15678|234 [2][2]		134567|28 [2][6]
135|678|24	35678|124		134568|27
137|258|46 [8][6]			134578|26
138|457|26			134678|25
157|368|24			135678|24
158|367|24			145678|23
138|257|46			
137|458|26			

TABLE 2   265

Table 2: *Complementary set partitions (contd.)*

13456|278 [2][15]
  13457|268
  13458|267
  13467|258
  13468|257
  13478|256
  13567|248
  13568|247
  13578|246
  13678|245
  14567|238
  14568|237
  14578|236
  14678|235
  15678|234

1345|2678 [2][10]
  1346|2578
  1347|2568
  1348|2567
  1356|2478
  1357|2468
  1358|2467
  1367|2458
  1368|2457
  1378|2456

1|2|3|4|5|6|7|8
12345678 [1]

References

Aigner, M. (1979) *Combinatorial Theory*, New York: Springer-Verlag

Amari, S.I. (1985) *Differential-Geometrical Methods in Statistics*, New York: Springer-Verlag.

Amari, S.I. & Kumon, M. (1983) Differential geometry of Edgeworth expansions in curved exponential family. *Ann. Inst. Statist. Math.* A **35**, 1–24.

Ames, J.S. & Murnaghan, F.D. (1929) *Theoretical Mechanics*, Boston: Ginn & Co.

Anscombe, F.J. (1961) Examination of residuals. *Proc. Fourth Berkeley Symposium* **1**, 1–36.

Atiqullah, M. (1962) The estimation of residual variance in quadratically balanced problems and the robustness of the F-test. *Biometrika* **49**, 83–91.

Bahadur, R.R (1971) *Some limit theorems in Statistics*, Philadelphia, SIAM.

Bahadur, R.R. & Ranga-Rao, R. (1960) On deviations of the sample mean. *Ann. Math. Statist.* **31**, 1015–1027.

Bahadur, R.R. & Zabell, S.L. (1979) Large deviations of the sample mean in general vector spaces. *Ann. Prob.* **7**, 587–621.

Barndorff-Nielsen, O.E. (1978) *Information and Exponential Families in Statistical Theory*, Chichester: Wiley.

Barndorff-Nielsen, O.E. (1980) Conditionality resolutions. *Biometrika* **67**, 293–310.

Barndorff-Nielsen, O.E. (1983) On a formula for the distribution of the maximum likelihood estimator. *Biometrika* **70**, 343–365.

Barndorff-Nielsen, O.E. (1984) On conditionality resolution and the likelihood ratio for curved exponential models. *Scand. J. Statist.* **11**, 157–170. corr: **12**, 191.

Barndorff-Nielsen, O.E. (1985) Confidence limits from $c|\hat{\jmath}|^{1/2}\bar{L}$ in the single parameter case. *Scand. J. Statist.* **12**, 83–87.

Barndorff-Nielsen, O.E. (1986) Strings, tensorial combinants, and Bartlett adjustments. *Proc R. Soc. Lond.* A **406**, 127–137.

Barndorff-Nielsen, O.E. & Cox, D.R. (1979) Edgeworth and saddle-point approximations with statistical applications (with Discussion). *J. Roy. Statist. Soc.* B, **41**, 279–312.

Barndorff-Nielsen, O.E. & Cox, D.R. (1984) Bartlett adjustments to the likelihood ratio statistic and the distribution of the maximum likelihood estimator. *J. Roy. Statist. Soc.* B **46**, 483–495.

Barndorff-Nielsen, O.E., Cox, D.R. & Reid, N.R. (1986) The role of differential geometry in statistical theory. *Int. Statist. Rev.*, **54**, 83–96.

Barnett, H.A.R. (1955) The variance of the product of two independent variables and its application to an investigation based on sample data. *J. Inst. Actuaries*, **81**, 190.

Bartlett, M.S. (1937) Properties of sufficiency and statistical tests. *Proc. Roy. Soc.* A **160**, 268–282.

Bartlett, M.S. (1938) The characteristic function of a conditional statistic. *J. Lond. Math. Soc.* **13**, 62–67.

Barton, D.E., David, F.N. & Fix, E. (1960) The polykays of the natural numbers. *Biometrika* **47**, 53–59.

Basu, D. (1955) On statistics independent of a complete sufficient statistic. *Sankhya* **15**, 377–380.

Basu, D. (1958) On statistics independent of sufficient statistics. *Sankhya* **20**, 223–226.

Basu, D. (1964) Recovery of ancillary information. *Sankhya* A **26**, 3–16.

Bates, D.M. & Watts, D.G. (1980) Relative curvature measures of nonlinearity (with Discussion). *J. Roy. Statist. Soc.* B **22**, 41–88.

Beale, E.M.L. (1960) Confidence regions in non-linear estimation (with Discussion). *J. Roy. Statist. Soc.* B **22**, 41–88.

Berger, J. & Wolpert, R. (1984) *The Likelihood Principle*, IMS Monograph No. 6: Hayward, California.

Bhattacharya, R.N. & Ghosh, J.K. (1978) On the validity of the formal Edgeworth expansion. *Ann. Statist.* **6**, 434–451.

Bhattacharya, R.N. & Rao, R.R. (1976) *Normal Approximation and Asymptotic Expansions*. New York: Wiley.

Bickel, P.J. (1978) Using residuals robustly I: Tests for heteroscedasticity, non-linearity. *Ann. Statist.* **6**, 266–291.

Billingsley, P. (1985) *Probability and Measure*, 2nd edition, New York: Wiley.

Breslow, N.E. & Day, N.E. (1980) *Statistical Methods in Cancer Research*, 1: *The Analysis of Case-Control Studies*. Lyon: I.A.R.C.

Brillinger, D.R. (1969) The calculation of cumulants via conditioning. *Ann. Inst. Statist. Math.* **21**, 215–218.

Brillinger, D.R. (1975) *Time Series: Data Analysis and Theory*, New York: Holt, Rinehart & Winston.

Buehler, R.J. (1982) Some ancillary statistics and their properties (with discussion). *J. Amer. Statist. Assoc.* **77**, 581–589.

Cartan, E. (1981) *The Theory of Spinors*, New York: Dover.

Cayley, A. (1885) A memoir on seminvariants. *Amer. J. Math.* **7**, 1–25.

Chambers, J.M. (1967) On methods of asymptotic approximation for multivariate distributions. *Biometrika* **54**, 367–383.

Chambers, J.M. (1977) *Computational Methods for Data Analysis*, New York: Wiley.

Chernoff, H. (1952) A measure of asymptotic efficiency for tests of a hypothesis based on the sum of observations. *Ann. Math. Statist.* **23**, 493–507.

Churchill, E. (1946) Information given by odd moments. *Ann. Math. Statist.* **17**, 244–246.

Cook, M.B. (1951) Bivariate *k*-statistics and cumulants of their joint sampling distribution. *Biometrika* **38**, 179–195.

Cordeiro, G.M. (1983) Improved likelihood-ratio tests for generalized linear models. *J. Roy. Statist. Soc.* B, **45**, 404–413.

Cornish, E.A. & Fisher, R.A. (1937) Moments and cumulants in the specification of distributions. *Revue de l'Institut Internationale de Statistique*, **4**, 1–14.

Cox, D.R. (1958) Some problems connected with statistical inference. *Ann. Math. Statist.* **29**, 357–372.

Cox, D.R. (1971) The choice between alternative ancillary statistics. *J. Roy. Statist. Soc.* B **33**, 251–255.

Cox, D.R. (1980) Local ancillarity. *Biometrika* **67**, 279–286.

Cox, D.R. & Hinkley, D.V. (1974) *Theoretical Statistics*, London: Chapman & Hall.

Cramér, H. (1937) *Random Variables and Probability Distributions*, Cambridge Tracts in Mathematics, No. 36.

Cramér, H. (1938) Sur un nouveau théorème-limite de la théorie des probabilités. *Actualités Sci. Indust.* **736**, 5–23.

Cramér, H. (1946) *Mathematical Methods of Statistics*, Princeton University Press.

Daniels, H.E. (1954) Saddlepoint approximations in statistics. *Ann. Math. Statist.* **25**, 631–650.

Daniels, H.E. (1960) Approximate solutions of Green's type for univariate stochastic processes. *J. Roy. Statist. Soc.* B **22**, 376–401.

Daniels, H.E. (1980) Exact saddlepoint approximations. *Biometrika* **67**, 59–63.

Daniels, H.E. (1983) Saddlepoint approximations for estimating equations. *Biometrika* **70**, 89–96.

Daniels, H.E. (1987) Tail probability approximations. *Int. Statist. Rev.* **54**, in press.

David, F.N. & Barton, D.E. (1962) *Combinatorial Chance*, London: Griffin.

David, F.N., Kendall, M.G. & Barton, D.E. (1966) *Symmetric Functions and Allied Tables*, Cambridge University Press.

Davis, A.W. (1976) Statistical distributions in univariate and multivariate Edgeworth populations. *Biometrika* **63**, 661–670.

Davis, A.W. (1980) On the effects of moderate non-normality on Wilks's likelihood ratio criterion. *Biometrika* **67**, 419–427.

Dirac, P.A.M. (1958) *The Principles of Quantum Mechanics*, Oxford: Clarendon Press.

Dressel, P.L. (1940) Statistical seminvariants and their setimates with particular emphasis on their relation to algebraic seminvariants. *Ann. Math. Statist.* **11**, 33–57.

Drucker, D.C. (1967) *Introduction to Mechanics of Deformable Solids*, New York: McGraw-Hill.

Durbin, J. (1980) Approximations for densities of sufficient estimators. *Biometrika* **67**, 311–333.

Dwyer, P.S. & Tracy, D.S. (1964) A combinatorial method for products of two polykays with some general formulae. *Ann. Math. Statist.* **35**, 1174–1185.

Eaton, M.L. (1983) *Multivariate Statistics*, New York: Wiley.

Edgeworth, F.Y. (1905) The law of error. *Trans. Camb. Phil. Soc.* **20**, 36–65, 113–141.

Edwards, A.W.F. (1972) *Likelihood*, Cambridge University Press.

Efron, B. (1975) Defining the curvature of a statistical problem (with applications to second order efficiency) (with discussion). *Ann. Statist.* **3**, 1189–1242.

Efron, B. & Hinkley, D.V. (1978) Assessing the accuracy of the maximum likelihood estimator: Observed versus expected Fisher information (with discussion). *Biometrika* **65**, 457–487.

Eisenhart, L.P. (1926) *Riemannian Geometry*, Princeton: University Press.

Ellis, R.S. (1985) *Entropy, Large Deviations and Statistical Mechanics*, New York: Springer-Verlag.

Esscher, F. (1932) On the probability function in the collective theory of risk. *Skand. Aktuarietidskrift* **15**, 175–195.

Esseen, C.V. (1945) Fourier analysis of distribution functions. *Acta Mathematica* **77**, 1–125.

Feller, W. (1971) *An Introduction to Probability Theory and Its Applications* **2**, New York: Wiley.

Fenchel, W. (1949) On conjugate convex functions. *Can. J. Math.* **1**, 73–77.

Fisher, R.A. (1925) Theory of statistical estimation. *Proc. Camb. Phil. Soc.* **22**, 700–725.

Fisher, R.A. (1929) Moments and product moments of sampling distributions. *Proc. Lond. Math. Soc.* Series 2, **30**, 199–238. Reprinted (1972) as paper 74 in *Collected Papers of R.A. Fisher*, vol. 2, (ed. J.H. Bennett. University of Adelaide Press, pp. 351–354).

Fisher, R.A. (1934) Two new properties of mathematical likelihood. *Proc. Roy. Soc.* A **144**, 285–307.

Fisher, R.A. (1936) Uncertain inference. *Proc. Amer. Acad. Arts & Sciences* **71**, 245–258.

Fisher, R.A. (1956) On a test of significance in Pearson's Biometrika tables (no. 11). *J. Roy. Statist. Soc.* B **18**, 56–60.

Fisher, R.A. & Wishart, J. (1931) The derivation of the pattern formulae of two-way partitions from those of similar partitions. *Proc. Lond. Math. Soc.*, Series 2, **33**, 195–208.

Folks, J.L. & Chhikara, R.S. (1978) The inverse Gaussian distribution and its statistical application – a review (with Discussion). *J. Roy. Statist. Soc.* B **40**, 263–289.

Foster, B.L. (1986) Would Leibnitz lie to you? (Three aspects of the affine connection). *Mathematical Intelligencer* **8**, No. 3, 34–40, 57.

Fraser, D.A.S. (1968) *The Structure of Inference*, New York: Wiley.

Gantmacher, F.R. (1960) *Matrix Theory* **1**, New York: Chelsea Publishing Co.

Gilbert, E.N. (1956) Enumeration of labelled graphs. *Can. J. Math.* **8**, 405–411.

Good, I.J. (1958) The interaction algorithm and practical Fourier analysis. *J. Roy. Statist. Soc.* B **20**, 361–372.

Good, I.J. (1960) The interaction algorithm and practical Fourier analysis: an addendum. *J. Roy. Statist. Soc.* B **22**, 372–375.

Good, I.J. (1975) A new formula for cumulants. *Math. Proc. Camb. Phil. Soc.* **78**, 333–337.

Good, I.J. (1977) A new formula for k-statistics. *Ann. Statist.* **5**, 224–228.

Goodman, L.A. (1960) On the exact variance of products. *J. Amer. Statist. Assoc.* **55**, 708–713.

Goodman, L.A. (1962) The variance of the product of K random variables. *J. Amer. Statist. Assoc.* **57**, 54–60.

Haldane, J.B.S. (1942) Mode and median of a nearly normal distribution with given cumulants. *Biometrika* **32**, 294–299.

Haldane, J.B.S. (1948) Note on the median of a multivariate distribution. *Biometrika* **35**, 414–415.

Halmos, P.R. (1948) The range of a vector measure. *Bull. Amer. Math. Soc.* **54**, 416–421.

Heyde, C.C. (1963) On a property of the lognormal distribution. *J. Roy. Statist. Soc.* B **25**, 392–393.

Hinkley, D.V. (1980) Likelihood as approximate pivotal distribution. *Biometrika* **67**, 287–292.

Hinkley, D.V. (1985) Transformation diagnostics for linear models. *Biometrika* **72**, 487–496.

Hooke, R. (1956a) Symmetric functions of a two-way array. *Ann. Math. Statist.* **27**, 55–79.

Hooke, R. (1956b) Some applications of bipolykays to the estimation of variance components and their moments. *Ann. Math. Statist.* **27**, 80–98.

Irwin, J.O. & Kendall, M.G. (1943–45) Sampling moments of moments for a finite population. *Ann. Eugenics* **12**, 138–142.

James, G.S. (1958) On moments and cumulants of systems of statistics. *Sankhya* **20**, 1–30.

James, G.S. & Mayne, A.J. (1962) Cumulants of functions of random variables. *Sankhya* **24**, 47–54.

Jarrett, R.G. (1973) *Efficiency and Estimation in Asymptotically Normal Distributions*, University of London PhD Thesis.

Jeffreys, H. (1952) *Cartesian Tensors*, Cambridge University Press.

Jeffreys, H. & Jeffreys, B.S. (1956) *Methods of Mathematical Physics*, 3rd edition. Cambridge University Press.

Johansen, S. (1983) Some topics in regression (with discussion). *Scand. J. Statist.* **10**, 161–194.

Johnson, N.L. & Kotz, S. (1970) *Continuous Univariate Distributions*, **2**. Boston, Mass: Houghton-Mifflin.

Kalbfleisch, J.D. (1975) Sufficiency and conditionality. *Biometrika* **62**, 251–259

Kaplan, E.L. (1952) Tensor notation and the sampling cumulants of k-statistics. *Biometrika* **39**, 319–323.

Karlin, S. & Studden, W.J. (1966) *Tchebycheff Systems:with applications in analysis and statistics*, New York: Wiley.

Kendall, M.G. (1940a) Some properties of k-statistics. *Ann. Eugenics* **10**, 106–111.

Kendall, M.G. (1940b) Proof of Fisher's rules for ascertaining the sampling semi-invariants of k-statistics. *Ann. Eugenics* **10**, 215–222.

Kendall, M.G. (1940c) The derivation of multivariate sampling formulae from univariate formulae by symbolic operation. *Ann. Eugenics* **10**, 392–402.

Kendall, M.G. & Stuart, A. (1977) *The Advanced Theory of Statistics* **1**, 4th edition. London: Griffin.

Kibble, T.W.B. (1985) *Classical Mechanics*, London: Longman.

Kiefer, J. (1977) Conditional confidence statements and confidence estimators. *J. Amer. Statist. Assoc.* **72**, 789–827.

Knuth, D.E. (1986) *The TEXbook*, Reading, Mass.: Addison-Wesley.

Kruskal, J.B. (1977) Three way arrays: Rank and uniqueness of trilinear decompositions, with application to arithmetic complexity and Statistics. *Linear Algebra and its Applications* **18**, 95–138.

Kruskal, W. (1975) The geometry of generalized inverses. *J. Roy. Statist. Soc.* B **37**, 272–283.

Lawden, D.F. (1968) *An Introduction to Tensor Calculus and Relativity*, London: Chapman & Hall.

Lawley, D.M. (1956) A general method for approximating to the distribution of likelihood-ratio criteria. *Biometrika* **43**, 295–303.

LeCam, L. (1986) The central limit theorem around 1935. *Statistical Science* **1**, 78–96.

Lehmann, E.L. (1981) An interpretation of completeness and Basu's theorem. *J. Amer. Statist. Assoc.* **76**, 335–340.

Leonov, V.P. & Shiryaev, A.M. (1959) On a method of calculation of semi-invariants. *Theory Prob. Applic.* **4**, 319–329.

Linnik, Y.V. (1968) *Statistical Problems with Nuisance Parameters*, Mathematical Monographs, No. 20. New York: American Mathematical Society.

Lloyd, C.J. (1985a) On external ancillarity. *Austral. J. Statist.* **27**, 202–220.

Lloyd, C.J. (1985b) Ancillaries sufficient for the sample size. *Austral. J. Statist.* **27**, 264–272.

Lugannani, R. & Rice, S.O. (1980) Saddlepoint approximation for the distribution of the sum of independent random variables. *Adv. Appl. Prob.* **12**, 475–490.

McConnell, A.J. (1931) *Applications of the Absolute Differential Calculus*, London: Blackie.

MacCullagh, J. (1855) On the attraction of ellipsiods, with a new demonstration of Clairaut's theorem. *Trans. Roy. Irish Acad.* **22**, 379–395.

McCullagh, P. (1984a) Local sufficiency. *Biometrika* **71**, 233–244.

McCullagh, P. (1984b) Tensor notation and cumulants of polynomials. *Biometrika* **71**, 461–476.

McCullagh, P. (1984c) Recurrence processes. *J. Appl. Prob.* **21**, 167–172.

McCullagh, P. (1985) On the asymptotic distribution of Pearson's statistic in linear exponential-family models. *Int. Statist. Rev.* **53**, 61–67.

McCullagh, P. & Cox, D.R. (1986) Invariants and likelihood ratio statistics. *Ann. Statist.* **14** , 1419–1430.

McCullagh, P. & Nelder, J.A. (1983) *Generalized Linear Models*, London: Chapman & Hall.

McCullagh, P. & Pregibon, D. (1987) k-statistics and dispersion effects in regression. *Ann. Statist.* **15**, (in press).

McCullagh, P. & Wilks, A.R. (1985a) Complementary set partitions. *AT&T Bell Labs Technical Memorandum* 11214–850328–07.

McCullagh, P. & Wilks, A.R. (1985b) Extended tables of complementary set partitions. *AT&T Bell Labs Technical Memorandum* 11214–850328–08.

Machado, S.B.G. (1976) Transformations of multivariate data and tests for multivariate normality. PhD Thesis, Dept. of Statistics, University of Chicago.

Machado, S.B.G. (1983) Two statistics for testing multivariate normality. *Biometrika* **70**, 713–718.

MacMahon, P.A. (1884) Seminvariants and symmetric functions. *Amer. J. Math.* **6**, 131–163.

MacMahon, P.A. (1886) Memoir on seminvariants. *Amer. J. Math.* **8**, 1–18.

MacMahon, P.A. (1915) *Combinatory Analysis 1, 2*, Cambridge University Press. Reprinted (1960) Chelsea Publishing Co. New York.

Malyshev, V.A. (1980) Cluster expansions in lattice models of Statistical Physics and the quantum theory of fields. *Russian Math. Surveys* **35**:2, 1–62.

Mardia, K.V. (1970) Measures of multivariate skewness and kurtosis with applications. *Biometrika* **57**, 519–530.

Michel, R. (1979) Asymptotic expansions for conditional distributions. *J. Multivariate Anal.* **9**, 393–400.

Moran, P.A.P. (1968) *An Introduction to Probability Theory*, Oxford: Clarendon Press.

Morris, C. (1982) Natural exponential families with quadratic variance functions. *Ann. Statist.* **10**, 65–80.

Mosteller, F. & Tukey, J.W. (1977) *Data Analysis and Regression*, New York: Addison-Wesley.

Murnaghan, F.D. (1951) *Finite Deformation of an Elastic Solid*, New York: Wiley.

Nelder, J.A. & Pregibon, D. (1986) An extended quasi-likelihood function. *Biometrika* in press.

Peers, H.W. (1978) Second order sufficiency and statistical invariants. *Biometrika* **65**, 489–496.

Peers, H.W. & Iqbal, M. (1985) Asymptotic expansions for confidence limits in the presence of nuisance parameters, with appplications. *J. Roy. Statist. Soc.* B, **47**, 547–554.

Pierce, D.A. (1973) On some difficulties in a frequency theory of inference. *Ann. Statist.* **1**, 241–250.

Plackett, R.L. (1960) *Principles of Regression Analysis*, Oxford: Clarendon Press.

Pukelsheim, F. (1980) Multilinear estimation of skewness and kurtosis in linear models. *Metrika* **27**, 103–113.

Rao, C.R. (1973) *Linear Statistical Inference and its Applications*, New York: Wiley.

Richtmyer, R.D. (1981) *Principles of Advanced Mathematical Physics* **2**, New York: Springer-Verlag.

Robinson, G.K. (1975) Some counter-examples to the theory of confidence intervals. *Biometrika* **62**, 155–161.

Robinson, G.K. (1978) On the necessity of Bayesian inference and the construction of measures of nearness to Bayesian form. *Biometrika* **65**, 49–52.

Robinson, J. (1982) Saddlepoint approximations for permutation tests and confidence intervals. *J. Roy. Statist. Soc.* B **44** , 91–101.

Rockafellar, R.T. (1970) *Convex Analysis*, Princeton University Press.

Rota, G.-C. (1964) On the foundations of combinatorial theory, I. Theory of Möbius functions. *Zeit. f. Warsch.* **2**, 340–368.

Seber, G.A.F. (1977) *Linear Regression Analysis*, New York: Wiley.

Shenton, L.R. & Bowman, K.O. (1977) *Maximum Likelihood Estimation in Small Samples*, London: Griffin.

Skovgaard, I.M. (1981a) Transformation of an Edgeworth expansion by a sequence of smooth functions. *Scand. J. Statist.* **8**, 207-217.

Skovgaard, I.M. (1981b) Edgeworth expansions of the distributions of maximum likelihood estimators in the general (non i.i.d.) case. *Scand. J. Statist.* **8**, 227-236.

Skovgaard, I.M. (1986a) A note on the differentiation of log-likelihood derivatives. *Int. Statist. Rev.* **54**, 29-32.

Skovgaard, I.M. (1986b) On multivariate Edgeworth expansions. *Int. Statist. Rev.* **54**, 169-186.

Skovgaard, I.M. (1986c) Successive improvement of the order of ancillarity. *Biometrika* **73**, 516-519.

Sokolnikoff, I.S. (1951) *Tensor Analysis:Theory and Applications*, New York: Wiley.

Speed, T.P. (1983) Cumulants and partition lattices. *Austral. J. Statist.* **25**, 378-388.

Speed, T.P. (1986a) Cumulants and partition lattices, II: generalized *k*-statistics. *J. Austral. Math. Soc.* A **40**, 34-53

Speed, T.P. (1986b) Cumulants and partition lattices, III: multiply indexed arrays. *J. Austral. Math. Soc.* A **40**, 161-182

Spivak, M. (1970) *Differential Geometry*, Berkeley: Publish or Perish.

Stigler, S.M. (1978) Francis Ysidro Edgeworth, Statistician. (with Discussion) *J. Roy. Statist. Soc.* A **141**, 287-322.

Stoker, J.J. (1969) *Differential Geometry*, New York: Wiley.

Synge, J.L. & Griffith, B.A. (1949) *Principles of Mechanics*, New York: McGraw-Hill.

Takemura, A. (1983) Tensor analysis of ANOVA decomposition. *J. Amer. Statist. Assoc.* **78**, 894-900.

Thiele, T.N. (1897) *Elementaer Iagttagelseslaere*, København: Gyldendalske. Reprinted in English as 'The Theory of Observations' *Ann. Math. Statist.* (1931) **2**, 165-308.

Thomas, T.Y. (1965) *Concepts from Tensor Analysis and Differential Geometry*, New York: Academic Press.

Tracy, D.S. (1968) Some rules for a combinatorial method for multiple products of generalized *k*-statistics. *Ann. Math. Statist.* **39**, 983-998.

Tukey, J.W. (1949) One degree of freedom for non-additivity. *Biometrics* **5**, 232-242.

Tukey, J.W. (1950) Some sampling simplified. *J. Amer. Statist. Assoc.* **45**, 501-519.

Tukey, J.W. (1956a) Keeping moment-like sampling computations simple. *Ann. Math. Statist.* **27**, 37-54.

Tukey, J.W. (1956b) Variances of variance components: I. Balanced designs. *Ann. Math. Statist.* **27**, 722-736.

Tweedie, M.C.K. (1957a) Statistical properties of inverse Gaussian distributions I. *Ann. Math. Statist.* **28**, 362-377.

Tweedie, M.C.K. (1957b) Statistical properties of inverse Gaussian distributions II. *Ann. Math. Statist.* **28**, 696-705.

Wallace, D.L. (1958) Asymptotic approximations to distributions. *Ann. Math. Statist.* **29**, 635–654.

Weatherburn, C.E. (1950) *An Introduction to Riemannian Geometry and the Tensor Calculus*, Cambridge University Press.

Wedderburn, R.W.M. (1974) Quasilikelihood functions, generalized linear models and the Gauss-Newton method. *Biometrika* **61**, 439–447.

Wichura, M. (1986) The PiCTeX manual. Technical Report No. 205, Dept. of Statistics, University of Chicago.

Williams, D.A. (1976) Improved likelihood ratio tests for complete contingency tables. *Biometrika* **63**, 33–37.

Willmore, T.J. (1982) *Total Curvature in Riemannian Geometry*, Chichester: Ellis Horwood.

Wishart, J. (1952) Moment coefficients of the k-statistics in samples from a finite population. *Biometrika* **39**, 1–13.

Zia ud-Din, M. (1954) Expression of the k-statistics k_9 and k_{10} in terms of power sums and sample moments. *Ann. Math. Statist.* **25**, 800–803. (corr.: Good, 1977).

Author index

Subject index